AROUND *the* PATIENT BED
HUMAN FACTORS AND SAFETY IN HEALTH CARE

Human Factors and Ergonomics Series

PUBLISHED TITLES

Around the Patient Bed: Human Factors and Safety in Health Care
Y. Donchin and D. Gopher

Conceptual Foundations of Human Factors Measurement
D. Meister

Content Preparation Guidelines for the Web and Information Appliances:
Cross-Cultural Comparisons
H. Liao, Y. Guo, A. Savoy, and G. Salvendy

Cross-Cultural Design for IT Products and Services
P. Rau, T. Plocher and Y. Choong

Data Mining: Theories, Algorithms, and Examples
Nong Ye

Designing for Accessibility: A Business Guide to Countering Design Exclusion
S. Keates

Handbook of Cognitive Task Design
E. Hollnagel

The Handbook of Data Mining
N. Ye

Handbook of Digital Human Modeling: Research for Applied Ergonomics
and Human Factors Engineering
V. G. Duffy

Handbook of Human Factors and Ergonomics in Health Care and Patient Safety
Second Edition
P. Carayon

Handbook of Human Factors in Web Design, Second Edition
K. Vu and R. Proctor

Handbook of Occupational Safety and Health
D. Koradecka

Handbook of Standards and Guidelines in Ergonomics and Human Factors
W. Karwowski

Handbook of Virtual Environments: Design, Implementation, and Applications
K. Stanney

Handbook of Warnings
M. Wogalter

Human–Computer Interaction: Designing for Diverse Users and Domains
A. Sears and J. A. Jacko

Human–Computer Interaction: Design Issues, Solutions, and Applications
A. Sears and J. A. Jacko

Human–Computer Interaction: Development Process
A. Sears and J. A. Jacko

Human–Computer Interaction: Fundamentals
A. Sears and J. A. Jacko

The Human–Computer Interaction Handbook: Fundamentals
Evolving Technologies, and Emerging Applications, Third Edition
A. Sears and J. A. Jacko

PUBLISHED TITLES (CONTINUED)

Human Factors in System Design, Development, and Testing
D. Meister and T. Enderwick

Introduction to Human Factors and Ergonomics for Engineers, Second Edition
M. R. Lehto

Macroergonomics: Theory, Methods and Applications
H. Hendrick and B. Kleiner

Practical Speech User Interface Design
James R. Lewis

The Science of Footwear
R. S. Goonetilleke

Skill Training in Multimodal Virtual Environments
M. Bergamsco, B. Bardy, and D. Gopher

Smart Clothing: Technology and Applications
Gilsoo Cho

Theories and Practice in Interaction Design
S. Bagnara and G. Crampton-Smith

The Universal Access Handbook
C. Stephanidis

Usability and Internationalization of Information Technology
N. Aykin

User Interfaces for All: Concepts, Methods, and Tools
C. Stephanidis

FORTHCOMING TITLES

Cognitive Neuroscience of Human Systems Work and Everyday Life
C. Forsythe and H. Liao

Computer-Aided Anthropometry for Research and Design
K. M. Robinette

Handbook of Human Factors in Air Transportation Systems
S. Landry

Handbook of Virtual Environments: Design, Implementation and Applications, Second Edition,
K. S. Hale and K M. Stanney

Variability in Human Performance
T. Smith, R. Henning, and M. Wade

AROUND *the* PATIENT BED

HUMAN FACTORS AND SAFETY IN HEALTH CARE

YOEL DONCHIN
DANIEL GOPHER

CRC Press
Taylor & Francis Group
Boca Raton London New York

CRC Press is an imprint of the
Taylor & Francis Group, an **informa** business

Book was previously published in Hebrew by Carta Jerusalem.

CRC Press
Taylor & Francis Group
6000 Broken Sound Parkway NW, Suite 300
Boca Raton, FL 33487-2742

© 2014 by Taylor & Francis Group, LLC
CRC Press is an imprint of Taylor & Francis Group, an Informa business

No claim to original U.S. Government works

Printed on acid-free paper
Version Date: 20130812

International Standard Book Number-13: 978-1-4665-7362-8 (Hardback)

This book contains information obtained from authentic and highly regarded sources. Reasonable efforts have been made to publish reliable data and information, but the author and publisher cannot assume responsibility for the validity of all materials or the consequences of their use. The authors and publishers have attempted to trace the copyright holders of all material reproduced in this publication and apologize to copyright holders if permission to publish in this form has not been obtained. If any copyright material has not been acknowledged please write and let us know so we may rectify in any future reprint.

Except as permitted under U.S. Copyright Law, no part of this book may be reprinted, reproduced, transmitted, or utilized in any form by any electronic, mechanical, or other means, now known or hereafter invented, including photocopying, microfilming, and recording, or in any information storage or retrieval system, without written permission from the publishers.

For permission to photocopy or use material electronically from this work, please access www.copyright.com (http://www.copyright.com/) or contact the Copyright Clearance Center, Inc. (CCC), 222 Rosewood Drive, Danvers, MA 01923, 978-750-8400. CCC is a not-for-profit organization that provides licenses and registration for a variety of users. For organizations that have been granted a photocopy license by the CCC, a separate system of payment has been arranged.

Trademark Notice: Product or corporate names may be trademarks or registered trademarks, and are used only for identification and explanation without intent to infringe.

Library of Congress Cataloging-in-Publication Data

Around the patient bed : human factors and safety in health care / [edited by] Yoel Donchin, Daniel Gopher.
     pages cm. -- (Human factors and ergonomics)
Includes bibliographical references and index.
ISBN 978-1-4665-7362-8 (hardback)
  1. Medical errors--Prevention. 2. Patients--Safety measures. 3. Medical care--Safety measures. I. Donchin, Yoel, editor of compilation. II. Gopher, Daniel, editor of compilation.

R729.8.A76 2014
610.28'9--dc23   2013030183

Visit the Taylor & Francis Web site at
http://www.taylorandfrancis.com

and the CRC Press Web site at
http://www.crcpress.com

Contents

Preface xi
The Editors xiii
Contributors xv

Chapter 1   Human Factors and Safety in Health Care 1
Daniel Gopher

Chapter 2   A History of Medical Errors 13
Yoel Donchin

Chapter 3   Types and Causes of Medical Errors in Intensive Care 23
Daniel Gopher and Yoel Donchin

Chapter 4   The Operating Room and Operating Process—Observations 29
Yael Einav and Daniel Gopher

Chapter 5   Mental Models as a Driving Concept for the Analysis of Team Performance in the Emergency Medicine Department 55
Shay Ben-Barak and Daniel Gopher

Chapter 6   Magnesium Sulphate Dosage—Analysis of Problems Involved in the Medication Administration Process 79
Efrat Kedmi Shahar and Yael Einav

Chapter 7   Human Engineering and Safety Aspects in Neonatal Care Units: Analysis and Appraisal 95
Yael Auerbach-Shpak, Efrat Kedmi Shahar, and Sivan Kramer

Chapter 8   Applying the Principles of Human–Computer Interaction to Improve the Efficiency of the Emergency Medicine Unit 119
Nirit Gavish

Chapter 9   Human Factors Contributions to the Design of a Medication Room ... 131
Zvi Straucher

Chapter 10 The User-Centered Design of a Radiotherapy Chart 139

Roni Sela and Yael Auerbach-Shpak

Chapter 11 Examining the Effectiveness of Using Designed Stickers for Labeling Drugs and Medical Tubing 163

Dorit Sheffi, Yoel Donchin, Nurit Porat, and Yuval Bitan

Chapter 12 The Emperor's New Clothes—Design of Garments for the Operating Room Staff 175

Anna Becker

Chapter 13 Thinking Patterns of Physicians and Nurses and the Communication between Them in the Intensive Care Unit 195

Yehuda Badihi and Daniel Gopher

Chapter 14 The Operating Room Briefing 207

Yael Einav, Yoel Donchin, and Daniel Gopher

Chapter 15 Analysis of the Rate of Interruptions during Physician Rounds 217

Yoel Donchin and Meirav Fogel

Chapter 16 How Does Risk Management Differ from Accident Prevention? 229

Yoel Donchin

Chapter 17 Reconstruction to Investigate the Sources of an Event in a Medical System 235

Yoel Donchin

Chapter 18 Development of a Human Factor Focused Reporting System for Hospital Medical Staff on Daily Difficulties and Problems in Carrying Out Their Work 245

Ido Morag and Daniel Gopher

Chapter 19 Patient Safety Climate: Development of a Valid Scale to Predict Safety Levels in Hospital Departments 263

Yael Livne and Dov Zohar

Chapter 20	Beyond Fatigue: Managerial Factors Related to Resident Physicians' Medical Errors 285
	Zvi Stern, Eitan Naveh, and Tal Katz-Navon
Chapter 21	Gentle Rule Enforcement 293
	Ido Erev and Dotan Rodensky
Chapter 22	Human Engineering and Safety in Health Care Systems—What Have We Learned? 299
	Daniel Gopher
Chapter 23	Summary: The Physician's Point of View 313
	Yoel Donchin
Index	 317

Preface

There has been a growing awareness among the general public and the medical professional community of the occurrence of failures and mistakes in health care, from primary care procedures to the complexities of the operating room. Medical personnel and policy makers are desirous for both an assessment and investigation of the problem in order to unveil the root cause to pinpoint the factors and guilty parties, and to create proposals for corrective measures and improvement of the situation.

This book examines the problem and investigates the tools to improve health care quality and safety from a human engineering viewpoint—the applied scientific field engaged in the interaction between the human operator (functionary, worker), the task requirements, the governing technical systems, and the characteristics of the work environment.

The editors' major claim is that the main cause for the multiplicity of medical errors is not lack of motivation or carelessness of care providers, rather it is the hostile and unfriendly work environment confronted by doctors, nurses, and other members of the medical team. The vast majority of health care working environments are not properly planned, nor are they appropriate to the tasks facing team members. They are considerably disadvantaged by the lack of a systemic thought approach enabling the system to carry out tasks in an efficient and safe manner.

The book's chapters are based on a theoretical and practical approach developed by the editors; Yoel Donchin, representing the medical profession, and Daniel Gopher, from the human factors engineering field; the two have cooperated over a period of approximately two decades. Students from the Center for Work Safety and Human Engineering at the Technion in Haifa participated in the research and application activities, together with nurses and doctors at medical centers throughout Israel.

The first two chapters of the book comprise an introduction and discussion of the general approach to the subject matter. The following 19 chapters describe case studies of human factors and safety in medical systems, and research work carried out in hospital wards, operating rooms, emergency units, and pharmacies. These studies, compiled and presented here for the first time, reveal a wide range of problems and weaknesses in contemporary health care, which impair its safety and quality, and increase the workload. Also presented are developed and implemented solutions based upon human engineering components and cognitive psychology, as well as their driving principles and methodologies. It is argued that this approach is a productive and efficient way to significantly reduce the number of errors, leading to the creation of a safe environment and improvement of the quality of health care.

The final two chapters of the book present discussions, concluding remarks, and directions for future activities.

The Editors

Yoel Donchin is a Clinical Professor of Anesthesia and Intensive Care at the Faculty of Medicine, Hadassah Hebrew University Medical Center in Jerusalem, Israel. Since completing his medical training, specializing in anesthesia and critical care, Donchin has held clinical and academic positions in Israel and the United States, where he was a fellow in Obstetric Anesthesia at the University of Florida in Gainesville. Donchin developed the prehospital emergency system in the Israeli defense forces as well as in the civil emergency services and served as the medical director of the national EMS (Mgen David Adom) services. He is the author of three textbooks on first aid and was among the first to propose and publish papers on the use of epidural narcotics for pain relief. As a senior physician in the Intensive Care Unit (ICU) of Hadassah, he noticed the high rate of medical mishaps. This led to cooperation with Daniel Gopher; they have since established a partnership between Hadassah hospital and the Technion Research Center for Work Safety and Human Factors Engineering. Donchin spent 2 years at Technion studying and initiating health care–related human factors research and application work. He then established and became the head of the Patient Safety Unit at the Hadassah Hebrew University Hospital. Donchin's research interests focus on human factors and safety in health care. He also serves as the chair of the Israel Society for the History of Medicine.

Daniel Gopher is a professor emeritus of cognitive psychology and human factors engineering at Technion—the Israel Institute of Technology. He is a fellow of the U.S. Human Factors and Ergonomics Society and the International Ergonomics Association. In 2013, Gopher was awarded the Hal Hendrick Distinguished International Scientist Award by the U.S. Human Factors and Ergonomic Society, for recognition of his outstanding contributions to the field of human factors and ergonomics. Since 1980, he has been the director of the Research Center for Work Safety and Human Engineering, an interdisciplinary research center. In 1996, he also established, together with Asher Koriat from Haifa University, the joint Technion–Haifa University Max Wertheimer Minerva Research Center for Cognitive Processes and Human Performance. Gopher joined the Technion Faculty of Industrial Engineering and Management in 1979, after serving 12 years in the Israel Defense Forces, during which he was a senior scientist and acting head of the Research Unit in the Personnel Division (1966–1970), and senior scientist and head of Human Factors for the Air Force (1970–1979). Gopher's research focuses on the study of human attention limitations, measurement of mental workload, training of complex skills and their applications to the design of aviation systems, medical systems (assessing the nature and causes of human error in medical work, and redesign of medical work environments to improve safety and efficiency), safety at work (developing methods and models for the analysis of human factors, ergonomics, safety, and health problems at the individual, team, and plant levels), and training of complex skills (development of computer-based cognitive trainers and virtual reality multimodal training platforms).

Contributors

In addition to the book editors, most of the chapters in the book were written by faculty members and students from the Center for Work Safety and Human Factors Engineering at the Technion and the Hadassah University Hospital, Jerusalem. Some chapters are based on joint projects with nurses, physicians, and members of the Ministry of Health who participated in the studies. Our thanks to all of our contributors and collaborators.

Yael Auerbach-Shpak
Research Center for Work Safety and Human Engineering
Technion
Haifa, Israel

Yehuda Badihi
Center for Work Safety and Human Engineering
Technion
Haifa, Israel

Anna Becker
Department of Industrial Design
Technion
Haifa, Israel

Shay Ben-Barak
Research Center for Work Safety and Human Engineering
Technion
Haifa, Israel

Yuval Bitan
Hadassah University Hospital
Jerusalem, Israel

Yael Einav
Research Center for Work Safety and Human Engineering
Technion
Haifa, Israel

Ido Erev
Industrial Engineering and Management
Technion
Haifa, Israel

Meirav Fogel
Hadassah University Hospital
Jerusalem, Israel

Nirit Gavish
Research Center of Work Safety and Human Engineering
Technion
Haifa, Israel

Tal Katz-Navon
Arison School of Business
The Interdisciplinary Center (IDC)
Herzelia, Israel

Efrat Kedmi Shahar
Research Center for Work Safety and Human Engineering
Technion
Haifa, Israel

Sivan Kramer
Research Center for Work Safety and Human Engineering
Technion
Haifa, Israel

Yael Livne
Department of Human Services
Yezreel Valley College
Israel

Ido Morag
Research Center for Work Safety and
  Human Engineering
Technion
Haifa, Israel

Eitan Naveh
Faculty of Industrial Engineering and
  Management
Technion
Haifa, Israel

Nurit Porat
Hadassah University Hospital
Jerusalem, Israel

Dotan Rodensky
Research Center for Work Safety and
  Human Engineering
Technion
Haifa, Israel

Roni Sela
Research Center for Work Safety and
  Human Engineering
Technion
Haifa, Israel

Dorit Sheffi
Hebrew University Medical School
Jerusalem, Israel

Zvi Stern
Hadassah University Medical Center
Jerusalem, Israel

Zvi Straucher
Research Center for Work Safety and
  Human Engineering
Technion
Haifa, Israel

Dov Zohar
Faculty of Industrial Engineering and
  Management
Technion
Haifa, Israel

1 Human Factors and Safety in Health Care

Daniel Gopher

CONTENTS

Health Care in the Age of Computers, the Information Revolution, and
Artificial Intelligence 1
Human Factor Engineering Components in Working Environments and
Medical Systems 3
Human Factor Characteristics in Contemporary Medical Systems 5
    Physical and Engineering Components of the Workstation and Its Surroundings ... 5
        Design of Devices and Unit Systems 5
        Design and Layout of Individual Workstations 6
        Design and Planning of Large Workspaces 6
        Recording of Information, Access to Information, and the Transfer of
        Information 7
    Work Procedures and Work Patterns 7
        Activities Are Multistage with Many Variables and Components,
        Operating in a Complex Technological Environment, with a Wealth of
        Information 7
        Effective Teamwork 8
    Data Collection for the Evaluation of Functional Problems and
    Performance of Medical Systems—Why Is It Insufficient to Report and
    Investigate Incidents, Malfunctions, and Accidents? 9
Cognitive Components, Role Perception, Mental Models, and Safety Climate 11
References 12

HEALTH CARE IN THE AGE OF COMPUTERS, THE INFORMATION REVOLUTION, AND ARTIFICIAL INTELLIGENCE

Modern medicine, in the second decade of the 21st century, is a powerful discipline and advancing rapidly, with the ability to diagnose, treat, and cure diseases that not long ago were considered incurable. An amazing revolution has taken place, thanks to modern diagnostic methods, long-term monitoring options, and development of analytical capabilities and intervention. But these capabilities have led to an extremely complex and costly system. Each year, modern industrialized nations invest a very high percentage of their gross national product (GNP) on health expenditure.

The 2010 Statistical Abstract of Israel provides health expenditure data as a percentage of gross domestic product (GDP) for various countries: the United States leads in the percentage of GDP spent on health (16%), followed by Switzerland (10.7%), and Austria and Germany (10.5% each). Moreover, this expenditure is steadily increasing. For example, between 2000 and 2004, health care costs in the United States rose from 13.3% to 15.2% and in Switzerland from 10.4% to 11.6%. At the same time, the number of medical mishaps increased and medical treatment safety began to decline. Lucien Leape and colleagues from the Harvard University School of Public Health wrote a seminal article published by the medical flagship journal *New England Journal of Medicine* (*NEJM*) in 1991, estimating the number of deaths per annum due to medical errors at between 98,000 and 120,000. This is a massive and sobering number, larger than the number of traffic fatalities and the number of deaths from heart disease and cancer. There are those who claim a lower number and those that claim a higher number, but no one disputes that the number of mishaps and fatalities due to mistakes is too high. Thus, the medical paradox was born, so-called in American medical jargon: a system where doctors and nurses have extensive knowledge and expertise, use the most modern and advanced technology available, and have an annual medical expenditure exceeding 15% of GDP and flourishing research—but a high number of mistakes and medical mishaps occur, most of which could be prevented! The question of questions is: What causes this paradox? And, perhaps more importantly—how can the rising rate of errors be prevented?

Another fact that should be taken into consideration: the high cost of advanced medicine prevents it from being available to all, making this a significant ethical dilemma for a modern and democratic society espousing equal opportunity and provision of welfare to the needy. The efforts to find a solution to this problem have been accompanied by an increasing workload on the system to treat a great number of patients in the public health sector and "cutting corners" to reduce costs. The burgeoning workload and limited resources create a conflict that increases the probability of the occurrence of adverse events and mishaps. This is a serious problem endangering the safety of patients and caregivers.

Another major development in the medical care system is the dramatic increase in the technological complexity of the system and the medical work environment.

Moreover, at all times and for all medical treatments, a rapidly expanding information base is created, in part because of the diversity of the medical team as well as the experts from different fields, which characterizes most medical treatment processes.

These facts turn every treatment process into a relay race, in which each baton transfer (patient transfer) must be accompanied by a transfer of responsibility and a transfer of appropriate information. A solitary failure would result in the failure of the whole process.

Referring again to the vast investments in medical systems, we find that the great majority of resources are directed toward investigation and ways of combating illness, toward development of new drugs, and toward development of technologies to improve the quality of life for chronic sufferers, while almost nothing is invested in creating user-friendly systems designed to assist the team in the use of developing technical capabilities, a system designed coherently and based upon human engineering factor principles and suited to the human operator's capabilities and limitations. Therefore, as mentioned above, from a human engineering viewpoint, those providing medical services are required to function in an unfriendly, poorly planned environment, which places a load on the shoulders of personnel, limits efficiency, and most importantly, is extremely susceptible to failures. In our view, these are the main causes for failures, not "medical carelessness" and not lack of motivation. Here is an opportunity for the intervention of human engineering personnel. The good news is the recent growing awareness of human engineering factors both as far as the general public is concerned and in terms of allocation of resources into the areas covered by this book.

HUMAN FACTOR ENGINEERING COMPONENTS IN WORKING ENVIRONMENTS AND MEDICAL SYSTEMS

The relationship between the worker and his tools or working environment—the essence of human engineering—did not happen in the latter part of World War I as claimed in many textbooks. Its beginnings can be found beneath the earth's surface in descriptions of the effect on miners of working in mines. In an extensive monograph from 1552, the mineralogist Georgius Agricola describes the many sufferings of mine workers in Italy. Eleven years later, the Swiss physician Paracelsus wrote a book about mine workers and described the toxic effects of inhalation of mercury vapor. In other words, there is a link between the workplace and the worker's health, well-being, and capability.

In 1700, the Italian doctor Bernardino Ramazzini (Photo 1.1) published the comprehensive "Lecture on Worker Diseases," describing almost all trades and professions existing at the time, beginning with diseases suffered by the workers, from lung diseases of silversmiths to a description of the more familiar insufferable wrist pains borne by writers using quills dipped in ink. In 1857, the Polish scientist Wojciech Jastrzębowski published a paper defining ergonomics, the science of work, detailing the advantages of applying science to improve the lot of the worker and his labors. The paper was published in Poland without any significant reaction at the time.

In the latter stages of World War I, it was found necessary to increase productivity. The workforce, who had been exploited to the limit, was unable to meet the tasks at hand. As a result, a committee was set up comprised of physiologists and psychologists, an initial task force named *The Committee for the Well-Being of Workers in the Arms Industry*—with the stated objectives to increase productivity without the loss of manpower. At the end of the war, the name was changed to *The Committee for Investigation of Worker Fatigue*, indicating that the root problem had already been identified (fatigue). The investigating team was complemented by experts from other fields, such as lighting experts and physicists, focusing on the working man and the interaction with the workplace. In fact, subsequent wars provided enormous momentum to the research and heightened the relationship to human engineering factors.

BERN. RAMAZZINI
In Patav. Archi-Lycæo Prof. Publ.
DE
MORBIS
ARTIFICUM
DIATRIBA.
ACCEDUNT
LUCÆ ANTONII PORTII
In Hippocratis librum
DE VETERI MEDICINA
PARAPHRASIS;
Nec non ejusdem
DISSERTATIO LOGICA.
EDITIO SECUNDA.

ULTRAJECTI,
Apud GUILIELMUM van de WATER,
Academiæ Typographum. 1703.

PHOTO 1.1 Bernardino Ramazzini's 1703 book on workers' diseases.

Human factors engineering is the scientific field engaged in the application of knowledge of human capabilities and limitations in the design of engineering systems, instrumentation, machinery, and working environments to meet the capabilities of the operator. The field also deals with the design and formulation of work processes, to enable efficient and safe operation of systems under controlled workloads. The key function of human engineering is the estimation of the measure of compatibility between the functional requirements and the ability of the worker charged with carrying out the function. This estimation is made by detailed analysis of the tasks, assessment, and definition of engineering design components, and the environment and work process for a given task, in terms of the requirements and effect upon the worker. The results of this analysis are compared with existing knowledge and assessment of the worker's capabilities and his level of knowledge and experience in handling the demands of the task.

Figure 1.1 illustrates this approach in a flow net that depicts the interaction between the various components: the task at hand, characteristics of the engineering system, and workplace features that together determine task requirements and the required level of implementation. The worker's capability, his skill, and his level of training will determine the measure of his ability and unsuitability to meet the demand of the task at hand. Unsuitability may be signified by a number of implications: deterioration in task implementation, delayed reactions, growing inaccuracy, and increasing workload due to failures, mistakes, and workplace accidents: the greater the mismatch, the more serious the problem.

Unsuitability can be treated in three main ways:

Human Factors and Safety in Health Care

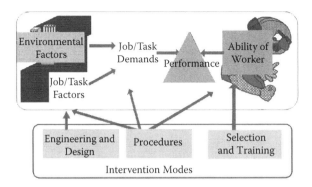

FIGURE 1.1 The interventions between system components and operator capabilities.

1. Engineering design changes or redesign in order to improve system suitability for the operator.
2. Changes to the work process and its formulation so that the worker can meet the work rate and deadlines and handle satisfactorily the task at hand.
3. Screening of workers and their training in order to match the task requirements.

Engineering design and work process and formulation planning are part and parcel of the work of human factor engineering. Screening and correct manpower selection, training, and practice come mainly under human resources development.

HUMAN FACTOR CHARACTERISTICS IN CONTEMPORARY MEDICAL SYSTEMS

From a human factor engineering viewpoint, modern medical systems are complex entities embracing engineering and human subsystems required to operate in unison with a very low level of allowable failure, where the price of a mistake, even a minor mistake, is liable to be very high. The general pattern in Figure 1.1 enables identification in the medical system, from a human factors engineering viewpoint, of a range of components that generate a complex array of worker requirements. The intervention paths shown in Figure 1.1 denote required operation directions: adoption of engineering steps and at the same time, creation of the framework and procedures applicable to functional implementation. The main components of each of the aforementioned viewpoints are detailed below.

PHYSICAL AND ENGINEERING COMPONENTS OF THE WORKSTATION AND ITS SURROUNDINGS

Design of Devices and Unit Systems

All medical staff members use a large number of health-support devices and systems to carry out their work: monitoring devices, surgical tools, various imaging devices, intravenous infusion systems, respirators, dialysis equipment, and so on. Design and planning of all these instruments impact work efficiency, operating speed, and response—level of awareness to mishaps. Correct design for easy usage and operation in the working environment at the patient's bedside reduces the workload. Poor design increases the load. For example, when the monitoring device literature or indicator is clearly visible even from a distance, discernment becomes easier and the burden on the employee is reduced.

Design and Layout of Individual Workstations

Medical systems feature many workstations where an individual member of staff provides treatment for an individual patient, while operating a large number of systems, such as the anesthesiologist's station in the operating room, the patient's bed in intensive care, as well as the dispensary and life support incubator for a premature baby. All these are personal workstations, typically encompassing a large number of custom designed instruments, not part of a complex system. In many cases, this leads to a lack of consistency and compatibility, and major workstation complications. The overall workstation system, including location of the various components, their arrangement and degree of compatibility, and the functioning of the employee in the given workstation is of vital importance. The best-known integrative and coherent workstation model is the flight cockpit, the result of considerable thought, both at a general level and from the planning perspective of human engineering factors. This begs the question, is it possible to design and install similar criteria in the planning of medical workstations?

Design and Planning of Large Workspaces

In many cases, personal workstations (Photo 1.2) are part and parcel of larger care units—for example, the operating room complex in large hospitals, recovery room, the emergency medicine department, and so forth. These units have a large variety of

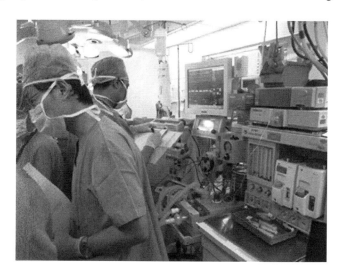

PHOTO 1.2 A medical workstation.

diverse equipment and staff who have to perform multiple tasks and take responsibility in different fields. For example, a hospital's ward has many beds surrounded by a given assembly of ancillary equipment, similar or different, in a particular order. The medical staff moves between these subunits to carry out their work. Major importance is attached to the planning of large units and their proper arrangement. Proper planning together with optimized and consistent internal arrangement increase work efficiency, reduce the load, and increase safety.

Recording of Information, Access to Information, and the Transfer of Information

Updated information is the most important component of today's medical work environment and its proper planning is one of the most important tasks facing human factor engineering in medical systems.

The information element holds a special status in today's medical environment—information is an integral part of diagnostic work, assessment of patient status, and the physician's decision making and that of other medical staff. Today, medical information (health informatics) is a major issue in every medical center and medical insurance system. However, although this is an ever-growing field, doctors and nurses are still struggling with the many requirements associated with making medical records and reports regarding the current patient. Staff are drowning in the proliferation of forms to be filled in, files to be organized, and computerized information systems that are not consistent with work claims in a medical department and they waste (or try to devote) precious time to delving through the various forms to extract essential information. An engineering solution, based on human engineering principles, could lead to a significant savings in time and enable the creation of a clearer situation report, enabling improvement in the quality of care and helping to prevent mishaps.

WORK PROCEDURES AND WORK PATTERNS

The importance of work procedures and work patterns in health care environments stems from two key features of these processes:

Activities Are Multistage with Many Variables and Components, Operating in a Complex Technological Environment, with a Wealth of Information

Multistage and multifunctional activities that are not well organized lead to confusion, a prevalence of stages, and many mishaps. Close supervision and constant monitoring are required and precious time is wasted in transition from stage to stage, especially in a tight and demanding schedule. This description fits the lion's share of medical activity and is similar to the functional stages of flight management. In both cases, health care and flight, the key to efficient performance lies in the possibility to split the entire function into a large number of intermediate tasks with definable objectives, definable substages, and a fixed order of measures to be implemented for each substage. The strength of this procedure is that it clarifies, defines, and organizes each of the functions unequivocally. The implementation format can be taught, learned, and assimilated until it becomes automatic, like changing gears when driving a manual transmission car—an action performed without any mental effort. Correct planning and design of processes and work patterns increase efficiency, enabling acceleration of execution of tasks, limiting the misleading options, and reducing the load. The aviation sector adopted these methods many years ago; the health care community, even if its tasks are more complex than those of a pilot, has yet to recognize this and has not yet adopted a significant number of these principles.

Effective Teamwork

Health care operations are mainly characterized by teamwork. A *team* may be on a simultaneous or serial basis in a trauma unit or a surgery department where the work of doctors, nurses, and other caregivers is coordinated simultaneously for the same patient. In many cases, one can define a team and teamwork in situations where staff do not work simultaneously, such as doctor rounds unaccompanied by nurses or shift changes. Incompatibility can occur when a patient due for an operation is hospitalized in one department, is operated on in another department, and comes into contact with other staff in the recovery room; each of these caregivers perform part of the treatment. In many cases, the same patient may be treated by many medical staff members who are not present simultaneously but are part of a single treatment process. Development of an appropriate framework is the most important challenge facing medical system designers. Key factors for such development are:

- Each team member must know the general purpose and the function's specific target or the whole procedure.
- Division of roles and areas of responsibility among team members should be clearly defined.
- Each and every staff member is committed to understanding his role and responsibilities, but must also be well aware of the activities of other members of the team and the definition of areas of responsibility.
- Each and every member of the team should be updated on the situation report of the patient: a current and correct image allowing an inclusive and extensive picture. This will enhance coordination between members of the team.
- If there are stages and no simultaneous caregivers, it is necessary to transfer information based on obtaining clear responsibility at each transition stage.

The five components of human engineering work featured in the engineering system and the physical working environment, and the two components of work procedure and pattern design are witness to the breadth of human engineering work in planning the human environment in general, and in the medical work environment, in particular, as well as the broad spectrum of issues involved. Naturally, the level and quality of design will have a crucial influence on the effective functional performance of the staff member, his safety, and the load level of the given role.

Representation of the scale of the issue and its components provides a guiding framework for data collection of the operational system and the quality of its performance. Appropriate data collection, evaluation, and lessons learned is one of the most important and at the same time, most problematic paths to the functional improvement of medical systems and the reduction of error rate in medical treatment. The following is a brief discussion highlighting the principal difficulties and their various aspects, the central theme in Chapter 18 deals with the development of complementary systems for reporting difficulties, risks, and problems in medical practice; the subject is also an important methodological constituent in many other chapters of the book.

DATA COLLECTION FOR THE EVALUATION OF FUNCTIONAL PROBLEMS AND PERFORMANCE OF MEDICAL SYSTEMS—WHY IS IT INSUFFICIENT TO REPORT AND INVESTIGATE INCIDENTS, MALFUNCTIONS, AND ACCIDENTS?

Recognition of the existence of safety problems in health care and the existence of medical mishaps, awareness of the fact, and growing public criticism led to a growing trend in the investigation of mishaps and accidents, as well as documentation, recording, and reporting of mishaps and events deviating from the norm. A growing number of countries and large-scale medical systems have adopted mandatory reporting and reporting systems have been established, requiring medical staff to report eventful incidents, accidents, or deviations occurring in the course of care activity. Investigative and test committees were established with growing frequency whose purpose was to inspect and investigate the cause of occurrence of incidents and accidental events.

Documentation of incidents and their investigation undoubtedly makes an important contribution to public awareness of the importance of the issue and places pressure on the decision makers and professional community to change the state of affairs. As a result, considerable resources are being invested in documentation of mishaps and report collection.

An important question we wish to pose at the outset: is the investigation of errors and accidents the only source of knowledge, or even the primary source, for data collection to satisfy the needs of human factors engineering? This question was the subject of a paper by the chapter author, published in 2004 in *Biomedical Instrumentation & Technology*.[3] The paper concluded that although major public importance is attached to gathering as much information as possible on errors and accidents, as a scientific database designed to steer and guide an orderly human factors engineering work program, there was little value to be gained. This conclusion stems from a number of key characteristics of this knowledge base:

> *Limited representation*: A major problem in relying on the investigation of errors and events deviating from the norm is limited reporting—a very small percentage of errors and events are reported, estimated between 5% and 7%, notwithstanding the large and growing investment in this area. This estimate does not differ from the estimated percentage of reported errors in civil aviation, despite the many years and the traditional care taken regarding flight safety.
>
> Even when there is a report, in many cases the report is incomplete or biased, this being an obvious fact. Complete and accurate reporting of

errors may be in the cause of "public service" but it also entails the possibility of discomforting personal results, imposition of personal liability, or the need to blame work colleagues or the reported employee's work unit or exposing them to trying procedures. Moreover, the report may be followed by an investigation or inquiry, taking up time and trouble, beyond the scope of regular reporting. Due to the low rates of reporting, sample reports are too small and biased, and therefore difficult to treat as representative scientific data to be used as the basis for a fundamental change in the system.

Quality of reported information and its completeness: Another problem also affecting the quality and validity of the reported material is that the reported errors are personal reports based on the reporter's memory. As in the case of eyewitness testimony in a trial, the quality of the reported material also depends on the quality of the reporter's memory and personal biases and preferences related to the reporter's personal perspective and involvement in the reported occurrence. A substantial and rich literature in cognitive psychology deals with biased reporting and impaired quality of eyewitness reporting and there is no reason to assume medically reported errors are not influenced by the same factors and biases.

Absence of reference base: Another obstacle to the ability to assess error and accident reports, to interpret and afford them proper weighting, is the absence of an appropriate reference base that would allow assessment of the relative frequency and severity of the reported problem. In most cases, if not all, reports of provision of incorrect medication, errors in filling out forms, incorrect medical procedure, calculation or incorrect reading of data (from a label or report) are not comparable to the frequency of similar operations performed properly and without error, as such a database simply does not exist.

For example, in a study carried out in a respiratory intensive care unit, Yoel Donchin, Daniel Gopher, and colleagues found a similar frequency of errors committed by doctors and nurses daily, even though the nurses were responsible for about 87% of the total activities in the unit. That is to say, the contribution of doctors to the number of errors was similar to the contribution of nurses, although the ratio of the scope of activity between the doctors and the nurses was about 1:7.[4]

This result clearly illustrates how the interpretation and the significance attributed to error reports in the absence of a basis of comparison capability may be flawed, and how impossible it is to achieve such a reference base for each individual case.

Wisdom in hindsight: One of the most difficult problems in investigating errors and relying on error reports as the basis for recommendations for changes and improvements is that interpretations and explanations of errors derived from their investigation result from wisdom in hindsight (*post factum*). This is one of the most serious risks inherent in drawing conclusions and their interpretation scientifically.

For example, a person who took office after working years or maintains lengthy contact with his/her spouse, if asked how he got his job or selected his spouse, will relate an orderly and logical story and recall, step by step, the sequence of events and decisions that led him to where he is today. This is wisdom looking backwards—wisdom in hindsight.

However, at each stage, if the same person was asked to predict the chances of eventually finding himself at the end of the process at the place where he is at a given moment (position, partnership), the predictive value would be reduced dramatically, if not disappear completely. For every step and point of decision the number of options is large and level of imprecision vis-à-vis all options is too high to allow a valid and reliable prediction. This is what is meant by wisdom in hindsight: the power to offer a logical and ordered explanation for the current state or occurrence but without any predictive ability, enabling prediction of the final outcome in the early stages.

This is an obstacle and a major difficulty in exploiting error investigation material as a basis for scientific work. Science aims at predicting future results from preconditions or basic characteristics. The ineffectiveness of such predictions negatively affects the ability of action and usefulness of the information collected.

Passive and reactive approach: Error and accident reports and the steps taken following their investigation characterizes a passive and reactive approach in attempting to make corrections and is not actively preventive or forward looking. Use of error and accident reports as a principal basis for system guidance in repairing a problem infers that the system changes the operation and performs fault correction after the occurrence, the "putting out fires" method, which does not take preventive action to reduce in advance the probability of accident occurrence. Relying exclusively on error reports and accident investigations as a principal source of information represents implicit adoption of this approach, minimizing active efforts to reduce errors and accidents by preventing their occurrence. For all the reasons listed above, it is absolutely clear that human factors engineering work and improving health care safety cannot rely on error reports and investigation as a central repository of information, as it is insufficient. The creation of an appropriate database, directed at the various issues discussed above, and building appropriate tools to collect, analyze, and interpret the information, are cornerstones and counted among the important challenges of human factors engineering activities in coping with improvement requirements and redesign of medical systems and methods. Various aspects of this topic will be discussed in later chapters.

COGNITIVE COMPONENTS, ROLE PERCEPTION, MENTAL MODELS, AND SAFETY CLIMATE

The main thrust of human factors engineering has been in the areas of engineering design, physical work environment planning, definition of procedures, and patterns of work, although the roots of the working group, whose work is described in various chapters of this book, is also fixed on expansive cognitive psychology, decision-making processes, and organizational psychology. These areas touch on the actual functioning skills of medical teams and the general behavioral style in the hospital wards or other medical work units. Earlier in this chapter, we discussed some issues regarding effective functioning of the team and work procedures. These issues are examined more broadly when considering the overall functioning of the unit; how different functionaries perceive their role and the major impact of their responsibilities on performance of their tasks and also on subjects of communication with other team members and ways and means of communication. Moreover, not only is role function perception important; but also it is important how functionaries perceive the role and responsibilities of others with whom they collaborate.

Communication between doctors and nurses in work teams and hospital departments is a painful and familiar problem. It can be shown that the origins of this problem can be found in role perception—each and everyone to their function and that of the other person.

Similarly, one can consider the mental model where the operator has a role in the process he is involved in. If the process is complex and multistage, devoid of work procedures and defined and clear work practices, it is likely that each functional subject will initiate its own mental model process, a model that organizes its work process in general and its work in particular. For example, doctors and nurses in the emergency department often differ in their perceptions of the patient flow process and stages of treatment. Investigation of the suitability of functional concepts and mental models of functionaries is an important issue in understanding the work processes and narrowing gaps. In any discussion on safety and safety behavior, general attitudes and general perceptions of team members on the critical nature of the subject and their willingness to invest in it are of major importance.

These variables complement the complex issues and variables that were discussed in earlier sections and contribute to overall safety and effective care. The book's chapters deal with different aspects of these issues.

REFERENCES

1. Leape L.L., Brennan T.A., Laird N., Lawthers A.G., Localio A.R., Barnes B.A., Hebert L., Newhouse J.P., Weiler P.C., and Hiatt H. 1991. The nature of adverse events in hospitalized patients. Results of the Harvard Medical Practice Study II. *N Engl J Med.* February 7;324(6):377–84.
2. Agricola G. 1912. *De re metallica*, trans. Herbert C. Hoover and Lou H. Hoove. London.
3. Gopher D. 2004. Why it is not sufficient to study errors and incidents: Human factors and safety in medical systems. *Biomed Instrum Technol.* 45:387–91.
4. Donchin Y., Gopher D., Olin M., Badihi Y., Biesky M., Sprung C.L., Pizov R., and Cotev S. 1995. A look into the nature and causes of human errors in the intensive care unit. *Crit Care Med.* February 23(2):294–300.

2 A History of Medical Errors

Yoel Donchin

To err is human.

Hieronymus

Then I shall be blameless
and innocent of any great transgression.

Psalms 19,13

Large hospitals are centers of day-to-day activities involving thousands of people, health requirements, and those striving to meet those requirements. Today, it is commonplace to replace the title of *hospital* with more positive sounding names, such as *Medical Health Center*. At the entrance to such a center one can find shops, cafes, as well as bank and post office branches. Numerous companies offer inpatients and care seekers home comforts in the form of a personal TV and private phone connection. Eye-catching signs direct newcomers to the many institutes, laboratories, and clinics, with slow-moving elevators to convey the needy to all parts of the hospital.

Many of these medical centers began life as small and simple structures. Over time, wings and new buildings were interconnected as one unit by means of complicated passages, staircases, and endless aisles.

A large hospital with a capacity of about 1,000 beds may employ a staff numbering some 2,000, from maintenance personnel, providing a constant supply of electricity and medical gases, to administrative personnel, who collect payments and manage the data collection system and prompt distribution among the various users—and in-between, nurses, doctors, and other relevant professional staff.

The medical center is to all intents and purposes an industrial plant in all respects: at the start of the "production line" are patients prior to medical diagnosis and finally (hopefully), after a series of manual and sophisticated instrumental tests, the appropriate diagnosis and treatment are determined.

This plant differs significantly from its early predecessors, that is, hospitals at the onset of the 19th century or institutions that served as poorhouses and charitable institutions rather than medical or healing care centers. In fact, the concentration of patients in these institutions increased the risks of mortality, from infections, joint use of contaminated tools, and poor hygiene. Giving birth in the home was a far safer proposition than in the maternity ward. The doctors in the selfsame "hospitals" were helpless against prevailing disease and suffering. Until 1846, for example, easing of pain after surgery was not possible, as anesthesia as we know it today, was unheard of. Moreover, surgeons at the time did not keep to the rules applying to disinfection, as the causes of infectious diseases had not yet been discovered and the patient's chances of surviving surgery were very slim.

Amazingly and paradoxically, it seems that a patient hospitalized in today's sophisticated medical center is prone to risks that are no less severe than the cause of the patient's hospitalization.

Medical systems have not paid sufficient attention to those factors that can cause mishaps, as they are immersed in the implementation of medical research successes. New surgical methods have been developed—but the team itself has never been tested, nor have heart surgeons been monitored, although they stand for hours on their feet, engrossed in a narrow upheaval-prone surgery; nurses and doctors have never been tested, although they are exposed to noise in intensive care: this goes as well for ambulance drivers, laboratory workers, floor cleaners (the most common accident in a hospital is an employee or a patient slipping on a polished floor), and cafeteria staff. A long period of time elapsed before consideration was given to hospital safety issues.

In 1940, a patient would receive one to three drugs, while today, the figure is closer to 10 to 15. Not all drugs are prescribed by the same doctor and some drugs may cancel out the actions of others. In order to be precise in the dosage of powerful drugs, use is made of supposedly precise instruments, but slight deviations in their calibration are likely to elicit unwanted or even dangerous patient reactions.

Diagnosis requires imaging of internal organs by various means—X-rays, injection of radioactive material, sound waves, magnetic fields, and so forth. Any deviation, even a small percentage, and the process could be harmful rather than beneficial. The gap between a harmful dose and a beneficial dose could be extremely narrow. The following is a recount of a medical event:

> Until 13:20 today, 57-year-old Mr. Haim Ratson has been in good health, feels good, and has never needed a doctor. Mr. Ratson is a passive sports fan (football, going to a match once a week). His parents are still alive. He is happy in his marriage and eagerly awaiting a third grandson due in a month. However, for a few hours after lunch, he felt pain in his chest. Initially, he blamed the spicy soup (this reaction of rejection of physical symptoms characterizes 60% or more of patients), then went to lie down and tried to forget the pain, but the pain stubbornly refused to disappear; on the contrary, it increased, causing undue distress. A few hours later, his wife intervened and despite his protests contacted MDA (Magen David Adom—Israel emergency medical services) and called for help.
>
> The person receiving the call addressed it with all seriousness, requesting Mr. Ratson to remain in bed, adding that help would arrive soon. Indeed, a few minutes later a doctor arrived accompanied by two men carrying bags and various instruments. They measured Haim Ratson's blood pressure, attached him to a device that recorded his heart activity and then, without further ceremony: "to the hospital!" They sat Haim on a chair, attached a greenish tube to his nose, tickling the inside of his nose with a thin stream of pure oxygen, descended the stairs that Haim had gone up and down thousands of times without any problem, carefully

lowered him into the ambulance, setting off immediately to the hospital, while honking and flashing to clear a path and not get stuck in traffic.

The hospital's emergency room staff was already waiting for Mr. Ratson, as the paramedic and MDA control center had announced their impending arrival with a "suspected heart attack." A sympathetic nurse was standing by to help Haim Ratson off the stretcher and on to the pre-prepared bed, which although very narrow was definitely more comfortable than the stretchered ride in the ambulance. The chest pain was less bothersome than before, probably thanks to the pill given to him in the ambulance ("Chew slowly," they said and placed in his mouth a small and somewhat bitter pill).

The nurse repeated blood pressure measurement, recorded the pulse rate, and took his temperature. Meanwhile, his clothes were removed and a needle injected in his arm. If that were not enough, he was asked to urinate into an odd-looking bottle from which a sample was taken and replaced in a test tube labeled with his name. The test tube was placed alongside other test tubes containing blood. A smiling and courteous male nurse attached his arms and legs to a device "for recording the heart's electrical activity" (Mr. Ratson never gave a thought to his heart's electrical activity, which until now had never disappointed him). When the test was complete, the smile disappeared from the male nurse's face, while muttering "I need to show the chart to Dr. Etgari."

When Dr. Etgari arrived, with an impressive stethoscope around his neck, he looked at the chart and proceeded to fire a volley of questions, one after the other: When did the pain start? Where did it progress? When were you last hospitalized? (Never been hospitalized.) What medications are you taking? (No medicines.) What are you sensitive to? (Not sensitive—only sensitive parts, LOL) What were the causes of your parents' death? (Both are still alive, wishing them long life.) Following the series of questions after auscultation of the heart and lungs, Dr. Etgari reaches a diagnostic assessment: myocardial infarction.

"Mr. Ratson," Dr. Etgari addresses Mr. Ratson, "All the signs indicate that you suffered a heart attack. At the moment you're in a good state of health and are receiving all the accepted treatment aimed at preventing any further damage. The arteries that supply blood and oxygen to the heart muscle are slightly clogged" (every sentence containing bad news is accompanied by soothing words, both for the patient and the doctor). "Today we can diagnose and rectify the situation in a 'single blow' (a really serious blow). To do so we will now transfer you to the Catheterization Institute for Angiography—or imaging of the coronary arteries, an uncomplicated test, lasting about an hour. Your heart will be injected with a dye (just hearing this can bring on a heart attack) to reveal the condition of the arteries. If there is a blockage, it can be unblocked and then you'll feel like a new person." Did he have a choice? Mr. Haim Ratson signed a consenting form, agreeing to the abovementioned medical procedure being carried out on his person. Before being transferred to the catheterization room he handed his wristwatch and car keys to his wife, as well as giving her precise instructions for arranging all outstanding personal matters. For whatever reasons, Mr. Ratson was calm and certain that everything would be fine (probably due to the morphine administered in the ambulance to relieve any pain). Now he's on his way to the second floor, the location of the Heart Institute and Catheterization Unit.

Mr. Ratson is admitted by a sympathetic nurse to a scene right out of a science fiction movie: huge machines and monitors everywhere. Dr. Lotem, the doctor responsible for the catheterization procedure, clad in a green robe with a mask covering most of his face, introduces himself and explains: "We shall first anesthetize the point of insertion of the catheter. You will momentarily sense a heating inside of you, but this will pass quickly. The test is not painful." Mr. Ratson was able to observe his pulsating heart on a monitor while receiving an explanation from Dr. Lotem at each stage of the procedure.

At the end of the test (not painful? not exactly), Dr. Lotem informed him in all solemnity: "Mr. Ratson, we have succeeded in opening up the blockage. You are healthy man and in a few days you will be able to return home. In fact you have not suffered myocardial infarction, but the start of myocardial infarction, and heart muscle was not damaged."

Haim Ratson was transferred from the Catheterization Room to the Recovery Room, where his wife was waiting for him. The next morning he was transferred to the Internal Medicine ward, where Dr. Lotem, gently and without any pain, removed the catheter that had been left overnight in his groin artery.

A week later, Haim Ratson underwent a checkup in the cardiology clinic. His file recorded that he follows a lifestyle appropriate to his condition, that is, to do everything necessary to prevent infarction: frequent checking of blood pressure, consumption of permissible foodstuffs only, otherwise, the rest—as usual.

A story with a perfect ending—a prompt rescue operation, efficient medical services, a well-oiled system, with all the links in the recovery chain operating in maximum harmony.

However, events can take a different course. The selfsame Haim Ratson, who was never ill or needed a doctor, is struck by a sudden chest pain and calls up Magen David Adom. The duty officer, loaded with work and dealing with cries for help, listens to the complaint. His reading of the situation is that this is not a serious condition, having participated in a brief 40-hour course, and dispatches an ambulance with a paramedic and driver to Haim Ratson's home. As the address given was not clear enough, the ambulance was delayed and 20 minutes elapsed before its arrival at the home of the patient. The paramedic realizes immediately that the situation requires summoning the intensive care unit ambulance and informs the MDA call center accordingly. Ten minutes later a doctor and intensive care team arrives. The ambulance's electrocardiogram (ECG) machine is not working due to a technical problem, however, based on clinical indications, the doctor concludes that Mr. Ratson is suffering from a heart attack and gives directions to transfer him to the duty hospital. After several attempts to contact the emergency room charge nurse failed, the doctor decides not to waste any more time and sends the ambulance on its way to the hospital. Mr. Ratson is taken to the overloaded emergency room. Fortunately, he is given a comfortable bed, where he must wait a long time for a doctor. About an hour later a nurse comes to him, holding a syringe, containing a yellowish liquid: "Mr. Bialik! Your shot is ready!"

"But I'm not Bialik."

"Oops…" She smiles apologetically and goes in search of the appropriate patient.

The chest pain intensifies. Mrs. Ratson went to arrange hospitalization; Haim Ratson lies alone in bed, unable to ask for help. Luckily for him, the replacement duty team arrived and a team of doctors, on arrival at his bed, were amazed to discover that an ECG test had not yet been carried out, that Mr. Ratson was experiencing chest

A History of Medical Errors

pain, that he was covered in a cold sweat and recording a rapid pulse rate. The team decides to act quickly. (At this point, time is of the essence!) Mr. Ratson is given an aspirin and intravenous pain relief ("What dedicated treatment and how good to be hospitalized here...") muses Mr. Ratson before his consciousness becomes hazy and a morphine-induced tranquility takes over his body.

Mr. Ratson's ECG chart shows clear signs of myocardial infarction and the team decides to transfer him immediately to the catheterization unit, but the room is currently undergoing maintenance operations and two patients are still waiting their turn. The catheterization procedure is set for later. Later that night the doctor informs Mr. Ratson that his blood vessels have opened up and there is no further risk to the heart muscle, however, due to the damage already caused, he will have to be hospitalized for several more days in the internal medicine department. Meanwhile, he can remain in the recovery room of the catheterization unit. Later, he is transferred to Internal Medicine. Several drugs are piped intravenously into his body via the groin artery. The catheter, when inserted, provides a system. At two in the morning, Mr. Ratson felt unwell and the bed was wet. After pressing the alarm bell and intervention of roommates, aroused from their sleep by his moaning and confused state, a nurse arrives and discovers that the catheter has removed itself from the groin and Mr. Ratson has lost two units of blood. Activity now moves into a higher gear, and so on and so forth. All events described are based on actual occurrences.

* * *

What is the probability that a patient undergoing hospital treatment will suffer harm due to error or medical negligence, due to mistaken medication, due to a respiratory device, or automatic syringe malfunction? Is it possible to check these questions? Is it possible to reach a quantitative calculation?

Dr. Lucian Leape, an epidemiologist from Harvard University, calculated the number of casualties from the mishaps and failures in U.S. hospitals.[1] His research examined thousands of cases of hospitalized patients in the State of New York. For every case in the records suggesting a failure or mishap, a thorough investigation was carried out of hospitalization procedures and of the incident itself.

In 15% of the cases, investigators found serious failures, including failures that led to fatalities. When Leape calculated the ratio of the number of hospitalized patients in the United States and the number of fatal mistakes, he found that each year 98,000 inpatients were likely to die in the United States. This number left a strong impression on the general public, especially as Leape calculated and discovered that this figure is equivalent to the number of victims of two large passenger plane disasters, including all the passengers!

With the publication of Leape's research, medical groups and associations awoke to the need to investigate and seek an explanation for why leading professionals—doctors, nurses, laboratory workers, and others—make mistakes.

An initial conference on this subject was held in 1996 at the Annenberg Center in Rancho Mirage, California. All those who had spoken openly about the occurrence of mistakes and investigated the phenomenon were invited to speak, about 160 researchers in total, most of them well informed. During a break in the proceedings, following the initial reviews and presentation of papers, a long table covered by a green tablecloth was placed on the stage displaying images of 7-year-old Joshua who died in the course of a simple operation to correct a defect in the inner ear.

When participants of the following discussion came onto the stage and their names and roles read out, a hush fell over the hall. They included the child's parents, the doctor who treated him, a representative from the insurance company, the family's lawyer, and nurses who initiated the investigation. The discussion was conducted by a well-known television broadcaster. To begin with, the realization of the mishap was discussed, what ensued, and what steps were taken as a consequence—a minor error in marking the drug drawn into the syringe and given to the surgeon resulted in the death of the child.

Each of the participants explained their part in the disaster. The nurse, whose job it was to prevent malfunctions, described how she entered the operating room after the disaster and gathered together all the syringes that were on the table, the needles, and surgical instruments, and how the empty syringes were transferred for examination to the central laboratories in Atlanta, and how the cause of the child's death was discovered—one of the syringes contained a high concentration of adrenaline that had not been properly diluted. The surgeon explained how he requested the appropriate syringe to anesthetize the area of the body to be operated on, the operating room nurse demonstrated how she held the flask containing the material, and the nurse who took part in the operation and received the syringe explained why she did not see the label. The father explained how he was dealt with in the hospital and who presented him with all the material relating to the investigation.

Such an event was unprecedented in the United States—mishaps were brushed under the carpet, as the legal ramifications hung heavily over all those involved in the mishap, and publication of research into the whole subject of mishaps was generally denied. The general public was made aware of the extent of the epidemic by specific cases involving celebrities, such as artist Andy Warhol, who died as a result of an overdose of fluids after gallbladder surgery. The most prominent case was that of Betsy Lehman, health columnist for the mass distribution daily newspaper, *The Boston Globe*.

> Betsy was diagnosed with breast cancer and hospitalized at the Dana Farber Cancer Institute, a facility specializing in these types of diseases, one of the leading and most famous ones in the United States. In the course of her treatment, Betsy was treated with "Taxol," a powerful drug; but because of a miscalculation, she received a dose 100 times greater than that required. No one at the hospital responded to her request to recheck the dose or listened to her husband who was telling people that Betsy's body was responding violently to the drug. In the end, Betsy succumbed and died. Friends who discovered the error brought the matter to the attention of the general public. Many leading newspapers in the United States published the story. Suddenly the public began asking questions: How could such a thing happen? Why does a leading hospital, engaged in research and promotion of treatment of cancer patients, slip up at the final stage of the treatment, submitting the medication to the patient?

A History of Medical Errors

At the second conference on the prevention of medical mishaps, Betsy's mother asked to say a few words to the conference. Her emotional speech was heard by hundreds of people, who meanwhile joined the circle of researchers and those striving to prevent mishaps. This was her message:

A Message from the Heart

As just one member of a family among many, many family members forced to cope with the loss or disability of a loved one due to medical error, I commend the convening of a national symposium on patient safety, an extremely important issue. The advancement of this goal, while still in its infancy, through cooperation and knowledge, is to be congratulated and has been wanting for a long time.

However, allow me to ask you to take a moment's pause from your important work. In departments away from your busy conference rooms can be heard the murmuring of patients who died due to a doctor's fatal error or were harmed at the hands of a trusted medical system. They cannot be present at your national symposia devoted to the welfare of patients. They include my young and brilliant daughter, taken from us suddenly six years ago due to an error in the medication she received from the medical staff at the Dana Farber Hospital Cancer Institute in Boston. Betsy, a mother of two, was ironically the health columnist for the popular *Boston Globe* newspaper.

Every day, modern medical treatment is saving the lives of the sick and people in pain. However, patients and medical practitioners alike understand that stringent consideration must be given to the actions taken in the struggle with death and suffering, which at the same time are both powerful and very delicate. The challenge is to improve the safety of future patients who are to be cured or saved.

Accusation of guilt as a remedy for errors ignores the duality of interests in our medical system. Unfortunately, our family became aware of this tension firsthand. I believe that your coming together at this symposium is an admission that patient safety must be top priority and a firm basis, for implementation in the system you seek to strengthen and in the hearts of caregivers. Certainly, systems to enhance patient safety systems are essential to our health system facilities, but the responsibility of each and everyone providing medical care is equally vital.

And thus, the movement for enhancement of patient safety gained momentum. The number of researchers increased by hundreds, dedicated journals were established, and the number of participants at conferences increased year by year.

In 2000, the National Institutes of Health (NIH) in the United States published the book *To Err is Human*,[2] some 300 pages long (one year before publication, it was already possible to download the book from the Internet free of charge).

After reviewing existing knowledge and the scope of the problem, the authors steadfastly maintained the many causes of errors and suggested ways and means of building a healthier system, reporting errors, and setting benchmarks for the safe operation of medical services.

It seemed that the book was read in every hospital and attempts made to apply the recommendations. Dozens of articles suggested methods for reporting errors. Many argued that: "Without a database we cannot familiarize ourselves with the problem and offer solutions." Thousands of medical errors were collected and documented at

Veteran Health Association (VHA) hospitals. Anonymous reporting of anesthesia and intensive care mishaps was possible on many Web sites and doctors in general were able to retrieve information regarding the scope of the problem. Patient safety units were established in most of the largest hospitals in America, in parallel and tangent to existing *risk management* units (a hospital *risk management system* is in essence a legal system responsible for minimizing the possibilities for negligence claims against the hospital, by ensuring proper registration, doctors' signing of documents according to the rules of protocol, as well as recording of events without doctor incrimination). The thirst for knowledge was huge and the book's authors were in great demand by Israeli hospitals and medical institutions to lecture on the subject. Many requests for seminars and workshops were also received from overseas.

Today, 10 years after the book was published and after thousands of research papers and articles, and many honest efforts to combat the problem, the number of mishaps and errors has not diminished and not only that, but even that which had been hidden and out of sight to the general public was now a "hot" subject in the press, reporting each case in bold print.

Along with the extensive publicity in the press, departments dealing with human engineering began to address the special problems of work in a medical environment.

Richard Cook, then engaged in human engineering, and his colleague David Woods began to investigate the medical system. Woods was among the researchers who examined the human engineering aspects following the major failure at the Three Mile Island facility in Pennsylvania.[3]

It was patently clear that failures can also be found in medical system staff training, pertaining to the link between the various teams and poorly designed systems. Later on, Richard Cook took up medical studies, initially specializing in anesthesia and he was one of the founders of the National Patient Safety Foundation in the United States, the selfsame organization that presented itself to the public at the first Annenberg Center conference.

Human engineering personnel were occupied with the man–machine interface relating to airplane disasters and flight safety. A few groups only, working in the medical environment, participated in a conference of the Human Factors Society, held in Denver, Colorado, in 1989. The aforementioned groups were engaged in investigation of the suitability of equipment for the disabled and those suffering from chronic illnesses. Our group presented its preliminary observations, conducted in an intensive care unit environment, and the high number of errors found. In 2004, at the New Orleans Society conference, more than 30 groups participated, with presentations and human engineering in medical environment workshops.

Researchers in human engineering published their findings in the professional press and outlined proposed procedures for prevention of errors. The doctors and nurses who collected and documented errors, and reported on abnormal events, published the results in medical journals. In essence, there were two parallel paths, two different languages and two different streams of thought. The main link between medicine and human engineering derives from comparisons made between civil aviation safety and operating room status. There were even those who copied into the medical environment the model of nonpunitive reporting and airplane teamwork.

Although it is obvious that this is not the end of the epidemic, the majority of those engaged in patient safety continue with reactive activities: collection and recording of errors, investigation of errors, and participation in conferences discussing one or another tragedy. Therefore, when it became clear that this in no way could bring a halt to the epidemic, voices began to be heard calling for a perceptional change, for a transition from waiting passively for an error to occur or reporting an error to preventive activity; an examination of the entire system through observation and identification of weak points in the system.

The current medical environment can be compared to a manufacturing production line built 50 years ago and occasionally supplemented by new machines, computerization, and procedures that are not an integral part of the original line. Accidents soon occur in this situation. A human engineering approach is more likely to reveal the potential error locations, expose weak points, and suggest a remedy. It is our hope that this be repeated in the medical setting.

Safety, according to Richard Cook, is similar to an infusion of adrenaline: so long as the drug is flowing intravenously, safety is maintained. Safety is not a commodity that can be bought and implemented; safety is a value. Once managers and decision makers make safety a top priority, not just an empty slogan but as a value to be fulfilled; with adoption of a program encompassing the culture of safety and environment of safety, as well as a working environment of taking responsibility but without punishment—maybe this will be the first step to eliminating the epidemic.

In this book, we present the results of the studies and the safety approach dating back more than a decade. We hope to have achieved full integration of medical terminology and terms used by the Center for Work Safety and Human Engineering.

And if by virtue of the book a single patient's life is spared—it has all been worth it.

REFERENCES

1. Leape L.L., Brennan T.A., Laird N., Lawthers A.G., Localio A.R., Barnes B.A., Hebert L., Newhouse J.P., Weiler P.C., and Hiatt H. 1991. The nature of adverse events in hospitalized patients. Results of the Harvard Medical Practice Study II. *N Engl J Med*. February 7;324(6):377–84.
2. Committee on Quality of Health Care in America. Kohn L.T., Corrigan J.M., and Donaldson M.S., eds. 2000. *To Err Is Human: Building a Safer Health System*. Washington, DC: Institute of Medicine, National Academy Press.
3. Operating at the sharp end: The human factors of complex technical work and its implication for patient safety. 2004. In: *Surgical Patient Safety: Essential Information for Surgeons in Today's Environment*, Manuel B.M. and Nora P.F. (eds). Chicago: American College of Surgeons, 19–30.

3 Types and Causes of Medical Errors in Intensive Care

Daniel Gopher and Yoel Donchin

Most patients hospitalized in internal medicine wards are there to undergo a series of tests in order to establish their diagnosis. Some patients receive medication, either orally, by intramuscular injection, or intravenously. Most of these patients are not in an immediate life-threatening situation, even though they are suffering from a serious illness liable to be fatal in the not-too-distant future. The patient's personal belongings are stored in a bedside cabinet. After hospitalization and preliminary treatment, patients are sent home or moved to another ward, according to the given diagnosis. Around 30 patients occupy a typical internal medicine ward and the medical team in attendance comprises nurses and doctors. Food is supplied from a central kitchen and patients eat by themselves. Blood samples for testing are taken the following morning. Test results are available within a day or at the most, a few days, on the basis of which an assessment is made of the situation.

On the other hand, the Intensive Care Unit (ICU) cares for patients who would have died within 24 hours had they not been admitted to the ICU. The number of patients receiving treatment in this unit is not large, 6 to 12 patients, one nurse taking care of one, two, or three patients at the most. In situations of distress, patients receive personal care by a single nurse. The patient's bed in the unit is surrounded by life support appliances, such as respirators, and continuous monitoring devices for measuring the patient's pulse, blood pressure, and many other vital indicators. The drugs, all very potent, are introduced intravenously using automatic syringes. All the indicators are recorded and any change must evoke a response immediately or within a short period of time. The intensive care unit is a very demanding environment, very noisy, and the staff does not have a moment's respite.

The abovementioned environmental conditions, where job requirements may exceed the ability of the staff to meet those requirements, may be a source of failure and mistake epidemics. Intensive care teams were among the first voices to be heard on this issue. The price to be paid for an error in the ICU is a high one. ICU patients do not have the reserves for and are unable to endure, for example, a fall in oxygen concentration due to incorrect knob rotation or medication that causes a fatal acceleration of the patient's pulse rate.

In fact, in our case, the implementation of human factors engineering methods (observation, interview, role analysis) at the Hadassah University Hospital ICU came about following an approach by intensive care unit teams to experts at the Technion Center for Work Safety and Human Engineering. Even the most experienced nurse was unable to offer a reasonable explanation for a number of errors and mishaps. Why, for example, was the incorrect drug taken from the medicine cabinet? How come two patients received adrenaline instead of morphine for pain relief? The team that discovered these mistakes was staggered by the banality of the error; a mistake caused not by lack of knowledge, negligence, or fatigue, but rather absentmindedness. On the very first visit by experts from the Technion Center to the hospital's operating rooms and ICU, it was patently clear that the medical team members in both working units were facing a hostile working environment, making it difficult for even highly qualified staff members to function properly. It was necessary to analyze the workstation from the point of view of human factors to discover its weaknesses; weaknesses that require intervention.

With all these reasons in mind, the ICU was selected to serve as a microcosm of the hospital. The unit selected for this study is the unit that the author served in as a senior physician. The ICU admitted patients after surgery or severe trauma, and patients requiring ventilation with recovery prospects. In the course of the study, the hospital maintained some 700 beds, while the ICU had only six beds, but due to overoccupancy for the most part, at least another four patients received intensive care in the hospital's recovery room. The department was headed by an experienced physician, a university professor, assisted by doctors from the anesthesia department and doctors and interns from various hospital departments. The ICU maintained a constant nurse–patient ratio of 2:1, regardless of the severity of the patient's condition. Nonetheless, the care level at the unit did not differ on formal indicators when compared to similar units in leading medical institutions worldwide (patient mortality rate was 12%, implying that even though the patients were in critical condition on admission to the unit, the majority left the unit in improved condition).

The morning shift receives information pertaining to the previous night—nurses from the night nurses, and doctors from the duty physician who was present in the unit from the previous afternoon to the changing of the shifts. Immediately after patients are admitted, the doctor is required to perform various actions, such as blood sampling, ordering an intravenous feed, and updating patients' records. Next begins the morning doctor's round with a lengthy discussion as to the treatment program for the remainder of the day. During the rounds, the ICU physicians are consulted by experts from other hospital departments, and they are responsible for discharging patients who are out of the critical stage and can be transferred to a regular ward or an interim unit. The doctor writes out orders for further treatment on the *Order Form* and later the nurse will read the form and follow the instructions therein to the letter. The entire body of information is recorded by the monitoring devices and sample summary recorded on a large chart at the patient's bedside. The nurse does not take an active part in the rounds; she is informed of the salient details of the information later. Introduction of a computerized sheet somewhat changed the nature of the activities in the ICU, but without relieving the workload!

Full cooperation between the ICU staff and the research team were assured by including medical and nursing staff at all stages of our research, from the planning to the publishing. The entire unit was inducted into the research program and undertook to report faults and fill out forms as required during the 4 months collecting data. After data collection, a severity fault assessment was performed by the ICU nurses. At joint staff meetings that took place outside the hospital, benchmarks for fault definition were set; a dedicated form was designed for reporting malfunctions and its location determined. In addition, the workers agreed to work under the watchful eye of observers, who recorded every activity performed at the patient's bed. Staff members were requested to fill in the form for reporting mishaps and to deposit the form in a box provided for the purpose in the unit. Forms could be submitted anonymously or by an identifiable member of the staff. The severity of the patient's condition was indicated by the number of lines attached to the patient (infusion, central line, etc.). The reporter was asked to describe the fault event in his own words. The observers also reported errors that they discovered. Observers were accompanied by a senior nurse, who did not intervene in the proceedings but, when requested, provided clarification regarding work procedures.

It was necessary to define an *error*. Mistaken diagnosis was not considered a human error; an error was defined as irregular deviation from routines or action not planned in advance, as well as missing actions which were supposed to take place. For example, if an infusion took place twice as fast as originally determined and not detected in a timely manner—it would be considered as an error. Medication not given, or given accidentally—incorrect dosage, wrong patient, and so forth—are all doubtlessly errors. In essence, an error is defined as an irregularity. If action A was planned and action B implemented—this is an error.

We documented care errors for 4 consecutive months, for all shifts, that is, 24 hours a day.

Many activities are carried out in the unit. In addition to the medical staff of the unit, the unit is visited by relatives, consultants from other departments, occupational therapists, physiotherapists, radiographers, and respiratory technicians. These activities were documented and measured by observing and recording each and every contact between patient and all those around him, whether during radiography, bathing, or catheter insertion. Any such contact was recorded as an *activity*, in order to obtain a benchmark for the scope of activities carried out in the unit and not just a record of errors. It may be assumed that increased activity leads to greater error probability. At night, when the scope of activity is reduced, one may assume a lesser error rate. Recording of activities encompasses all the relevant information, such as length of time and nature of activities. Activities were sorted into three groups: planned activity (routine activities of the unit); planned activities deviating from routine; and reactive activity, that is, prompt response of a nurse or doctor to a change in the patient's condition.

The observers recorded 8,178 activities during 24 hours monitoring 46 patients, an average of 178 activities per patient per day (Figure 3.1). During this time period, 78 errors were discovered (Figure 3.2). In this way, through observation, it was possible to obtain a global situation map of the unit and patient treatment. Another analysis of these findings showed that just 4.7% of the activities were carried out by a single physician and 84% was carried out by nurses: 3.8% of the activities were carried out by a doctor and nurse working in tandem. The detailed results are presented in Donchin, Gopher, Olin et al.[1]

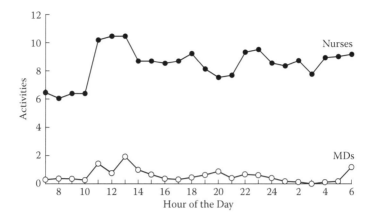

FIGURE 3.1 Diurnal distribution of physician and nurse activities in the intensive care unit. (From Donchin Y., Gopher D., Olin M. et al., 1995, The Nature and Causes of Human Error in the Intensive Care Unit, *Critical Care Medicine,* February 23(2): 294–300.)

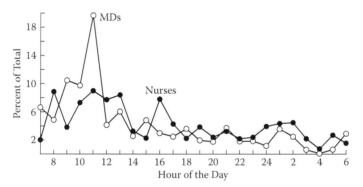

FIGURE 3.2 Diurnal distribution of commitment of errors. (From Donchin Y., Gopher D., Olin M. et al., 1995, The Nature and Causes of Human Error in the Intensive Care Unit, *Critical Care Medicine,* February 23(2): 294–300.)

Data collection continued over a period of 4 months, during which time 554 errors were recorded on purpose-made forms—476 by staff and 78 by the observers. Everything that the observers discovered was also discovered by the team, thus providing verification of the reported material.

The errors themselves were classified at five levels of severity, from typing errors to errors, which, if not discovered in time, could cause the patient to die. The error severity classification was individually judged by nurses and doctors without being aware of how others classified identical errors. The resultant symmetry lent credence to the uniform perception of error severity.

The general listing of patient care activities, during all three shifts, enables clarification of the prevalence and nature of errors relative to the overall number of care activities per patient and of the person responsible for the error. A total number of 78 errors were revealed in the 24-hour observations, which constitutes less than 1%

(0.95%) of the overall number of care activities observed. In other words, the vast majority of activities took place without errors and mishaps. Nevertheless, 78 directly observed errors in a sample of 46 patients under observation implies 1.7 errors per patient per day, errors that if not discovered in time, could aggravate the patient's medical condition or even cause death.

This calculation illustrates, on the one hand, the significant cost of an error even when the number of errors is small, and on the other hand, the calculation highlights the importance of the base frequency of overall activities for the purposes of error interpretation and attributed error significance.

Another important example is evident when comparing the relative contribution of the doctors and nurses to the number of errors discovered. The overall rate of the doctors' activities from the total number of unit activities amounted to 4.7%, while the nurses' share was 84%. However, of all the errors observed and documented, 45% were attributable to doctors and 55% attributable to nurses. This fact underscores the importance of the overall number and distribution of activities as a reference point. Although the number of errors attributable to doctors was less that of nurses, their relative *contribution* to the error rate, compared to the scope of activities, was ten times as great! This significant difference would not have been evident without a comparison to the overall frequency of activities of each type, including correct and erroneous performance.

Another important finding showed that more errors were made and discovered close to shift changeovers. Shift changeovers entail the transfer of responsibility and information from the departing to the incoming staff member. The increase in the frequency of error detection can be explained by the control and inspection processes conducted during the changeover. The increase in the number of mishaps occurring at this juncture can be explained by the fact that some of the information was not transferred at the shift changeover, or is passed on informally and not documented and well understood. Incomplete transfer or unclear information will increase the probability of occurrence of errors close to the time of shift change. These results provide objective evidence of problems and failures that may occur when transferring responsibility between team members in the health care process.

To summarize, an analysis of the findings exposes the ICU as a multirisk work environment, even though 75% or more of patients treated in ICUs live to tell the tale and if they had not been treated in the ICU they would not have survived. Important findings emerging from this study revealed that the unit is indeed beset by problems at all levels, but all of them can be solved or improved. In the work environment, general organization level, and the communication between doctors and nurses—transfer of information and update of patient medical status map between shifts is not structured in such a way that all important details are indeed transferred. The patient's bedside layout is disorganized and unstructured, from the intertwined tangle of cables all connected to the same power point or gas supply point, the arrangement of monitors and information display positions, to actual physical obstacles preventing access to the patient's head in need of resuscitation.

The approach and methods adopted for this study is the format we are proposing for those proactively seeking the root problems of faults and a plan for change. The first step is to motivate staff, acutely aware of the frightening possibility that any one of them could make a mistake. The ability to recognize human shortcomings when operating a system is an important step toward a solution. The next step is a comprehensive analysis of the system and working environment from a human engineering perspective. Posing questions such as: what is the role and the tasks of each staff member? How is the workstation organized (or not organized)? What is the nature of the interrelationship between the medical team (doctors) and nursing staff (nurses)? How is information transferred? What is the workload? Such an analysis allows for detection of weaknesses in the system and finding possible solutions (emphasizing practical solutions). Suggestions such as: "We have to rebuild the entire unit and replace the entire crew" are not feasible or logical.

Following this study, several solutions were proposed for the ICU staff. Some of the solutions were accepted (form changes, reorganization of the patient's bedside environment, determination of a unique procedure for recording medication dosage), some, unfortunately, did not stand the test of hospital constraints, either because of shift changeovers determined by the hospital's shuttle system or because of the overcrowded wards.

However, planning the construction of a new intensive care unit taking into account all the research findings is indeed more user friendly to staff and patients alike.

REFERENCE

1. Donchin Y., Gopher D., Olin M., Badihi Y., Biesky M., Sprung C.L., Pizov R., and Cotev S. 1995. The nature and causes of human error in the intensive care unit. *Critical Care Medicine.* February 23(2): 294–300.

4 The Operating Room and Operating Process—Observations

Yael Einav* and Daniel Gopher

CONTENTS

Description of the Process 33
    Numerous Staff 36
    Numerous Locations 37
    Multistage Process 37
    Information-Rich Process 38
Results of Observations 38
    General Data 38
    Classification of Abnormal Events 40
        Teamwork 40
        Knowledge and Its Management 41
        Lack of Knowledge and Expertise 41
        Procedures 41
        Timetables 42
        Human Engineering and Safety 42
        Deviations from the Norm 42
        Equipment 42
        General Notes 42
    Management Problems versus Safety Problems 43
Pelvic Fracture Fixation—Observations and Interpretation 43
Troubleshooting 51
    Human Engineering Problems 51
Acknowledgments 53
References 53

---

* This research project was conducted in partial fulfillment of the requirements for a doctoral degree, and was supervised by Daniel Gopher and Yoel Donchin. The observations described were performed in the Hadassah Ein Kerem Hospital in Jerusalem operating rooms, with the cooperation of the directors and staff of the gynecology and orthopedics operating rooms (Orna Ben Yosef and Margaret Lun), assisted by Yitzchak Karah, Director of Nursing Services of the Operating Rooms Division. Ilan Palatine, Dr. Ella Miron-Spector, and Maya Peyman-Etzion participated in the observations and analysis of the findings. Dr. Abishag Sphilinger and Yael Auerbach-Shpak participated in the analysis of the results.

Surgical practice was almost unknown before 1864. Most of these activities were not carried out in a properly equipped operating room, but rather in the patient's home or at a location with the appropriate equipment—a table and ropes to restrain the unfortunate patient. Only after the discovery of the anesthetic properties of ether and the beginning of methodical anesthesia did surgery take off, a historical process entitled *Allocations of Science*. Thanks to the discovery of the importance of washing one's hands and disinfection of devices, along with a broadening knowledge of physiology, surgeons dared to penetrate areas they had not dreamed of beforehand and at the same time, sophisticated devices were developed allowing access to anatomical layers that no one had previously attempted. Technical capability did not go along with a corresponding assessment of the cognitive ability of the surgeon and assessment of the physical exertion involved in having to remain standing up under difficult conditions. Mishaps and disastrous errors were soon in evidence—initially, fatalities due to excessive use of ether for anesthetic purposes and later, many surgical complications during the operation itself. Today, anesthesia is safe and mortality due to anesthesia is a rare event. Furthermore, the incidence rate of complications as a result of surgery is low, but occasionally errors are detected that in retrospect appear as disastrous. Some errors are not perceived at all, not among the general public and not by the surgeon himself: How come the operation was carried out on the wrong side of the body? How come a surgical instrument remained inside the abdomen? How come an appropriate surgical instrument was not available even though it was needed?

These are serious questions and the answers to these questions should emerge from the study of human behavior, namely cognitive psychology.

Current surgical procedures are complex and risky, performed in a frequently changing environment, often requiring a program change and prompt intervention, analogous to the constantly changing conditions on a battlefield, and the level of certainty and predictability is limited. Therefore, work efficiency and safety can only be planned through the combined and coordinated efforts of the operating room (OR) staff: a team effort.

What is the OR team? Who is who on the team? How does the team operate? Does it operate as a team or as a tribe? Is it possible to influence the behavior of the team members?

This chapter will describe the research carried out to find the answers to these questions.

* * *

The medical environment in which patients are most prone to injury is the operating room. About 50% of injuries occur when patients undergo surgical procedures. Doctors and anesthesiologists were the first to study and propose interceptive methods for the prevention of mishaps in the operating room, but precious little research time was devoted to examination of the entire team, as a single unit working in unison. Most studies were carried out by medical personnel; human engineering personnel have only recently entered the hospital scene—shedding new light on the picture from a somewhat different angle.

The operating room is a complex work environment run by a multidisciplinary medical team who must work in strict coordination. Investigative research, based on questionnaires and observations of the interaction between OR personnel revealed discrepancies between the desired interaction and the status quo. Observations made in teaching hospitals in Europe revealed failures in meeting regular work requirements, such as delays due to noncompliance with planned work schedules, as well as misunderstandings and poor communication between teams, between nurses and surgeons, and surgeons and anesthesiologists. The researchers listed the failures observed, according to behavioral categories:

- Irregularities in communication or in the decision-making process.
- Errors in the preparatory stage, planning, or preparedness.
- Workload distribution.

The researchers also drew attention to the hostility between team members. This research program focused on interpersonal and systemic aspects contributing to medical errors, but without quantifying the frequency of occurrence of all types of events and without proposals for concrete improvements. That is to say, there is a fundamental problem in the operating room of working in a tribal fashion rather than as a team, and despite the widespread assumption that the patient's well-being is the primary consideration, in practice things are different.

We wished to examine the social structure and the dynamics of the entire operating room using methods of human engineering factors analysis, thereby enhancing and broadening the knowledge of the various factors liable to lead to mishaps and OR malfunctions.

The basic assumption was that failures do not occur due to medical activities alone, that is, anesthesia or surgery, but that there is another system component, related to all the activities carried out in the operating room and not directly related to the operation itself, such as the interaction between teams (e.g., anesthesia staff and surgical team); shaping of the work environment; procedures and their implementation; transfer of information between members of staff themselves and between the team and outside parties (OR centers or hospital wards).

The modern medical treatment process is compared to a relay race, where each runner hands over the baton to the next runner. However, the medical runners are in the habit of dropping the baton, thus leaving a gap, noncontinuity of treatment, stemming from a whole spectrum of causes: shift changeovers, treatment by more than one member of the team, and so forth (Photo 4.1).

These discrepancies are a natural outcome of the modern treatment process, but in contrast to the approach that perceives human beings as the weak link whose influence should be minimized, however, there are those who see professionals (the experts) as the most likely individuals to bridge the existing gaps in the system. Gaps spawn mishaps, and in order to help professionals to bridge these gaps and as a first step, the proposal is to address the gaps in the system. The research carried out by the team from the Research Center for Work Safety and Human Engineering was designed to describe the OR work process and discover the inherent discrepancies in the process, the weak points that set off the fault process and ultimately harm the patient (or therapists).

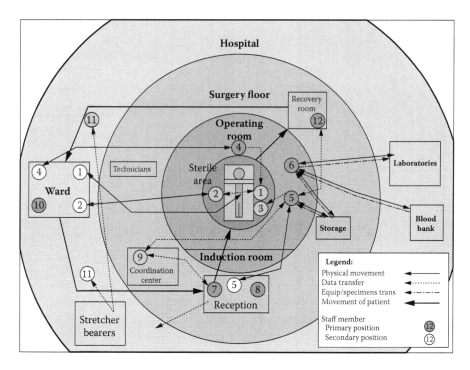

PHOTO 4.1 A global overview of the surgery procedure system.

The research comprises three parts:

1. Charting a formal description of the processes in the operating room, especially identifying the weak points.
2. Classification and quantification of abnormal events in the operating room.
3. Proposed improvements and assessment of said improvements in terms of safety enhancement.

In order to describe what happens in the operating room, observations were made by four observers from the Center for Work Safety and Human Engineering at Technion and by two medical students who participated in the project in the framework of their studies. Two departments in the hospital gave us free reign to carry out our observations of operations carried out by their doctors: the orthopedics and gynecology departments.

Observation dates were chosen at random, independent of the surgical program, in order to obtain a representative situation picture of OR activities. One to three observers were assigned to each operating room and for every day that observations were carried out, simultaneous activities in one to three rooms were documented. Observations began with the arrival of the patient to the OR reception room or when the patient was already in reception. From reception, the observers accompanied the patient to the anesthesia preparation room and then to the OR. Observations ended

The Operating Room and Operating Process—Observations

with the arrival of the patient to the recovery room. During surgery, the observers remained in the OR and carried out their work without interfering with the normal OR activities. Each observer noted down the events on the Observation Form, specifying the event time frame, the functionary linked to the event, and the activities carried out.

The observers were directed to focus on activities taking place in the room and not on abnormal events or mishaps, and to concentrate upon the means by which information was transferred within the room or between the room and the outside world, to listen attentively to the interaction between the different members of staff, and interaction between them and the physical environment. The focus of observations was on activities in the operating room and not on activities outside the room, even surgery-related activities.

A report was made of each operation, consisting of all the observations made of the given operation. After a large number of reports had been accumulated, it was possible to filter out abnormal events, that is, events that may put patients at risk or endanger the safety of team members, or events that disrupted or delayed the process. The observations and reports facilitated characterization of the system that we call the *operating room*.

DESCRIPTION OF THE PROCESS

Operating room workers include surgeons, anesthesiologists, anesthesia technicians, OR nurses (sterile and *circulating*), on-call nurses, reception room nurses, auxiliary staff (sanitary workers), cleaners, and maintenance personnel. All team members remain in the operating rooms at all times during their shift, except for the surgeons.

The order of operations for the following day must arrive at the OR office by 2:00 P.M. The schedule of operations (timetable for operations for all the hospital's ORs) is printed and posted in all areas of OR activity. Planned nonurgent operations are entitled elective operations. The program changes from time to time due to the inclusion of an unplanned emergency operation. The tendency is to complete the planned program and if an urgent operation is to be performed it is postponed to the end of the working day. However, in an emergency situation, an ad hoc change in the program is always in the cards.

During the day, nurses arrive at work for the interim duty shift (11:00 A.M. or 12:00 P.M.). The purpose of the interim duty shifts is to provide assistance to the morning team to complete the elective operations. Elective surgery is set for earlier in the day so as not to stretch into the evening hours, leaving the evening team free to perform emergency operations. The morning duty shift is from 7:30 A.M. to 3:00 P.M.

We will track the process from the beginning; from the moment the patient is summoned until surgery has been completed and the patient transferred to the recovery room (Figure 4.1 and Table 4.1).

As seen from the above description, we are addressing a complex, multistage process, in which many medical personnel are participating. From the description that appears in the flowchart and characterization of personnel classification, it is possible to direct one's attention to a few focal points that are potentially hazardous.

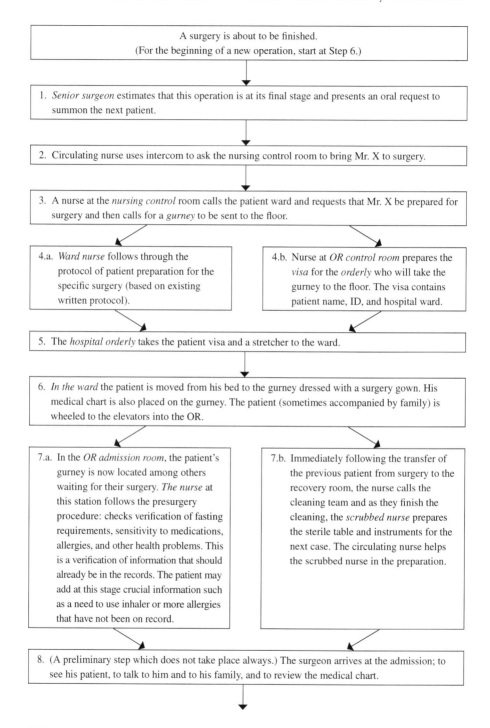

FIGURE 4.1 Procedure flowchart: From admitting the patient to the OR to the recovery room.

The Operating Room and Operating Process—Observations

9. *When OR is ready*, the senior nurse calls the surgical team via intercom or phone. If they do not respond, she asks the control room to locate them. Circulating nurse/an aide/surgeon (single or together) arrive at the admission room and are directed by the admission nurse to the patient, who is rolled into the OR.

10. Patient arrives at the OR and meets his anesthesiologist. This step may take place in the induction room or in the OR. Anesthesiologist reviews the chart and converses with the patient. Common questions—Allergy? Did you undergo anesthesia before? He then starts an IV and checks BP. He is helped by a nurse or aides.

11.a. In the OR, the patient is moved and placed on the operating table. The team (anesthesia plus nursing) put him to sleep. He is undressed; the skin in the location to be operated is cleaned and prepared. The patient is postured according to surgery (body, legs up, etc.).

11.b. Concurrently with the preparation of the anesthesia, surgeons arrive and review the chart and discuss the surgery plan. They then enter the scrub room and scrub. They return to the OR and with the assistance of the circulating nurse put on their sterile gowns.

12. Surgery is performed and during the procedure the nurse also has to perform fetching, delivering, and counting of instruments; recording and sending pathological samples; injection of special drugs, and more.

13. At the time of the skin closure, the surgeon requests next patient. The circulating nurse also checks if there is a free position in the recovery room.

14. Before final skin closure, the surgeon asks for a finishing of the counting process, which is performed jointly by the scrubbed and circulating nurses.

15. Surgeon writes orders for the next hours, anesthesiologist wakes the patient, surgeon dictates or writes a summary. Nurses start to clean the room.

16. The patient is now awake and is accompanied to recovery by his surgeon, anesthesiologist, and nurse. Information is transferred to the recovery room nurse.

FIGURE 4.1 (Continued) Procedure flowchart: From admitting the patient to the OR to the recovery room.

TABLE 4.1
Process Team Members and Their Main Roles

Job Description	Where	Role	No. in Figure
The decision to operate, planning the surgical schedule, follow-up	Surgical ward, around the operating table	Senior surgeon	1
Take care of patients in the postoperative period	Surgical ward, help the senior surgeon during operation	Surgical resident	2
Prepare instruments, deliver surgical items to the surgeon, in charge of counting items before and after surgery	Within the sterile field in the OR	Scrubbing OR nurse	3
Add items upon request, arrange equipment, support scrubbing nurse	Nonsterile field in the operating room	Circulating OR nurse	4
Responsible for maintaining vital signs during surgery; induce anesthesia and analgesia; in charge of the recovery room	Preoperative evaluation, in the OR, near patient head	Anesthesiologist	5
Help scrubbing nurse, help position patient, help with POP	Nonsterile area in OR	Orderly	6
Identify patient, check for denture, signatures on forms and other formalities	Admitting area of OR	Reception nurse	7
Inform families on surgery process	Admitting area, recovery room, operating room	Communication nurse	8
Control operations, schedule, substitute for a break	HQ office in OR	OR chief nurse	9
Prepare patient for surgery	Surgical ward	Ward nurse	10
Get an order from OR HQ and go to fetch the patient from the ward	Between ward and OR	Hospital orderly	11
Admit patient postsurgery and begin monitoring	Recovery room near OR	Recovery room nurse	12

NUMEROUS STAFF

A team should work in a coordinated fashion in order to achieve its goals. A harmonious team is vital to the success of the mission, especially under pressure. The three main factors affecting teamwork, as opposed to the work of an individual, are shared information, shared resources, and allocation of responsibility. As staff numbers grow, control over these three factors becomes more complex and difficult.

Some 11 to 19 staff members participate in both major and peripheral surgical processes (Table 4.1). The doctors and nurses comprise the main team, working alongside other staff members, who do not participate throughout the whole process, such as an anesthesia technician and senior anesthetist who may join forces with a junior anesthetist during surgery; and a cleaner, who is not part of the team itself although his contribution definitely affects the efficient conduct of the process. Taking into account that members of the staff fulfill different roles and those in authority are assigned different responsibilities, while some of them perform the procedure for a number of patients in parallel, we can conclude that teamwork and coordination are complex tasks, a veritable minefield with potential failure spots.

NUMEROUS LOCATIONS

The process takes place in seven main locations: inpatient department, reception, nursing center, anesthesia preparation room, operating room (divided into two areas, sterile and nonsterile areas), and a recovery room. In addition to these locations there are secondary locations, such as laboratories and a blood bank. Staff members are not positioned in a given location at all hours of the day, but circulate in a certain location or even throughout the entire hospital (cafeteria, other departments, laboratories, etc.). Part of the staff's time is spent looking for one another in order to transfer important information. Communication is problematic, as each team member deals with a great deal of information and is required to commit to memory: What to transfer? To whom? Which member of staff has not received information that the member was supposed to receive? In other words, intercommunication is complex and complicates the process.

MULTISTAGE PROCESS

According to our findings, the entire surgical process consists of 16 main stages. Transition between stages may involve the transfer of information and responsibility for the patient to another responsible party, with the transfer of the patient to another location or transfer to another stage in the process (without any change of patient and staff location). These transfers give rise to a significant probability of error occurrence—errors due to incomplete transfer of information relating to the patient. Sometimes staff will rely on memory or on information that only they have the whereabouts of. In fact, in many instances staff members are not at all aware that they are relying on their memory while treating a patient. When responsibility for treatment passes from one member of staff to another, important pieces of information can go astray; there is no *memory transfer*, and whatever is saved to memory at any time is not passed on to staff at the next stage. Stage interfaces are error prone areas.

Sometimes transition between stages does not entail the physical transfer of the patient to another responsible party but transfer of information, that is, multiplicity of stages in the process and within each procedure may result in incomplete transfer of information between team members and ultimately—to error occurrence.

INFORMATION-RICH PROCESS

Information relating to every individual patient may encompass thousands of details, not one of which cannot be ignored, from initial hospitalization, related diseases, anatomical changes, and the like, occurring in the operating room, such as materials added to the infusion bags. This is vital information that if not passed on to the next staff member in the chain, may put the patient at serious risk.

At this juncture, we present the results of detailed observations of 48 gynecological and orthopedic operations. These observations will enable the reader to appreciate the significance of the four focal points and how they influence the issue.

RESULTS OF OBSERVATIONS

GENERAL DATA

In the course of the research program, observations were made over 8 days of 34 operations in the gynecology department and 14 operations in the orthopedic department. In each operating room, continuous observations were made from the start of the daily program of operations to their completion. Observations were made in two ORs; one OR of the orthopedic department and gynecological department, respectively. Types of commonplace gynecological surgery observed included: tumor removal from the womb, cesarean section, abortion, cervical stitching, endoscopy, and hysterectomy. Orthopedic surgery observed included: extension of the Achilles tendon, removal of metallic ankle stabilizer, screw removal, the introduction of platinum, hip replacement, foot correction, and so forth.

A total of 204 abnormal events were documented (Table 4.2). These events are not errors but potential risk situations or potential errors if things had turned out differently. These are events that reflect safety compromises in the workplace or process inefficiency, communication, or coordination failure.

TABLE 4.2
Distribution of Abnormal Events Observed during Operations, According to Length of Time and Type of Operation

	Short-Time Operations (Less Than 1 Hour)		Long-Time Operations (More Than 1 Hour)		Total	
	Number of Operations	Number of Abnormal Events	Number of Operations	Number of Abnormal Events	Number of Operations Observed	Number of Documented Events
Gynecology	15	15	19	117	34	132
Orthopedics	3	12	11	60	14	72
Total	(31%) 18	(13%) 27	(69%) 30	(87%) 177	(100%) 48	(100%) 204

TABLE 4.3
Distribution of Abnormal Events Observed by Category and Subcategory

General Category	Subcategory	Frequency	Relative Frequency
Teamwork	Coordination between staff members	12	30%
62 nonroutine events	Unaddressed requests	6	
	Communication between staff members	13	
	Staff changes	7	
	Staff member discipline	6	
	Staff absenteeism	8	
	Coordination with personnel outside the room	10	
Procedures and Practices	Sustaining the activity sequence	16	17%
36 nonroutine events	Complying with practices	16	
	Omitting practices	4	
Human Engineering and Safety		17	8%
17 nonroutine events			
Deviations from Routine	Special procedures	6	6%
13 nonroutine events	Unexpected changes	7	
Scheduling	Delays	13	10%
19 nonroutine events	Schedule changes	4	
	Work arrangements	4	
Equipment		19	9%
19 nonroutine events			
Existence and Management of Knowledge	Availability of information or incorrect information	13	9%
18 nonroutine events	Informal transfer	3	
	Recording and documenting	2	
General Notes		11	5%
11 nonroutine events			
Lack of Knowledge and Skills		7	3%
7 nonroutine events			
		204	Total: 97%

The distribution of abnormal events varies with different types of operations—in certain operations, the occurrence rate of abnormal events was higher than average (normal average is four per operation) while in other operations no abnormal events occurred, except for the isolated instance. From this finding we learn that various factors affect the probability of error occurrence during surgical operations. These factors are discussed in greater detail below.

From Table 4.3 it can be deduced that the number of short-time operations in the orthopedic department is low (21% of all operations in this department) relative to the numbers in the gynecological department (35% short-time operations, of less than an hour's duration).

CLASSIFICATION OF ABNORMAL EVENTS

According to the conceptual framework that is presented and the general description of the process, as noted in the previous sections, the observations focused on abnormal events from the problem groups discussed in the previous chapter. These problems guided the observations and the analysis of the data obtained from the observations.

Although the observations focused mainly on the operating room itself, the overall process was also reviewed in a less systematic and more informal manner as summarized in previous sections of this chapter.

General categories under observation were:

- Teamwork
- Knowledge and its management
- Lack of knowledge and expertise
- Processes and procedures
- Schedules
- Human engineering and safety
- Deviations from routine
- Equipment
- General

A brief description of these categories and their scope are given below.

Teamwork

Coordination—Staff act without coordinating with each other. For example, nurses and surgeons change the position of the patient without notifying the anesthesiologist, or regulate room temperature without prior notice; the anesthesiologist repositions the operating table without notifying the surgeons; the surgeon does not inform the anesthesiologist that an incision has been performed, or the abrupt completion of the operation.

Communication—Staff are not communicating with each other or communicating in such a manner that does not allow for efficient transfer of information. For example, changeover of nursing personnel during surgery without informing the surgeons; a failed attempt to locate a team member; a team member uses terms unfamiliar to other team members; lack of familiarity between staff.

Inexpert requests—A team member requests assistance without addressing a specific team member. In this way, the responsibility for fulfilling the request falls upon all staff members who heard the original request lessening the likelihood of the request being fulfilled (diffusion of responsibility). For example, the surgeon asks for the lights to be turned off in the OR, the anesthesiologist asks for the air conditioner temperature to be changed, and so forth.

Team changeover—A staff member leaves and is replaced by another staff member. Problems may arise under these circumstances such as failure to update the senior surgeon regarding the impending changeover (and the surgeon does not know with whom he is working) or communication problems in the transfer of information between members of staff during the changeover.

Absenteeism—A team member is missing from his post. For example, a circulating nurse is not present in the OR; a senior nurse is not present in the reception room in order to make changes in the operation schedule; a junior surgeon was absent, and so forth.

Team discipline—Actions taken are not in accordance with the procedure, *shortcuts* and *cutting corners*, without any reference to the team.

Coordination with outside parties—Absence of mutual coordination between outside parties and the OR.

Knowledge and Its Management

Know-how availability—Missing vital information, such as blood test results or the patient's weight.

Unofficial transfer of information—Vital information is passed along off the record.

Records and documentation—Incorrect recording of information (patient's name misreported on the patient's form), missing documentation (such as nonreporting of additional dressing or needle during the operation immediately after the addition), or information that was not recorded.

Lack of Knowledge and Expertise

For example, a team member is without the appropriate expertise for the role assigned to him (e.g., unfamiliarity with the device that he is supposed to operate and not knowing the whereabouts of the information on the device, etc.).

Procedures

Compliance with actionable sequence—A specific sequence of actions are not taken in the correct order. For example, stitching up the patient before checking the number of swabs and needles; a patient is brought into the OR in plaster cast, even though the plaster cast should have been removed before entering the OR.

Compliance with procedures—Members of staff are taking actions that are not in accordance with the existing procedure. For example, the departmental nurses are required to provide the patient with tranquilizers as instructed by the anesthesiologist exactly as recorded. In practice, the nurses provide the tranquilizer before the patient arrives for surgery.

Lack of procedures—There is no procedure that members of staff perform where they are forced to act in a way that is not systematic or structured. For example, there is no procedure for dealing with information relating to a patient's blood tests obtained shortly before surgery; there is no procedure for communication between orthopedic surgeons as they adjust the X-ray equipment ("move it toward me... not toward him, toward me").

Timetables

Delays—Delays at the onset or during surgery for various reasons; for example, the cleaners were not summoned in time to clean the operating room. Many delays are caused by absence of the surgeons from the operating room during operation preliminaries with the patient already anesthetized.

Changes in the operations program—The order of operations is changed for various reasons, such as postponement to a later date or advancement of operations.

Work plans—Events related to the work plans of nurses in the operating room, at the end of their shifts, at shift deployment, and so forth: very relevant when an operation continues beyond the accepted working hours, triggering a discussion on the team as to who will remain in the OR and whether or not to continue with further operations; other instances relate to the surgeons' work plans; on occasion the surgeon is obliged to leave the OR before the end of surgery (for example, if he has to give a lecture at the medical school), and a replacement must be found.

Human Engineering and Safety

Failures found in the design of medical equipment used by the medical team or in the team's working environment, such as problematic use of X-ray equipment.

Deviations from the Norm

Special procedures—Team members perform special medical procedures, requiring particular preparation. For example, brain surgery anesthesia demands assistance from anesthesia technicians and high levels of concentration and high awareness of the anesthesiologist himself during surgery.

Unexpected changes—Changes to the order of surgical operations introduced at short notice.

Equipment

Dysfunctional equipment unsuitable for the required procedure, and so forth.

General Notes

Events that were difficult to classify in one or other of the above categories but were nevertheless important to make note of.

The incidence of abnormal events in each category was determined by the frequency of event recurrence in the observation records.

This method does not attach weight to event importance and understanding the impact on system performance and probability of failure associated with events. For example, human engineering problems, as reflected in some of the orthopedic surgical operations, especially operations using X-ray equipment: human engineering factors related to the X-ray equipment and its suitability carried significant weight in the surgical chain of events.

The Operating Room and Operating Process—Observations 43

MANAGEMENT PROBLEMS VERSUS SAFETY PROBLEMS

The procedure is characterized by two types of abnormal events: management events and safety events. A management event is defined as an event that will affect the carrying out of the procedure and its effectiveness, without harming the patient's health or that of the team (for example, an event causing a delay). A safety event is an event that may affect the safety of the patient or OR team or their well-being.

The following table (Table 4.4) details individual abnormal events recorded, possible results stemming from these events, and corrective actions taken by team members to prevent these results. The abnormal events are detailed according to the occurrence stage in the procedure (according to Figure 4.1), and also if the event is managerial or safety (or both together).

PELVIC FRACTURE FIXATION—OBSERVATIONS AND INTERPRETATION

The purpose of the surgery is the mending of pelvic bone fracture, fractures caused by road accidents. Surgery was performed by four doctors: two orthopedic surgeons and two interns.

The patient entered the anesthesia preparation room at 9:00. He was the first patient in the surgical program that day. The anesthetist asked the patient a few questions as she had not interviewed him the previous evening, as required.

At the same time the nurse charged with bathing the patient is waiting, wearing a lead apron (for X-ray radiation protection). *The apron is very heavy relative to the weight of the nurse. As the nurse is forbidden to sit*, she must wait while standing (*problem #1*), wait for a long time (*problem #2*), but remain in good humor.

Three minutes after the patient had been injected with immobilizing material, the senior anesthetist realized that he had no baseline (*problem #3*). This should have been verified before injecting the immobilizing material, to know later if the injected material has already taken effect. In order to solve the problem the senior anesthetist left the operating room and after a short while, returned with the peripheral nerve stimulating device, whose purpose is to define the degree of patient muscle immobilization. The anesthetist performed the test on the patient's foot and concluded that the patient was in an immobilized state. *The anesthetist withdrew a drug from an ampoule without checking the label (problem #4). The anesthetist sprinkled artificial tears into the patient's eyes without checking the label on the bottle (problem #5).*

The anesthetist realized that he had not checked the coagulation functional status (problem #6).

This surgery was the first time that a new device for securing the patient's legs was used. It emerged that none of the surgeons had ever used this device or practiced using it. A debate ensued among junior doctors on how to use the device. *It was not clear whether or not the leg should be raised or lowered (problem #7).*

Later, it was unclear whether the senior physicians wanted the patient's legs to be secured in the required position for surgery before they entered the OR or should

TABLE 4.4
Mishaps Stemming from Weak Points in the Various Surgical Procedures (See Figure 4.1)

Stage[a]	Nonroutine Event	Results	Corrective Action
1. Summoning the next patient	The request to summon the next patient in line only occurs once the current surgery is complete. *Managerial event*	The next patient is not ready on time—delays the start of the next surgery, or perhaps necessitation of the cancellation of the surgeries scheduled at the end of the list.	A nurse coordination center inquires as to the progress of the current surgery—which reminds the operating room staff to prep the next patient in line.
2. The circulating nurse contacts the coordination center	A communication problem between the operating room nurse and the nurse at the center—the surgical nurse had in mind a particular patient, but the nurse at the center takes in a different name, or neglects to identify the patient by name. *Managerial and safety event*	The wrong patient is sent in for the operation—delays the start of the surgery for the correct patient. The patient that was mistakenly prepped becomes distressed.	There is no corrective action and thus a different patient arrives to the admission room.
3. The nurse from the center contacts the department and reception	The center is notified that the next patient should be sent, but the message is not transmitted to the department or to the reception staff. *Managerial event*	The next patient is not ready in time—delays the start of the surgery and surgeries at the end of the list may have to be cancelled.	An operating room nurse calls reception to ask if the next patient has arrived, and reception reports that the center did not notify them about this patient.
3. The nurse from the center contacts the department and reception	The nurse in the center notified the department but forgot to notify reception. *Managerial event*	The patient in the department is prepared but a gurney does not arrive to transport him—delays the start of the surgery.	An operating room nurse calls reception to ask if the next patient has arrived, and reception reports that the center did not notify them about this patient.

The Operating Room and Operating Process—Observations

3. The nurse from the center contacts the department and reception	Poor communication between the center's nurse and the department nurse (similar to stage 2—miscommunication of the patient's name, or the patient is not identified by name). *Managerial and safety event*	The department nurse prepares the wrong patient. The correct patient is not ready for the surgery—delays the start of the surgery.	The gurney does not arrive to transport the patient and the department personnel realize that something has gone wrong. On the other hand, the gurney that arrives to take the correct patient will notify staff that the correct department did not prepare the patient.
4. The procedure of prepping the patient in the department	The procedure of preparing a patient for surgery is not performed or incompletely performed—for example, tranquilizing medication is administered at the wrong time, or presurgical fasting is not ensured. *Managerial and safety event*	The patient is not ready for surgery. The patient is not prepared as anticipated by the staff members (for example, they had expected, incorrectly, that the patient's cast would have been removed). If they see that the patient isn't prepared, surgery will be delayed or postponed to another day. If this goes unnoticed and the staff begins to operate, it may pose a problem to the patient being operated on.	
5. Orderly transports the patient to surgery	Shortage of gurneys. *Managerial event*	Delay in the patient's arrival to the admitting room—delay in the start of the surgery.	
7a. Procedure for receiving a patient to admitting	The admitting nurse does not complete the entire procedure of admitting the patient for surgery. *Managerial and safety event*	As in stage 4: The patient is not ready for the surgery and his/her condition is not as anticipated by the staff (especially the anesthesiologist and the surgeons). This poses a risk for emergent problems during the course of the surgery, and may pose a risk to the patient's health.	Discussion between the patient and the anesthesiologist or the operating room nurse.

(Continued)

TABLE 4.4 (Continued)
Mishaps Stemming from Weak Points in the Various Surgical Procedures (See Figure 4.1)

Stage[a]	Nonroutine Event	Results	Corrective Action
7b. Preparation of the operating room for the next operation	The head nurse arranges the equipment in the room according to the type of surgery about to be conducted. If there is a change in the surgical schedule and the chart schedule was not updated so that the nurse is not aware of the change, the room will not be properly equipped. *Managerial event*	Delay in starting the operation until the room is rearranged, or delays during the surgery while the appropriate equipment is obtained.	The surgeons update the nurse as to the schedule change immediately before the surgery.
7a. Procedure for receiving a patient from admitting	A patient in critical condition is transported to the reception room and is not being treated according to his/her condition since the nurses are unaware that the patient is in critical condition (e.g., a woman with preeclampsia giving birth). *Safety event*	The patient's condition deteriorates.	Revealed accidentally by the anesthesiologist who passes by the woman giving birth.
9. A patient on the way to the operating room	The surgeons are not in the area of the operating rooms and the nurse cannot find them. *Managerial event*	Delay in starting the operation.	
10. A patient arrives at the induction room	Poor communication between the anesthesiologist and the patient, for example, if there are language difficulties or if the patient is in a coma. *Safety event*	The anesthesiologist relies solely on what is recorded in the chart. The anesthesiologist does not consider all the necessary parameters because they are not recorded in the chart.	Peruse the patient's file, ask the department physicians or the surgeons, if they are familiar with the patient.

10. A patient arrives at the induction room	There is no verbal communication with the patient (does not speak Hebrew, is unconscious or a child) and the information was not received from the family or staff who possess it. *Safety event*	The physicians and the anesthesiologist are unaware of critical information that affects the patient's health (such as allergies or previous surgeries) that may result in harming the patient and the success of the operation.	Further clarification if possible.
11a. Initiating anesthesia and preparation of the patient for surgery	The patient is laid down in a certain position without consulting the anesthesiologist or other parties who should be consulted (such as an orthopedist). *Safety event*	The patient is placed in a position that is not healthy for him/her while being anesthetized, or not healthy for him/her for other reasons (for example, if the patient has had orthopedic surgeries).	The anesthesiologist hears beeps from the instruments or discovers problems in the patient's vital signs.
12. The surgery	The patient's blood work results do not arrive prior to the initiation of the surgery. *Safety event*	Providing anesthesia that is unsuitable given the patient's condition, the risk of adverse effects from various substances being injected into the patient during surgery.	The anesthesiologist notices this immediately before the surgery, when the patient is in the induction room (from studying the patient's chart).
12. The surgery	The screening equipment in the room is unsuitable. *Managerial and safety event*	Doctors find it difficult to identify the relevant bones for the surgery.	Replace the screening equipment with the appropriate equipment.
12. The surgery	The patient's X-ray film is not current. *Safety event*	Between the date of the X-ray and that of surgery, other procedures were performed (a screw was removed) and the patient can be injured since these factors were not considered.	The surgeon searches for the missing screw manually.

(Continued)

TABLE 4.4 (Continued)
Mishaps Stemming from Weak Points in the Various Surgical Procedures (See Figure 4.1)

Stage[a]	Nonroutine Event	Results	Corrective Action
12. The surgery	Transfer of the patient from the operating table to the gurney without coordination between the staff performing the transfer. *Safety event*	Injury to the patient's body parts (e.g., as a result of the neck falling).	There were no corrective actions.
12. The surgery	Injection of a substance into the infusion bag without recording it on the bag. *Safety event*	The patient is injected with a double dose of the substance.	Clarify with the staff member who performed the injection.
12. The surgery	The sheet of stickers referring to the patient whose surgery was completed is left in the operating room and they are used for the next patient being operated on. *Safety event*	The results of the blood tests and samples sent for testing arrive for the wrong patient. Drug therapy is given to the wrong patient.	The anesthesiologist calls out the name of the patient out loud and the surgeon realizes that it is not the name listed on the sticker on the surgical order sheet.
16. Exit to the recovery room	Some of the data is not transmitted to the recovery room nurses. *Safety event*	The patient does not receive the appropriate treatment.	

Note: 50% of events observed were safety events, 21% were management events, and 28% were defined as safety and management events.
[a] See Stages, Figure 4.1.

staff wait for their arrival before securing the patient's legs. Even on this issue, there was no agreement between the orthopedic interns: one intern claimed that he was told to secure the leg, while another, who entered the operating room after talking on his cell phone, stated that one should stand by. Eventually, the left foot was secured by the device (compromise?).

The senior surgeon, #2, enters the room first. He called the orthopedic surgeons' attention to a light table where they could review the pelvic X-ray results, after which they moved to another side of the room to another light table to review the CT results. It is unclear why two light tables are positioned at two ends of the room (*problem #8*).

A few minutes later a call is received over the intercom. Apparently, it was senior surgeon #1, wanting to pass on a few guidelines to senior surgeon #2. However, the intercom line quality or the quality of the speaker was poor and therefore, possibly due to the noise produced by the anesthesia machine beeps, *senior surgeon #2 could not understand the guidelines (problem #9)*. After several attempts, he finally received an accurate message.

The senior surgeon #2 then addressed those present, stating that they had no experience working with the new orthopedic device at the foot of the table (*the new pump?*).

Senior surgeon #1 enters the room, while at the same time a few members of staff are engaged in updating their Palm Pilots (*problem #10*). At the request of the senior anesthetist, the patient's left leg was removed earlier from its secured position, so that an intravenous line could be injected in the patient's foot. At this juncture, with the arrival of the senior orthopedic surgeon, the moment came to once again secure the leg to the device. Senior surgeons #1 and #2 put on lead aprons before scrubbing down. Only at this stage is the X-ray machine brought into play and only at the request of senior surgeon #1. Incidentally, it transpired that the scrub nurse had scrubbed down about 40 minutes too early and had to wait around all that time in a standing position. Nevertheless, to be on the safe side, the nurse scrubbed down again.

The orthopedic surgeons examined the X-ray image. The X-ray image angles are not defined, and a side view image is approximately side viewed, as the X-ray machine axes are not graduated and have no other accurate means of obtaining specific image angles. One problem observed illustrated the failed attempt by doctors to reproduce what seemed to them to be appropriate camera angles: in order to obtain the desired angle they conducted a series of X-rays, repositioning the X-ray machine between one X-ray and another, until they obtained the appropriate angle (*problem #11*).

When the X-ray machine was turned on for this surgery, the monitor displayed a blurred image, and even the senior orthopedic surgeons had difficulty identifying different anatomical reference points.

At this stage, one of the patient's legs was moved in order to set the leg in a natural position. The rod connecting the screws inserted in the fractured bones was removed, after which considerable energy was needed to move the leg in order to move the pelvic bones. After removing the rod, expert effort was enlisted in the form of an able-bodied doctor. He took hold of the leg tightly, like a wrestler, and climbed up onto the operating table to better his position and summon up more energy to manipulate the leg. The other orthopedic surgeons ordered him off the table for fear that the table would collapse. All of a sudden, the question arose, not addressed to anyone specifically: has the patient been immobilized? (A patient whose muscles are slack may be problematic, especially when a broken bone is under tension.)

Senior surgeon #1 goes to the nearby orthopedic operating room to request another X-ray machine with a larger display, in the hope that the new machine displays clearer information that is vital for carrying out surgical manipulation (problem #12). Meanwhile, the other orthopedic surgeons unscrew the temporary fixation screws from the body of the patient.

At this stage, instructions are given to bathe the patient and the orthopedic surgeons also scrub down. *At the same time, the X-ray machine with the larger display is brought into the operating room from the adjacent room. Due to the narrow passageway the transfer of the machine disturbs the doctors' scrubbing routine (problem #13).* Once the machine was in the operating room, maneuverability was made difficult due to the tangled tubes of the anesthesia machine (*problem #14*).

When the orthopedic surgeons return from scrubbing and are wearing sterile garments (robe, hat, gloves), an orderly washes the bed areas around the operated patient. Next, the patient is covered with sterile sheets. At this point, the surgeons are standing around the operating table, with their feet in water that dripped to the floor while the patient was being washed. *The orderly decided to wipe up the spill, at least on one side of the operating table. He bent down behind the orthopedic surgeons and wiped up the spill with a sheet. The orthopedic surgeons were not paying attention and did not notice that they were still standing in small spills of water (problem #15).*

The operating room was occupied by 12 personnel (*problem #16*).

Blood sample test results dispatched at around 9:45 were received at 10:33 and handed over to the anesthetist. She did not check the results of the test. According to her, there was no urgency.

The first surgical incision was performed at 10:38; 1 hour and 38 minutes after the patient entered the operating room. As before, an intern held tight to the leg of the patient, probably to enable senior surgeon #2 to drill into two parts along the bone. Senior surgeon #2 manually placed the drill where he intended to drill. A series of X-rays from different angles was carried out in order to check that the drill was correctly positioned. Next, senior surgeon #2 tapped the drill lightly with a hammer (to fix it in the bone) and finally connected the drill to a pneumatic drill for drilling. After a second series of X-rays, senior surgeon #2 had to admit that the drilling operation was unsuccessful and withdrew the drill.

Any problem identified during surgery involves communication between the orthopedic surgeons and the orthopedic surgeons with the remaining members of the team. Despite the intensive teamwork, there is no formal framework for colleagues to address one another. Problems of communication that arose during this surgery are:

- Members of the team were not identified when addressing one another. The orthopedic surgeons were not face to face as they are focused on the body of the patient and therefore any one of the team members may assume that he or she was being addressed, or alternatively ignore what was said without comment.

- The orthopedic surgeons interact informally. They use expressions such as "move over a little," without any clear indication of the manner or the extent of movement. For all that, as far as X-rays are concerned, the situation was far better. The orthopedic surgeons ask each other to move the X-ray machine to obtain angles defined as: AP, lateral, inlet. However, as mentioned above, X-rays are not carried out exactly at these angles, but at approximate angles, making X-ray orientation more difficult (for example, the exact spatial positioning of the drill in the patient's body).
- There was no formal means for requesting instruments from the scrub nurse (an extreme example of this: "Give me that") (*problem #17*).

During surgery, a senior anesthetist replaced the female anesthetist (problem #18). At the same time the female anesthetist's log was updated to 10:30. The changeover was summed up by the statement, "the patient is stable." The senior anesthetist did not inspect the monitors but went to check up on the actions of the orthopedic surgeons. He then went to check the anesthesia form. In the course of the surgery it became clear *that a screw-tap was missing from the sterile kit (problem #19)*.

A member of staff arrives from an adjoining operating room, requesting that the X-ray machine be returned. Half an hour later, someone is again sent from the orthopedic operating room to request that the X-ray machine be returned.

In order to transfer the patient from the operating table to the bed from the induction room, a number of personnel are required to support and transfer the patient in a coordinated fashion. The transfer is made by physically supporting the patient's body or by gripping the sheet on which the patient is lying. At the end of the operation in question, some members of the team took hold of the patient's body, lifted him up and transferred him to the bed. No one verified whether or not the anesthetist, positioned next to the head of the patient, was supporting the patient's head. Furthermore, no one adjusted the bed or the operating table so that both were at the same height. As a result, the difference in height between the table and the bed was almost 10 cm. *While transferring the patient, his head bounced up and down, finally coming to rest on the bed*. Confusion among the participants in the room, both senior and junior, was very evident (*problem #20*).

An intern and senior surgeon #2 verified that the patient was able to move his fingers and his legs. In total, the patient's sojourn in the operating room lasted 240 minutes, of which the surgical procedure lasted 183 minutes.

TROUBLESHOOTING

HUMAN ENGINEERING PROBLEMS

The study documented a number of human engineering problems, mostly in the operating room. These problems are related to defined processes that require specific solutions. During surgery and afterwards an inventory was carried out to verify that no instruments or foreign substances were forgotten in the patient's body. A detailed examination of the inventory revealed error sensitivity stemming from the disparity between the requirements made of the nursing staff and their ability to meet those requirements under existing conditions.

The current practice of ignoring the medical chart and overreliance on memory and informal sources of information was noted. We assume that one of the reasons for this is that formal information sources (such as the patient's medical chart) are not readily available or that the source information is not clear, poorly written, or not updated, and therefore members of the team prefer more accessible information.

A cogently organized patient's chart, accounting for all team members who may access the chart, and the redesign of the charts (assisted by team members) is likely to encourage the search for location of information in the chart and encourage its use. A simple solution such as using transparent binders (the patient's chart is actually a portfolio containing all the patient's charts and labeled with the patient's name and title) should go a long way to emphasizing vital information (for example, if the first page of the binder contains vital information such as allergies, identification details, type of surgery, the side to be operated on, blood type, etc.).

Other medical information, such as substances injected into infusion bags, should also be given prominence. Availability of white sticky labels and indelible, thick writing implements in locations where drugs and infusion bags are being prepared should encourage appropriate listing. Thus, dependence upon the presence of team members who carried out the actions will be neutralized and it will not be necessary to search for them (or in the worst case scenario—to carry out an action on the basis of assumed actions that were or were not carried out). The observations showed that there was no uniform and complete situation picture for all team members, thus giving rise to coordination problems. Every team member was aware of his or her role and his objective. Each member has ability and knowledge that differs from the ability and knowledge of other team members. The general, overall situation was not always clear and familiar to the team overall. For example, we believe that if the departmental nurse were aware of the effects of a delay in the department to continue the process, she could provide advance notice of the delay to the operating rooms (or prevent the delay at the outset). Every member of the team participating in the process should have a clear picture of the process—how to conduct themselves, the character of the different stages, identities of those involved, and ways to communicate with them. Work coordination will be more effective under these conditions.

Almost no operating room procedure is performed by one person only. The processes—from positioning a patient on the operating table, through preparation of drugs for injection to equipment inventory and listing of treatments—are all carried out through the cooperation of several team members. For each process to be carried out efficiently, communication and coordination are needed between team members. Each team member must know exactly how and when to fulfill their respective roles and what to expect from other team members, and so forth. Lack of coordination at the most basic level, such as taking action without vocally reporting to the rest of the team, may result in serious safety incidents and cause harm to the patient or medical team. Today, surgery is conducted without management. Team members operate independently, performing at a rate and in a sequence that they are accustomed to, without adapting to each other. We propose that prior to surgery, the surgeon briefs the entire team who are to be involved in the surgery (including the orderly, anesthesia technicians, medical students, nurses, junior surgeons). This method and its application in practice is described in Chapter 14.

ACKNOWLEDGMENTS

The authors wish to thank Ilan Paltin for his part in organizing the observations and their formal description.

REFERENCES

1. Bonger S.B. (1994). Introduction. In: S.B. Bonger (ed.), *Human Error in Medicine*. Hillsdale, NJ: Lawrence Erlbaum, 1–11.
2. Cook, R.I., Render, M., and Woods, D.D. (2000). Gaps in the continuity of care and progress on patient safety, *British Medical Journal,* 320, 791–794.
3. Copra, V., Bovill, J.G., Spierdijk, J., and Koornneef, F. (1992). Reported significant observations during anesthesia: A prospective analysis over an 18-month period, *British Journal of Anesthesia*, 68, 13–17.
4. Gaba, D.M. (1994). Human error in dynamic medical domains. In: S.B. Bonger (ed.), *Human Error in Medicine*. Hillsdale, NJ: Lawrence Erlbaum, 197–224.
5. Galletly, D.C., Mushet, N.N. (1991). Anesthesia system errors, *Anaesthesia and Intensive Care*, 19, 66–73.
6. Helmreich, R.L. and Schaefer, H.G. (1994). Team performance in the operating room. In: S.B. Bonger (ed.), *Human Error in Medicine*. Hillsdale, NJ: Lawrence Erlbaum, 225–253.
7. Kohn, L.T., Corrigan, J.M., and Donaldson, M.S., eds. (2000). *To Err Is Human: Building a Safer Health System*. Washington, DC: Institute of Medicine, National Academy Press.
8. Kumar, V., Barcellos, W.A., Metha, M.P., and Carter, J.P. (1988). An analysis of critical incidents in a teaching department for quality assurance—A survey of mishaps during anaesthesia, *Anaesthesia*, 43, 879–883.
9. Leape, L.L. (1994). The preventability of medical injury. In: S.B. Bonger (ed.), *Human Error in Medicine*. Hillsdale, NJ: Lawrence Erlbaum, 13–25.
10. Runciman, W.B., Sellen, A., Webb, R.K., Williamson, J.A., Currie, M., Morgan, C., and Russell, W.J. (1993). Errors, incidents and accidents in anaesthetic practice, *Anaesthesia and Intensive Care*, 21, 506–519.
11. Vincent, C. (1993). The study of errors and accidents in medicine. In: C. Vincent, M. Ennis, and R.J. Audley (eds.), *Medical Accidents*. Oxford: Oxford Medical Publications, 17–33.
12. Wilson, M.W., Gibberd, R., Hamilton, J., and Harrison, B. (1999). Safety of healthcare in Australia: Adverse events to hospitalized patients. In: M.M. Rosenthal, L. Mulachy, and S. Lloyd-Bostock (eds.), *Medical Mishaps*. Buckingham: Open University Press, 95–106.

5 Mental Models as a Driving Concept for the Analysis of Team Performance in the Emergency Medicine Department

Shay Ben-Barak and Daniel Gopher

CONTENTS

Introduction 56
An Industrial Engineering System Approach 56
The Mental Model 58
Mental Models of the Team 59
An Index for Gap Assessment between Different Mental Models 61
Process Flowchart Operations Similarity Index 63
    Operations Similarity 63
    Transition Similarity 63
    General Similarity 63
Use of Similarity Index for Performance Assessment 64
    Description of the Department of Emergency Medicine Selected for the
    Field Experiment 64
    Management Policy Model 64
    Mental Models of the Medical Team in the Department of Emergency
    Medicine 66
    Calculation of Similarity Indices between the Team and Management
    According to the Proposed Approach 69
    Calculation of the Similarity Indices among the Team Members under the
    Proposed Approach 72
Discussion 73
Appendix A: Describing and Quantifying Mental Models 74
    Comparing Process Charts 75
    An Example of Comparing Two Process Charts 76
References 77

INTRODUCTION

Until quite recently, the main entrance to a hospital was called the *Emergency Room*, that is, the place where prompt decisions relating to treatment or organization were made. The name was changed to the *Department for Emergency Medicine*, that is, not only for *screening* patients but also for providing preliminary, lifesaving procedures. Both young and old enter via the hospital entrance, including chronically sick patients whose condition has flared up, and speeding ambulances bringing in victims of road accidents. A system was devised to respond quickly to the aforementioned intake—to replace lost blood volume to the wounded and balance the blood sugar level of a diabetic patient. Unlike other hospital departments, the department for emergency medicine operates under difficult conditions: the workload can vary between zero and an indescribable overload; doctors and nurses have to treat a large number of patients simultaneously, patients with whom they are not familiar as in the internal wards; but patients who are admitted on the spot and whose medical problems require immediate attention. However, this is a difficult service to provide, in part because of the dependency on many auxiliary agents in the hospital and because of the need to move patients from place to place and obtain advice from physicians within a reasonable time period, which is usually not reasonable. The department for emergency medicine at times serves as an intake regulator to prevent occupancy of other hospital wards reaching bursting point. The time spent in the department for emergency medicine may be prolonged even though the department itself is short of available beds, that is, beds that have been evacuated where the patient's problem has already been diagnosed and a treatment program already mapped out.

Yoel Donchin

AN INDUSTRIAL ENGINEERING SYSTEM APPROACH

A system is defined as a set of physical components operating together in order to convert selected input to desired output, through a series of work processes governed by predefined rules and conventions. The system concept is used in many areas, such as computerized systems, air conditioning systems, control systems, and automated production, and so forth. Despite the basic differences between these systems, all of them can be defined as complex, high-technology systems, operating in a dynamic environment where the human factor has a minor role, if at all. For example, climate control in a multistory office building can be described as follows:

> If the temperature is lower than 22°C, switch the air conditioner to heating mode. If the temperature is higher than 26°C, switch the air conditioner to cooling mode. In any other situations, do not switch the air conditioner on.
>
> More complex systems also monitor the humidity and switch the air conditioner on or off according to a heat index based on temperature and humidity.

Since these systems are fully configured, most work processes can be described by a series of mathematical equations, or a set of precise analytical definitions. Their operation is relatively easy with respect to performance, control, and monitoring.

However, with more complex systems, particularly systems incorporating human performers, performance analysis is more complicated. For example, cash withdrawals from the bank by personal check can be described as follows:

Verify customer ID, check account status. If the account is positive or does not deviate from the credit limit, update account, and defray the required sum. Otherwise, inform the customer that the withdrawal is not possible.

In the case of structured processes, such as the above process, where employees and operators have a low level of control over the work mode, pace, and style, there are successful quantitative approaches in industrial system engineering to predict, design, optimize, and adjust variables such as capacity, quality, safety, utilization of resources, and so forth, usually on the basis of functional indices, representing system performance.

Alongside these systems, many systems operate wherein the human factor constitutes a central stratum, also as part of the operating system resources and also in control and decision making that affects system performance. With these systems, employees, controllers, and managers exercise broad-based discretion in determining the style of work, the process flow, and implementation sequence. For example, patient admittance to the department of emergency medicine:

Admitting a patient depends on a combination of factors such as age, complaint severity, and personal impressions of the patient's behavior. These factors will direct the medical team member charged with evaluating the patient's condition, categorizing the patient as *mild* and referring him or her to walkabout areas, categorizing the patient as *moderate* and referring him or her to bed areas, or categorizing the patient as *severe* and in need of urgent and immediate treatment.

In all these cases there is considerable dependence between the individual staff member engaged in the implementation process and the actual quality of implementation, that is, the nature of work process implementation depends upon individual perception of the system and understanding the situation at any given moment. Perception and understanding are themselves dependent on an inner subjective global concept that the employee has of the system, its components, how it works, the immediate environment, and his or her performance amid all these factors. This perception is also known as *The Mental Model of the Employee*.[8]

Due to the pivotal role of the human factor in these systems, accepted work processes are fuzzy and not completely understood and therefore it is much more difficult, as opposed to structured systems, to describe and model their performance in accordance with conventional engineering methods, such as queuing theory and linear programming. From the literature[1] the accepted argument is that system performance errors cannot be defined, except by comparison with the accurate and normative behavior of that system. Hence, the lack of an accurate and detailed model describing system operation makes system response to changes, internal or environmental, difficult to predict, and it also makes it difficult to search for points of potential failure. In the absence of appropriate analytical skills, it is far more difficult to make the functioning of the system more efficient and improve it.

In the absence of an engineering-quantitative path to the description of these complex processes, we shall present in this chapter a description of the work processes in nonstructured systems through an approach based on process flowcharts. The power of these charts is to describe the process as perceived by the management, the operators, and by the controllers. The management model, referred to as Management's Operation Policy (MOP), presents the work process as perceived and understood by management, according to how it should be managed, including the utilization of different system resources, staff roles, and relationships between team members operating in the system. This, in fact, is the management mental model. Since management adopts a broad view of the system components in total, including a comprehensive understanding of events in the environment in which the system operates, the work process described must be a reference model for comparing models of the individual items operating in a system. The claim found in the literature[2-6] perceives a common process for employees and similarity between this approach and management policy creating adjusted and more efficient performances, limiting the probability of failures and mishaps.[7] Diverse implementation or deviation from a defined norm, due to different perceptions or situation pictures or incompatible perceptions or situation pictures, may harm the efficient and coordinated workings of the system and thus cause system failures, which may well claim a heavy toll on human life.

THE MENTAL MODEL

In the natural environment we are required to deal with different systems, some simple, such as an operator and a machine, and some much more complex, requiring coordination, control, decision making, and teamwork. When human operators interact with systems, a mental model is created—a subjective internal model of the system and required role. The mental model is usually a simplification of reality, by breaking down the system—in abstract terms.[8] The mental model differs from the objective model of the system in that human beings tend to unreasonably assess the probability of occurrence of different events[9]; they err in assessing cause and effect relationships and take wrong turns in understanding how different systems operate. Moreover, over time, mental models are not consistent—they are affected by the frequency of recourse to them and develop on the basis of experience. Notwithstanding all these drawbacks, the mental model of the system provides tools to describe and explain the reality and tools for predicting according to the given situation. There is a connection between the mental model of a worker and his or her level of operation in the work environment: as the mental model converges on the objective model, which describes the behavior of the system, improvement of system assessment, selection of actions to be taken, and prediction of system behavior is more than evident.

Professional literature dealing with mental models has not ignored the medical environment; however, the focus of most research studies has been on physiological understanding and clinical conclusions (diagnostic) and has not touched on environment mapping or management of medical procedures. A number of studies have dealt with comparisons between specialist doctors and trainee doctors in terms of the respective approach to problem solution, reasoning, and decision making. The conclusion[10] was that experienced doctors speculate diagnostically earlier than trainees and tend to arrive at a more accurate evaluation of the hypotheses on the additional and cumulative information. Apparently, the mental models of specialist doctors[11] lead to fewer errors than those made by trainee doctors, at different levels, and to realistic and better assessments than trainee doctors. In most cases, there is a linear relationship between the level of the doctor's formal expertise and accuracy of his mental model (in terms of predictive and explanatory attributes). From trial studies in the respiratory ICU,[12] a connection was found between the mental models of the nurses and doctors and the probability of error occurrence (clinical). In addition, a major difference was found between doctors and nurses in terms of *ways of thinking*: doctors process more information but tend to focus on additional information search processes and on existing information enrichment processes, while the nurses manage to better organize existing knowledge and are therefore more focused.

MENTAL MODELS OF THE TEAM

In a team framework, team members share common resources and similar information and work toward a common goal. It is therefore patently clear that the team should be well aware of role allocation and respective responsibility, that is, who does what and who is responsible for issuing orders. Research and development teams, whose members are charged with generating original thoughts and unique ideas, are also required to maintain proper communications and appropriate levels of coordination and synchronization. This means that a common mental model to everyone on the team is an important and central factor in determining the level of performance of the team. Other examples of teamwork in a medical system are given below.

A trauma unit and a team is comprised of at least two physicians, nurses, on-call physician, trauma coordinator, and other experts. All these individuals are responsible for the safety of the injured whose life may be in danger. Under these circumstances, the physical location and responsibility borne by each individual are defined. Moreover, the required sequence series of actions to assess and stabilize the patient is defined by procedures issued by the appropriate medical associations, such as trauma protocol.

Operating rooms are where surgeons, anesthetists, nurses, and logistic team personnel operate together to bring about successful surgery. In the course of the surgery, the leadership mantle passes between individual roles, according to a defined procedure.

The department of emergency medicine operates with a team of physicians drawn from different disciplines, nurses, and logistical personnel, all working together to provide care for a wide range of patients at varying levels of severity. Their goal is problem classification, severity assessment, and routing of patients to the appropriate location (ward, operating room, or discharge from hospital and clinic referral). Meanwhile, a few patients undergo physical evaluation and the initial care in the department of emergency medicine. In other words, in addition to clinical and nursing decisions that have to be taken with respect to all of the patients in the system, managerial/operational decisions must also be taken regarding routing of patients (and their related information) promptly and expediently, as well as correct utilization of the department's resources. Routing patients between the different departmental functions and decisions regarding each patient's processing constitute the work of the team of doctors and nurses, who serve as an integral part of system resources.

To perform the tasks described above, team members are required to operate, each and everyone, according to a predefined scenario. For example, the surgeon is required to perform a clinical procedure defined by the type of surgery and the anesthetist is required to perform a different clinical procedure, defined by the patient and type of surgery; nurses as well, scrubbed and nonscrubbed. Regardless of the complexity of the different scenarios that team members are required to act upon, the level of performance of surgical operations is dependent upon the individual skills and experience of each team member, namely the complexity of his or her personal mental model.

Over and above individual capability, the three examples given above describe the required processes with respect to teamwork. It resembles a relay race where the winner is not necessarily the team with the fastest runner but rather the most skillful, more experienced, and better balanced team, capable of transferring the baton from hand to hand with greater efficiency and speed. *Hence, the quality of process execution is not only dependent on the individual ability and skill of each member of the team, but on coordination and communication between team members.* Mission success depends on all members of the team understanding the process, in other words, the presence of a shared mental model, based upon coordination, synchronization, collaboration, and order of execution, differing from the individual mental model focused on the individual's scenario, however complex it may be.

The personal scenario required from the physicians in the shock/trauma unit and operating room is more complex than the personal scenario required from them in the department of emergency medicine. In contrast with other studies dealing with mental models, mainly the physicians' physiological understanding and their clinical reasoning,[11,12] this chapter focuses on the systemic concept and shared mental model required for correct, safe, and efficient implementation of the various stages of the process. The importance of the shared mental model and its centrality depend upon the degree of comprehension of each of the three processes described above. The first situation is structured and well defined, as the order of work and procedures are dictated and clear to all concerned and the degree of freedom of team members for independent action is lower, therefore the lesser the importance of the mental model, even if the order of work and operational procedures in the department of emergency medicine are much more complex. Patient information emanates from several sources: patients themselves, their families, results of clinical tests, and samples sent to the laboratory and imaging results. The medical team is required to take all these factors into a single account. However, this combination is not at all simple, since immediately following admittance the patient becomes two separate entities: the individual and the individual's information. For example, a sample taken from the patient's body becomes an item of information that flows along in the system independently of the patient. This information should coincide with the patient at specific points in the process enabling the medical team in the department of emergency medicine to make appropriate decisions on routing the patient. Therefore, the department of emergency medicine is a much less defined and structured environment, hence the shared mental model, coordinated between the various team members is crucial and has an immediate effect on the quality of performance of the department of emergency medicine.

From what has been said thus far, it is clear that the method of implementation of unstructured processes is dependent upon individual mental models of the process that applies to each and everyone in the team. Therefore, team members working together in the department of emergency medicine retain differing perceptions and insights regarding the system, and uncoordinated implementation is liable to interfere with the process, disrupt, or prevent the transfer of data and eventually lead to loss of information, damaging patient treatment. From the literature,[13] these events are referred to as gaps or mismatches in the mental model of the team. A single or several mismatches and gaps are not necessarily decisive evidence of an inefficient or an unresponsive system, since most of them are foreseeable, detectable, and treatable by team members spearheading the system.[14] However, a single gap in the team member mental models increases the probability of error occurrence.

Many of the events referred to as *human error* actually stem from multifactor system failure. That is to say, the cause of human error is not the level of performance of team members at the forefront of the system but a combination of circumstances comprising many factors. The localized failure referred to as *human error* is a necessary, but not generally a sufficient, condition.[13,15] For a localized failure to become a systemic failure specific conditions are required and at times, additional localized failures may appear together or correlate with one another. The greater the number of gaps in mental models of the team members, the higher the probability that these gaps will arrange themselves in such a way that a localized event becomes a systemic failure. Therefore, in order to raise the level of safety in medical systems and improve the level of performance, the gaps and their means of formation should be investigated, and the gaps located through training should be bridged, with changes in procedures and other organizational changes.[14]

AN INDEX FOR GAP ASSESSMENT BETWEEN DIFFERENT MENTAL MODELS

This chapter presents a quantitative index that allows the differences between various mental models to be estimated and therefore a comparison to be made between team members operating in the system (physicians and nurses) and an evaluation of the level of coordination between the subjective perceptions of the staff and the model describing management policy, MOP. The role of the index is to assist the prediction of system behavior by detecting gaps between the different mental models. Detection will enable focusing upon issues that require administrative intervention to improve coordination, teamwork, and overall system performance.

The engineering literature presents different tools for describing the processes, some graphic, some mathematical/logical, and some based on imaging. The process flowchart was selected from these tools—simple-to-apply engineering graphic tools able to clearly display all key factors in the process: the necessary resources, activities, decisions to be taken, the logical relationship and probabilities of referral, and routing. This tool is general enough to describe a wide range of processes covering different areas, is highly valid, readily understood, and can be learned and put to use quickly and easily. For example, consider a complex process consisting of nine A–9 operations, exploiting five resources. The hypothetical flowchart below sets out the different process operations, according to resources, according to their performance, and according to logical relationships between the operations. The process starts with A; in 30% of cases the process ends at D and in 70% of cases, at B. We assume that for the process described in Figure 5.1, the division of labor between the various resources is as follows: α performs the tasks A and D, β performs operation B, γ performs operations C, E, and H, δ performs operations F and I, ε performs operation G. This allocation of responsibility can be displayed in the process flowchart.

From the resulting process flowchart, a graph of vertices and arcs may be derived—vertices represent operations and arcs represent links and routing between operations, implying that each and every process flowchart can be represented by an adjacency matrix, showing the relationship between the various graph vertices (1 if there is a connection, 0 if there is no connection; cross-references between vertices are probabilistic and represented numerically from 0 to 1). For example, the process flowchart shown in Figure 5.1 can be represented by the adjacency matrix shown in Figure 5.2. The resulting matrix describes the individual's cognitive map whose process flowchart, from the literature,[16] was used to compare perceptions of different details about the systems they operate and the processes they perform.

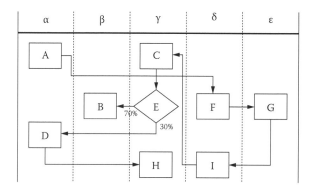

FIGURE 5.1 Illustration of a hypothetical flowchart.

	A	B	C	D	E	F	G	H	I
A	—	0	0	0	0	1	0	0	0
B	0	—	0	0	0	0	0	0	0
C	0	0	—	0	1	0	0	0	0
D	0	0	0	—	0	0	0	1	0
E	0	0.7	0	0.3	—	0	0	0	0
F	0	0	0	0	0	—	1	0	0
G	0	0	0	0	0	0	—	0	1
H	0	0	0	0	0	0	0	—	0
I	0	0	1	0	0	0	0	0	—

FIGURE 5.2 Association matrix of the flowchart elements depicted in Figure 5.1.

PROCESS FLOWCHART OPERATIONS SIMILARITY INDEX

OPERATIONS SIMILARITY

When comparing two process flowcharts (physician—a nurse, physician A—physician B, etc.) the similarity index between process descriptions can be evaluated by comparison from a group of three values:

1. Number of joint operations in two process flowcharts.
2. Number of activities listed in the first flowchart, but not in the second flowchart.
3. Number of activities listed in the second flowchart, but not the first flowchart.

These groups allow the calculation of an average operation similarity index, as detailed in Appendix A at the end of this chapter.

TRANSITION SIMILARITY

Calculation of the transition similarity index of operations and interoperation transitions (see Appendix A) is similar to calculation of the operation similarity index.

GENERAL SIMILARITY

The overall score of similarity between two process flowcharts is the average of the similarity scores of operations and interoperation transitions (see Appendix A).

USE OF SIMILARITY INDEX FOR PERFORMANCE ASSESSMENT

The department of emergency medicine constitutes a nonstructured working environment that provides extensive freedom of action and freedom for decision making to the team of physicians and nurses working in the department and coordination between team members is critical to the functional goals of the department. Therefore, significant importance is attached to efficient assessment of the work of the team and its performance. This assessment should be made at frequent intervals to enable performance of staff to be monitored and changes and trends in patterns of work and performance to be identified in a timely manner while in their formation. Team performance should be evaluated after new procedures and patterns of work have been decided upon and after the introduction of new technologies. The approach presented here for calculating the similarity between process flowcharts representing subjective descriptions of work processes in an appropriate system is particularly suitable for evaluating gaps between mental models of team members (physicians and nurses) and comparing these models to the model representing management policy. To test the proposed approach, a field experiment was conducted in one of the local hospitals as part of a Master's degree thesis at the Faculty of Industrial Engineering and Management at the Technion.[17]

DESCRIPTION OF THE DEPARTMENT OF EMERGENCY MEDICINE SELECTED FOR THE FIELD EXPERIMENT

A hospital with 444 beds and 18 wards was selected for the experiment. The hospital serves a population of half a million people and some 70,000 patients per year are seen in the department of emergency medicine. The department of emergency medicine was divided into three primary function areas. Patients are referred to the different areas according to medical considerations, taking into account the urgency of the situation. Patients admitted to the first area are those diagnosed with medical problems at a level requiring immediate hospitalization and they make up about 20% of all those seen in the department of emergency medicine. Patients admitted to the second area are those requiring surgical or orthopedic treatment and constitute about 60% of those seen in the department of emergency medicine. The third area deals with patients whose condition does not require hospitalization.

Department of emergency medicine physicians operate in different areas according to their areas of expertise: internists in areas one and three; surgeons and orthopedists in area two. The nurses operate in all areas according to the work schedule.

MANAGEMENT POLICY MODEL

In order to demonstrate the management policy model, interviews were held with the senior staff of the department of emergency medicine: the head of the department of emergency medicine (who is also the physician responsible for areas 1 and 3), the physician responsible for area 2, and the head nurse. As a result of the interviews, seven characteristic patient types were identified: internally ill mobile, internally ill, complex internally ill (requiring care by two nurses), light trauma, minor surgery, complex surgery, and orthopedic patient.

After a brief explanation of the nature of the process flow diagrams and possibility for translation of the reality to a flowchart, senior team members were requested to expand the process that all categories of patients identified in the department of emergency medicine were required to go through. The final stage entailed a combination of team members' individual perceptions of the one and only management policy. This process was performed a few times and encompassed updating of the flowcharts until final approval by all senior team members. The process flowcharts of typical patients are shown in Figures 5.3 through 5.7.

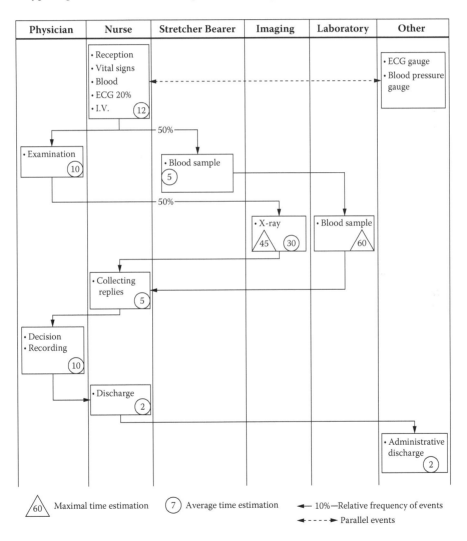

FIGURE 5.3 Process chart: Internal patient—fast track.

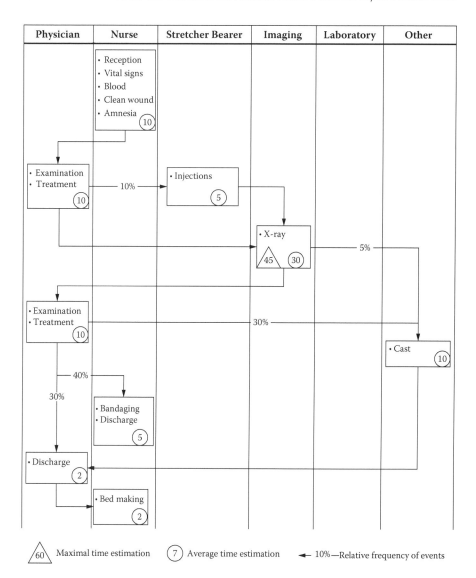

FIGURE 5.4 Process chart: Trauma patient—minor trauma.

Mental Models of the Medical Team in the Department of Emergency Medicine

After demonstrating the mental model of management (management policy) we moved on to demonstration of the mental models of members of the medical team. The study encompassed five physicians and four nurses, all members of the permanent team. The physicians, the professional core of the department of emergency medicine (apart from the management team), operate only on the day shift, from 8:00 A.M. to 4:00 P.M., while other shifts operate across all hours of the day and

Mental Models of Team Performance

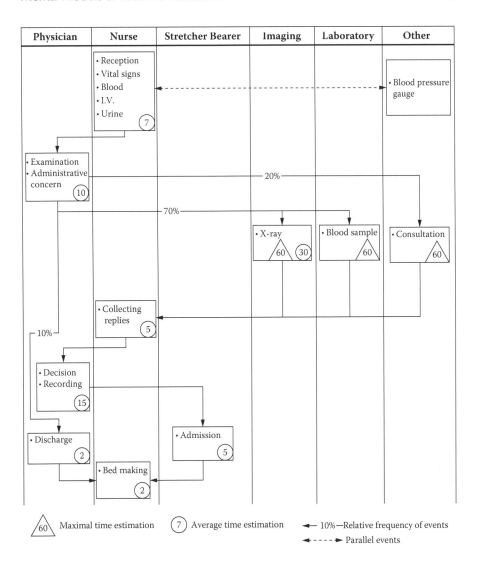

FIGURE 5.5 Process chart: Trauma patient—minor surgical.

night. The nurses are fully aware of the daily operational nature of the department of emergency medicine, as well as evening and night shifts.

The five physicians and four nurses were asked to identify the patient characteristic types. The comparison shows that a number of the doctors and nurses did not identify all typical patient types relevant to their medical expertise as defined by management. For example, one physician does not indicate any difference between patients he meets but defines them all as "internally ill," contrary to the opinions of two other physicians with similar training and roles; one nurse identifies five types of patients, while another nurse identifies three types only (Table 5.1).

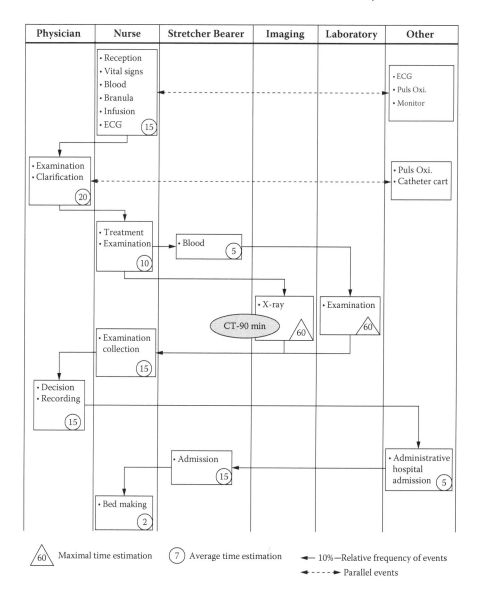

FIGURE 5.6 Process chart: Trauma patient—surgical.

In the second stage, members of the medical team were asked to describe the process flowcharts corresponding to each typical patient type as noted. Extraction of the mental models was carried out in a similar fashion to the extraction of senior management models, from interviews to approval of personal process flowcharts of each and every one of the physicians and nurses.

Mental Models of Team Performance

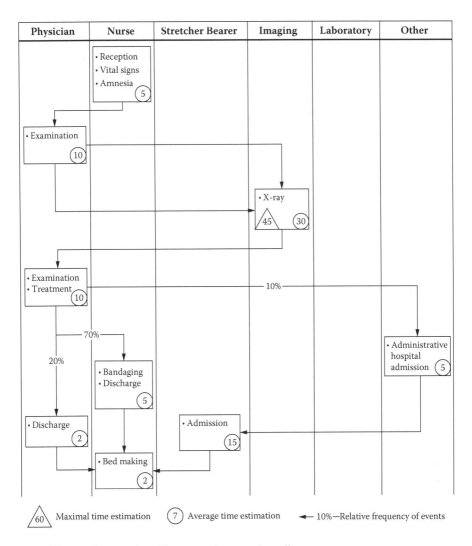

FIGURE 5.7 Process chart: Trauma patient—orthopedic.

CALCULATION OF SIMILARITY INDICES BETWEEN THE TEAM AND MANAGEMENT ACCORDING TO THE PROPOSED APPROACH

After extracting the mental models of the physicians and nurses who participated in the study, similarity indices were calculated according to the proposed approach between these models and the models representing management policy. Since doctors and nurses are crucial resources to the success of the process, the weighting attributed to them was 2, while weighting 1 was attributed to the remaining

TABLE 5.1
Patient Categorization According to Members of the Medical Team Participating in the Study

Management Categorization	Minor Trauma	Minor Surgery	Complex Surgery	Orthopedic	Internally Ill–Mobile	Internally Ill	Internally Ill–Complex
Physicians							
MD1					√	√	√
MD2						√	
MD3					√	√	√
MD4	√			√			
MD5	√	√					
Nurses							
N1	√				√	√	√
N2	√				√	√	
N3	√			√	√		√
N4	√	√			√	√	√

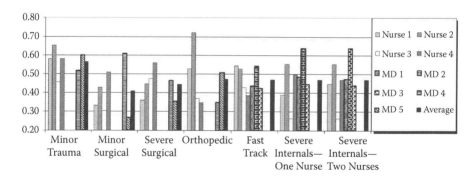

FIGURE 5.8 Overlap with reference model.

resources allocated to the process. Where members of the team did not identify the characteristic patient type as defined by management, the nearest process flowchart of the same team member was used for comparison purposes. Figure 5.8 shows the calculated similarity index values between the physicians' and nurses' models and management models.

Comparing process flowcharts as described by physicians and nurses to management descriptions indicates an average value 0.47 for the similarity index of 0.47; the maximum similarity (0.56) is obtained in the case of a minor trauma patient and minimal similarity in the case of a minor surgery patient (0.41). These similarity values indicate incompatibility between the medical team's perception of the accepted work processes and management policy.

Figure 5.9 shows the process flowchart of an orthopedic patient as perceived by Physician #4. The similarity index between this process and the management model is 0.35.

Figure 5.10 shows a flowchart of a fast track medical patient as perceived by Nurse #4. The similarity index between this process and the management model is 0.38.

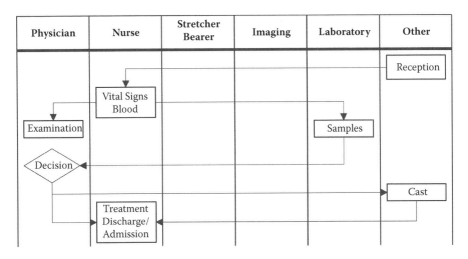

FIGURE 5.9 Process chart of an orthopedic patient as captured from MD 4.

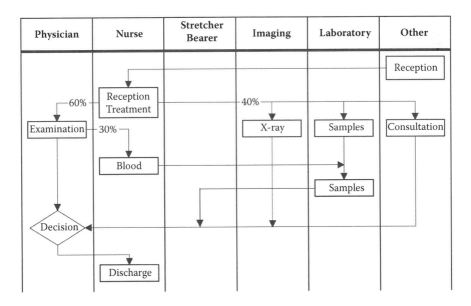

FIGURE 5.10 Process chart of a fast-track patient as captured from Nurse 4.

These similarity indices are very low, but not surprising when taking into account the fact that Nurse #4 points out that in the case of some 40% of patients defined as *mobile internally ill* the nurse is the one to decide on the need for X-rays, specialist advice, or laboratory tests, without consulting the physician.

CALCULATION OF THE SIMILARITY INDICES AMONG THE TEAM MEMBERS UNDER THE PROPOSED APPROACH

The professional literature reveals a close link between the quality of procedure performance and the degree of agreement that exists between team members who carried out the procedure. To test the degree of agreement, the similarity index within the team was calculated for each identifiable patient type. Table 5.2 presents four similarity indices: similarity among physicians, similarity among nurses, similarity between physicians and nurses, and similarity taking into account all the relevant team members (the number of comparisons that make up the average is in parentheses).

According to Table 5.2, the best agreement is obtained for the required process to admit a mobile internally ill patient to the department of emergency medicine. The effect of the result is significant as it is based on 21 comparisons and also correct for comparison among nurses and also for comparison between physicians and nurses. Furthermore, it should be noted that six out of seven members of the medical staff identified this type of patient. And at the same time, the highest degree of overall disagreement was obtained for the required processing for admittance of an orthopedic patient to the department of emergency medicine. This result also has a significant effect as it is based on 15 comparisons and is correct for comparison among nurses and comparison between physicians and nurses. This result is not surprising given that only two team members of the six members of the relevant medical team identified this type of patient, even though approximately 28% of all patients arriving at the department of emergency medicine are defined as orthopedic patients. The agreement among doctors regarding the process that a patient defined as *minor trauma* goes through and the disagreement regarding the process that a patient defined as *minor surgery* goes through are not significantly valid as they are based on one comparison only.

TABLE 5.2
Similarity Indices among Medical Team Members

	Minor Trauma	Minor Surgery	Complex Surgery	Orthopedic	Mobile Medical	Medical	Complex Medical
Physicians	0.73 (1)	0.34 (1)	0.58 (1)	0.38 (1)	0.52 (3)	0.60 (3)	0.54 (3)
Nurses	0.49 (6)	0.44 (6)	0.43 (6)	0.42 (6)	0.72 (6)	0.47 (6)	0.53 (6)
Physicians and Nurses	0.52 (8)	0.46 (8)	0.47 (8)	0.38 (8)	0.58 (12)	0.57 (12)	0.55 (12)
Total	0.52 (15)	0.44 (15)	0.46 (15)	0.40 (15)	0.61 (21)	0.55 (21)	0.54 (21)

DISCUSSION

The results obtained at the department of emergency medicine under testing provide highly significant insights regarding the quality of departmental teamwork and existing coordination between members of the medical team and the departmental management. The strength of the index is the ability to point out gaps in perception and weak links that require outside intervention, such as training, guidance, regulatory changes, or tighter team control. Naturally, the similarity index should also be examined after outside intervention in order to evaluate its effectiveness.

The index of agreement between the medical team and the management policy in the department of emergency medicine was relatively low (0.47), but this does not mean that the clinical and diagnostic performance of the department of emergency medicine is incorrect or invalid. It cannot be concluded from the findings that management policy is the optimal policy and the correct and only way to care for patients. In fact, even the management models are a generalization and simplification of the reality. An objective process flowchart of an orthopedic patient, which was based on observations and data collected from information systems at the hospital, is shown in Figure 5.9. The flowchart represents only one patient type out of all the different types of patients arriving at the department of emergency medicine, but it is clear from the degree of complexity of the departmental work process and the degree of coordination required among the members of the medical team working together and between the medical team and management, making operational decisions that affect the department of emergency medicine performance. The obvious conclusion: despite the differences between the flowcharts, management is responsible for the routine operation of the department of emergency medicine, as well as being required to decide on issues such as resource inventory, allocation, and priority setting. This is done on the basis of a systemic policy model. If policy is uncoordinated with the medical team's perception of its role, management decisions may hinder the routine operation of the department of emergency medicine.

Past studies have argued that workers' perspective is narrower and more personal from the perspective of management. The example cited above strengthens this argument. Members of the medical team, physicians and nurses, identify correct types of patients less than management and process flowcharts of physicians and nurses are less detailed than management flowcharts—compared to management models, an average 18% less operations and 13% fewer links for physicians' flowcharts and an average of 16% less operations and 7% fewer links for nurses' flowcharts.

In the department of emergency medicine, the subject of the study, the lowest agreement level among medical team members was obtained with respect to the process describing orthopedic patients. This result can be explained objectively: the physical space in which patients were treated is shared by minor trauma patients and surgery patients, and the orthopedic patients are not treated differentially. Moreover, the physician responsible for the orthopedic patients had no clinical expertise and also, management concerned with making efforts into many other areas, was not active in this field.

The greatest level of agreement among members of the medical team was with respect to the processing of mobile internally ill patients. It seems that in recent years, management devoted special attention to this process and hence, the process underwent updating and changes, including setting up a special treatment area for these patients, with physicians treating them, all specialists in internal medicine. This being the case, it is clear that the proposed index was successful in detecting these differences and indicating the most ordered and most efficient process.

In conclusion, the proposed index can be used as a tool by an organization in which unstructured processes are prevalent. The index allows for detection of ineffective processes or processes with a higher than average probability of negative and problematical events. The organization can determine the processes worthy of resource allocation and order of priority of optimal measures to be taken. Under certain conditions the index may also be used to test the effectiveness of outside intervention in the system, even before receiving the actual findings regarding performance following the implementation of changes. If the editors of the training system change procedures or introduce engineering changes, it is possible to carry out the process of extracting mental models among a sample of employees, to compare models to a standard model (in terms of overlapping of process flowcharts), and measure the effectiveness of processes, their safety, and continuity.

APPENDIX A: DESCRIBING AND QUANTIFYING MENTAL MODELS

In order to assess and quantify the difference between mental models (MMs) of the individuals in a work unit and the MOP model of the work processes in this unit, a common method to describe the MMs has to be used. Such a description should represent resource consumption, causality, and probabilistic transitions of stages within the processes. The process description method chosen in this study is the common Process Chart (PC). This simple method meets all requirements—it enables simple tracking of resource use and consumption, it represents very clearly the precedence dimension of the process while at the same time it has high face validity and is simple to learn and understand even by nonprofessionals. Finally, PCs are generic enough to represent a wide array of processes ranging from health care to industrial systems. A few simple process charts are illustrated in Figure A5.1. For example,

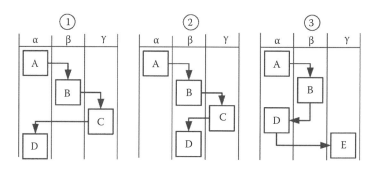

FIGURE A5.1 An example of three different process charts.

Mental Models of Team Performance

all three process charts contain three resources α, β, γ. In PC1 event D, performed by resource α follows event C, performed by resource γ. In PC2 event D by resource β follows again event C performed by resource γ, and so forth.

COMPARING PROCESS CHARTS

As indicated earlier, key factors in representing processes are resource activities and resource transitions (flows). Let us consider the process chart as a graph, which is comprised of nodes (activities) and arcs (resource transitions). Henceforth, any similarity/dissimilarity measure between two process charts has to be based on these two components, denoted hereafter as activities (a) and relationships (r), which make up the graphs (process charts). The first component represents the activities (nodes) of the different resources in the charts, for example, A, B, C, D, and E in Figure A5.1. The activity similarity measure a^{ij} can be obtained using Equation (5.1).

$$a^{ij} = \frac{e^{ij}}{e^{ij} + b^{ij} + b^{ji}} \qquad (5.1)$$

e^{ij} denotes the number of identical activities in process chart i and j (in the coding phase identical activities that carry different names should be recognized). b^{ij} denotes the number of activities that exist in PC i while do not exist in PC j (see that $b^{ij} \neq b^{ji}$). It is clear that $1 \geq a^{ij} \geq 0$. In the case both PCs are identical in terms of their activities (not necessarily their relationships) $a^{ij} = 1$ while $a^{ij} = 0$ if no common activities exist. By definition $a^{ij} = a^{ji}$.

The second component represents the relationships between activities (arcs) in the chart. A relationship is defined by the activities it connects (may be more than one connection between activities) and by the direction of the connecting arc, for example, the arcs that connect activities AB, BC, and CD in PC1 in Figure A5.1. The first step in calculating the relationship similarity measure is to calculate the relationship intensity matrices H^i between each pair of resources k and l for each of the process charts. The relationship intensity is a function of the number of arcs f_{kl}^i between each pair of resources and the weights[*] associated to each resource ω_k, ω_l. This value can be calculated as follows:

$$h_{kl}^i = f_{kl}^i \times \omega_k \times \omega_l \qquad (5.2)$$

In the case of probabilistic relationships the value of h_{kl}^i may be a noninteger number.

Based on the relationship intensity the sum of all the common arcs c^{ij} and the sum of all exclusive arcs d^{ij} between any two process charts i and j can be calculated as shown in Equations (5.3) and (5.4), respectively. Finally, the relationship similarity measure r^{ij} can be obtained using Equation (5.5).

---

[*] Setting the resource weights is not a trivial issue. The weights can be set based on their relative cost or on how critical the relationships between the resources are to the safe completion of the entire process or even based on management wishes. Nonetheless, this issue is beyond the scope of this chapter.

$$c^{ij} = \sum_k \sum_l \min\{h^i_{kl}, h^j_{kl}\} \tag{5.3}$$

$$d^{ij} = \sum_k \sum_l |h^i_{kl} - h^j_{kl}| \tag{5.4}$$

$$r^{ij} = \frac{c^{ij}}{c^{ij} + d^{ij}} \tag{5.5}$$

It is clear that $1 \geq r^{ij} \geq 0$. In the case both PCs are identical in terms of their relationships (arcs) $r^{ij} = 1$ while $r^{ij} = 0$ if no common relationships exist between the two process charts. By definition $r^{ij} = r^{ji}$.

Finally, based on the activity and relationship similarity measures, a combined measure s^{ij}, which defines the overall similarity level between any two process plans i and j, is calculated as the average of the individuals similarity measures.

$$s^{ij} = \frac{a^{ij} + r^{ij}}{2} \tag{5.6}$$

In this model the absence of any resource in one of the process charts has the same effect on the a^{ij} measure regardless of which resource it is. In the case not all resources are of the same importance, thus there are those which are more critical than others, means not all dissimilarities are of equal importance. To incorporate this into the model a weight has to be associated to each resource to reflect its relative importance.

AN EXAMPLE OF COMPARING TWO PROCESS CHARTS

In order to illustrate the calculation procedure of the similarity measure, three process charts, PC1–3 presented in Figure A5.1, are used.

Choosing PC1 and PC2: the following values are obtained $e^{12} = 3$, $b^{12} = 1$, $b^{21} = 1$. Using these values in Equation (5.1) results in an activity similarity measure of $a^{12} = 0.6$. It should be noted, that in the case activity D is performed by different recourses, it should be considered as two different activities.

Choosing PC1 and PC3: $e^{13} = 3$, $b^{13} = 1$, $b^{31} = 1$, and $a^{13} = 0.6$

Choosing PC2 and PC3: $e^{23} = 2$, $b^{23} = 2$, $b^{32} = 2$, and $a^{23} = 0.333$

Using the vector weights (2,1,1) for each of the resources illustrated in Figure A5.1 (α, β, γ), respectively, the relationship intensity matrices H^i for each of the three process charts PC1–3 can be calculated using Equation (5.2) as follows:

$$H^1 = \begin{pmatrix} 0 & 1\times 2 \times 1 & 0 \times 2 \times 1 \\ 0 \times 1 \times 2 & 0 & 1 \times 1 \times 1 \\ 1 \times 1 \times 2 & 0 \times 1 \times 1 & 0 \end{pmatrix} = \begin{pmatrix} 0 & 2 & 0 \\ 0 & 0 & 1 \\ 2 & 0 & 0 \end{pmatrix}$$

$$H^2 = \begin{pmatrix} 0 & 2 & 0 \\ 0 & 0 & 1 \\ 0 & 1 & 0 \end{pmatrix} \quad H^3 = \begin{pmatrix} 0 & 2 & 2 \\ 2 & 0 & 0 \\ 0 & 0 & 0 \end{pmatrix}$$

Using these values in conjunction with Equations (5.3) and (5.4), the common and exclusive relationship values can be calculated as follows:

$$c^{12} = 0+2+0+0+0+1+0+0+0 = 3, c^{13} = 2, c^{23} = 2$$

$$d^{12} = 0+0+0+0+0+0+2+1+0 = 3, d^{13} = 7, d^{23} = 6$$

Based on these values and Equation (5.5), the relationship similarity measure can be calculated as follows:

$$r^{12} = 3/(3+3) = 0.5, r^{13} = 0.222, r^{23} = 0.25$$

Finally, the overall similarity measure between the three process charts illustrated in Figure A5.1 can be calculated using Equation (5.6) as follows:

$$s^{12} = (0.6+0.5)/2 = 0.55, s^{13} = 0.41, s^{23} = 0.29$$

The calculation shows that the highest similarity is between PC1 and PC2 while the lowest similarity is obtained when comparing PC2 and PC3.

REFERENCES

1. Sanders, J.W. and Moray, N.P. (1991). *Human Error—Cause, Prediction and Reduction*, Hillsdale, NJ: Lawrence Erlbaum.
2. Cannon-Bowers, J.A., Salas, E., and Converse, S. (1993). Shared Mental Models in Expert Team Decision Making, In: Castellan, N. and John, Jr. (eds.), *Individual and Group Decision Making: Current Issues*, Hillsdale, NJ: Lawrence Erlbaum, pp. 221–246.
3. Cannon-Bowers, J.A., Tannenbaum, S.I., Salas, E., and Volpe, C.E. (1995). Defining Team Competencies and Establishing Training Requirements, In: Guzzo, R. and Salas, E. (eds.), *Team Effectiveness and Decision Making in Organizations*, San Francisco, CA: Jossey-Bass, pp. 333–380.
4. Kraiger, K. and Wenzel, L.H. (1997). Conceptual Development and Empirical Evaluation of Measures of Shared Mental Models as Indicators of Team Effectiveness, In: Brannick, M.T., Salas, E. (eds.), *Team Performance Assessment and Measurement: Theory, Methods and Application*, Hillsdale, NJ: Lawrence Erlbaum, pp. 63–84.
5. Stout, R.J., Cannon-Bowers, J.A., Salas, E., and Milanovich, D. (1999). Planning Shared Mental Models and Coordinated Performance: An Empirical Link Is Established, *Human Factors*, Vol. 41, No. 1, pp. 61–71.
6. Webber, S.S., Chen, G., Payne, S.C., Marsh, S.M., and Zaccaro, S.J. (2000). Enhancing Team Mental Model Measurements with Performance Appraisal Practice, *Organizational Research Methods*, Vol. 3, No. 4, pp. 307–322.

7. Rouse W.B., Cannon-Bowers, J.A., and Salas, E. (1992). The Role of Mental Models in Team Performance in Complex Systems, *IEEE Transactions on Systems, Man, and Cybernetics*, Vol. 22, No. 6, pp. 1296–1308.
8. Moray, N. (1999). Mental Models in Theory and Practice, In: Gopher, D. and Koriat, A. (eds.), *Attention and Performance XVII: Theory and Application*, Cambridge: MIT Press.
9. Tversky, A. and Kahneman, D. (1974). Judgment under Uncertainty: Heuristics and Biases, *Science*, Vol. 185, pp. 1124–1131.
10. Patel, V.L. and Arocha, J.F. (1998). Expertise and Reasoning in Medicine: Evidence from Cognitive Psychological Studies, In: Singh, I. and Parasuraman, R. (eds.), *Human Cognition—A Multidisciplinary Perspective*, Thousand Oaks, CA: Sage, ch.13.
11. Kaufman, D.R. and Patel, V.L. (1998). Progressions of Mental Models in Understanding Circulatory Physiology, In: Singh, I. and Parasuraman, R. (eds.), *Human Cognition—A Multidisciplinary Perspective*, Thousand Oaks, CA: Sage, ch.14.
12. Badihi, Y., Gopher, D., and Arnan, U. (1993). Analysis of Human Errors in Intensive Care Unit, A Technical Report Research Center for Work Safety and Human Engineering HEIS-4-93, Haifa, Israel.
13. Cook, R.I. and Woods, D.D. (1994). Operating at the Sharp End: The Complexity of Human Error, In: Bogner, M.S. et. al. (eds.), *Human Error in Medicine*, Hillsdale, NJ: Lawrence Erlbaum, pp. 255–310.
14. Cook, R.I., Render, M., and Woods, D.D. (2000). Gaps in the Continuity of Care and Progress on Patient Safety, *BMJ* 320, pp. 791–794.
15. Woods, D.D. and Cook, R.I. (1999). Perspectives on Human Error: Hindsight Biases and Local Rationality, In: Durso, F.T., Nickerson, R.S., Schvaneveldt, R.W., Dumais, S.T., Lindsay, D.S., and Chi, M.T.H., *Handbook of Applied Cognition*, New York: John Wiley & Sons, pp. 141–171.
16. Langfield-Smith, K. and Wirth, A. (1992). Measuring Differences between Cognitive Maps, *Journal of the Operation Research Society*, Vol. 43, No. 12, pp. 1135–1150.
17. Sinreich, D., Gopher, D., Ben-Barak S., Marmor Y., and Lahta, R. (2005). Mental Models as a Practical Tool in the Engineer's Toolbox, *International Journal of Production Research*, 43, pp. 2977–2996.

6 Magnesium Sulphate Dosage—Analysis of Problems Involved in the Medication Administration Process

Efrat Kedmi Shahar and Yael Einav[*]

CONTENTS

The Study Objective 81
Method 81
Agenda 81
Tools 82
    Human Factor Engineering Method to Detect and Analyze Work Safety
    Problems 82
    Cognitive Function Analysis 82
    Analysis and Prediction of Errors and Analysis of Factors Affecting
    Performance 82
Findings 83
    Nurse Task Description—Administration of Magnesium Sulphate 83
Analysis of Task Requirements and Possible Errors at Each Stage 83
Discussion 86
    Accuracy of Drug Dosage Calculations 86
    The Nurse's General Knowledge of the Drug and Dosage Calculations ... 87
    The Ability to Obtain Information on the Drug from Outside Sources 87
    Labels on Drug Packages 88
    Magnesium Ampoule Volume 90

---

[*] "The Analysis of the Process of Administering Magnesium Sulfate to Illustrate the Problems Related to Drug Administration Procedures" by Efrat Kedmi Shahar, Yael Einav, Zvi Straucher, and Daniel Gopher, Center for Work Safety and Human Engineering, Technion. Yael Appelbaum, Director of the Assessment Department, Quality Assurance Division, Ministry of Health.

Conclusions, the Effects on Drug Administration Processes, and
Recommended Solutions 90
    Engineering Solutions 91
    Procedural–Organizational Solutions.. 93
    Training Improvements 93
References... 94

A Hospital Erred in Drug Dosage and the Baby Was Left Paralyzed

DRUG DOSAGE ERROR

The baby ▇▇▇▇ suffered from cancer and needed chemotherapy treatment. She was referred to the hospital for treatment. The claim, submitted by Attorney-at-Law ▇▇▇▇, stated that during the baby's treatment in June 2001 in ▇▇▇▇ Hospital, the nurse injected a drug containing 100 times the Magnesium dosage prescribed by the physician treating the baby.

This story, as reported on the Internet, discusses an event of mistaken drug dosage, just one event of many similar events. The process of drug administration in a hospital is an example of a complex work situation, multistaged and risky, and subject to human factor involvement at all stages. Every day, hundreds of thousands of drugs are administered in hospitals, and incorrect drug dosages are the most common among medical mishaps.[1] Studies have shown that abnormal events associated with drugs are the main causes of mortality among hospitalized patients.[2]

The U.S. Food and Drug Administration (FDA) reported 426,109 abnormal drug administration events in 2004 in the United States. Another study showed that in the United States, adverse events, both in and out of the hospital, resulted in more than 7,000 deaths per year.[3] However, the effect of these types of mishaps may still have a significant influence even if the issue is not a matter of life or death. A study conducted at a university hospital shows that if two out of every hundred drug administration events are abnormal, the average cost is $4,700, that is, 2.8 million dollars per year in a hospital with 700 beds.[4]

The situation is no better in Israel. A recent study carried out in two internal medicine departments at the Bnei Zion Hospital in Haifa revealed 13.4 drug-related errors for every 100 days of hospitalization, of which 5.2 errors were related to the drug database (drug name spelling errors, dosage errors, drug interaction errors) and 7.2 errors related to the patient's database (list of drug allergies, drug-disease interaction, and drug-laboratory interaction).

The frequency of adverse events in the administration of drugs and their aftereffects is no trivial matter. As mentioned above, drug administration is one of the most frequent medical processes, a process that every physician and nurse performs dozens of times a day. In order to minimize, as far as possible, the probability of occurrence of adverse events, in depth studies of the drug administration processes should be conducted in order to pinpoint the inherent failings in the process and understand the role that the human factor plays in the process.

In 2000, the Quality Assurance Department at the Israel Ministry of Health requested a team of human factors engineers from the Technion Research Center for Work Safety and Human Engineering to investigate the failures inherent in the process of administering magnesium sulphate. The request emanated from accidents in Mg administration, which will not be detailed in this chapter; the process and adverse events as presented in this chapter are based on general protocols and resultant incidents that were observed during this study.

THE STUDY OBJECTIVE

The purpose of this study was to map and analyze the process of administration of magnesium sulphate, to locate the foci of high probability for errors, and to find practical solutions for reducing the error probability.

Magnesium sulphate (hereinafter *magnesium*) is administered in hospitals for the treatment of a wide range of problems in different departments: prevention of premature labor, treatment of malnutrition (especially with respect to oncology patients), and in emergency cases the drug is used to treat preeclampsia. Magnesium is administered in a mix intravenously. It reduces muscle tension and blocks peripheral nerve transfer. High dosage or incorrect infusion rates may cause serious disturbances to heart rate and severe muscle spasms, even proving fatal.

Magnesium is administered by a nurse, compliant with the directions of a physician. Reported studies focus on the role of the nurse in administrating the drug (not on the physician's written prescription). The role of the nurse in the process begins with reading the physician's instructions for administration of the drug and continues until monitoring the patient's condition after receiving the drug has run its course.

This chapter will present a case study of the magnesium sulphate drug administration process, attempting to demonstrate a broader framework of the nature of the problems that characterize administration of drugs in general. Firstly, the work method is presented, then the findings relating to error risks and their consequences, and finally proposed solutions for the reduction of error risks.

METHOD

Administration of magnesium is performed by a nurse, according to a physician's instructions, as described above. The task study follows the nurse's actions from the moment the instructions for administering the drug are read by the nurse through the stages of calculating and preparing the drug and administration to the patient and finally, monitoring the patient's condition after the start of the infusion. To understand the nurse's role (mental and physical demands) and the nature of the nurse's interaction (human factor) with the physical work environment (devices, materials, accessories, information, workplace, spatial structure, etc.) and the human working environment (medical staff and patients), several tools were brought into use, as described below.

AGENDA

Mapping of the process of magnesium sulphate administration is based on visits to six departments in which the drug is administered, at two local hospitals: gynecology department, obstetrics department, general intensive care unit, pediatric intensive care, department of internal medicine, and surgical department. Magnesium sulphate is administered at different frequencies in all these departments. During visits to these departments, the form of magnesium administration was requested and provided (by departmental nurses) and a practical demonstration given, as well as interviews with the nurses.

TOOLS

After collecting the data, the nurse's role was analyzed using a combination of the following tools.

Human Factor Engineering Method to Detect and Analyze Work Safety Problems[5]

This model is based on the principles of analysis of human factor engineering. The model shows the relationship between environmental characteristics, the characteristics of employee capability and job characteristics, and job demands and manner of execution. The model enables recognition of the existence of gaps between role demands and worker capability, assuming that these gaps can lead to human errors. The model detects these gaps by cognitive function analysis (as described later) and identification of the role demands of the worker.

Cognitive Function Analysis[6–8]

The role analysis encompassed a hierarchical tree-task diagram of the role of the nurse, and breakdown of the task to its cognitive subtasks. The role analysis performed is related to the work environment and specific working systems that the nurse comes into contact with.

Analysis and Prediction of Errors and Analysis of Factors Affecting Performance[9,10]

The purpose of these methods is to understand how role mishaps occur. The study analyzes the role stage by stage and at each stage checks the errors that are likely to occur during the stage execution and identifies the factors that affect the likelihood of error and the impact of the error itself. More can be learned in this way; if and how to recover from errors identified during role performance, in a way that does not involve negative effects.

* * *

As already mentioned, the study encompassed a combination of three working tools. During the study the nurse's attributes were analyzed, role attributes and working environment attributes—to identify situations where there is a gap between the demands of the job and the nurse's ability to implement those demands (i.e., role demands exceed the nurse's capability to fulfill those demands in a given situation). Delineation was made of the linkage between each subtask relating to the nurse's role and errors that may occur at the same stage, and the factors that may affect error probability were evaluated.

After identifying the gaps, error probabilities, and the factors affecting implementation, solutions were proposed aimed at the reduction of the gaps and error probabilities—solutions in the engineering, management-organizational, and training fields.

FINDINGS

This section describes the preparation and administration task and its stages, followed by a description of the requirements made of the nurse at every stage and the possible errors.

Nurse Task Description—Administration of Magnesium Sulphate

The nurse administering magnesium sulphate to a patient begins by reading the physician's orders from the Instructions Form. According to the written instruction the nurse calculates the volume of the drug to withdraw from the original package of the drug. After making the calculation, the nurse withdraws the calculated drug volume from the bottle and dissolves the drug in the IV bag to obtain a solution at the required fluid concentration. A further calculation is then made—the number of drops of solution per minute that the patient should receive (drip rate). All these operations are carried out at the nurses' station.

The nurse approaches the patient, attaches the IV tubings to a flow regulator or syringe pump, sets the regulator to the calculated drip rate, and starts the infusion. Then, the next step is to monitor the effects of the drug on the patient.

The task is divided into five stages (see Figure 6.1):

1. Reading orders for administration of magnesium.
2. Calculating the dosage.
3. Drug withdrawn from Mg original bottle is dissolved.
4. Calculate drip rate: Number of drops per minute.
5. Drug administration: Set the regulator and attach infusion tubing to patient.

ANALYSIS OF TASK REQUIREMENTS AND POSSIBLE ERRORS AT EACH STAGE

Task performance imposes on the nurse a complex set of requirements. Figure 6.2 displays the task requirements and the possible errors at each stage. The role requirements are summarized below:

Sensory requirements—Absorption of visual information (physician's instructions, reading calculations, reading the label, etc.) and tactile information (drug container, use of syringe, IV kit, etc.).

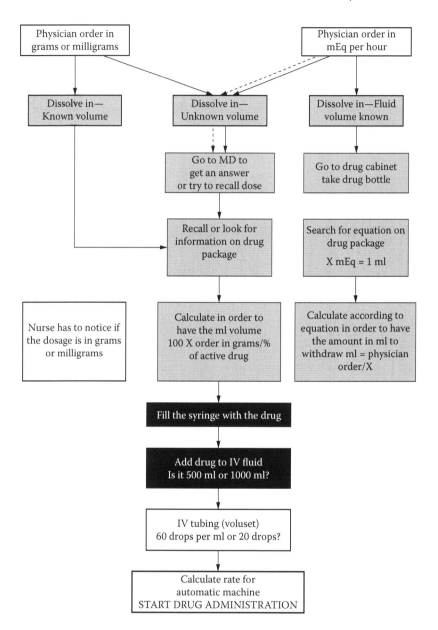

FIGURE 6.1 Magnesium administration flowchart.

Knowledge requirements—The nurse needs to be familiar with the drug to be administered, in all respects—possible side effects, excessive dosages, means of administration, and so on. In order to make the correct calculations the nurse must have a thorough knowledge of pharmacological calculation (for example, expressing 50% concentrated solution in grams per millimeter). The nurse is also required to

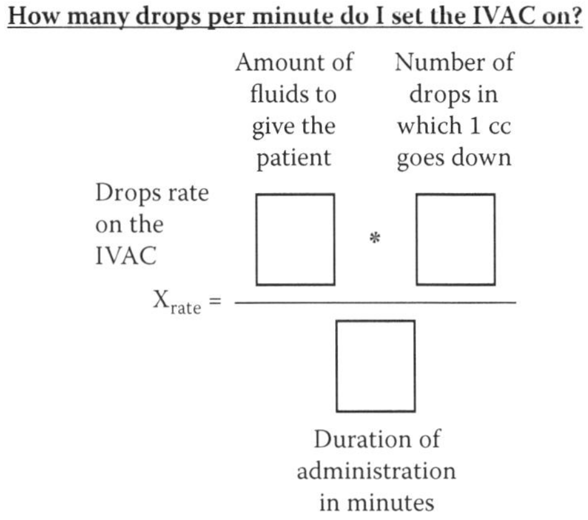

Magnesium Sulphate Dosage

How many drops per minute do I set the IVAC on?

$$X_{rate} = \frac{\text{Amount of fluids to give the patient} * \text{Number of drops in which 1 cc goes down}}{\text{Duration of administration in minutes}}$$

(Drops rate on the IVAC)

FIGURE 6.2 Calculation of magnesium drops per minute.

be intimately familiar with stages of implementation and various work procedures, and know how to deal with errors or uncertainties at each stage. The nurse must have the knowledge to be able to fully understand the physician's instructions. For example, an instruction such as "1 gram per hour in a solution of 1000 cc for a period of 10 hours" signifies that "the patient should receive 1 g every hour," *not* that "each gram should be diluted in 1000 cc." The nurse must know that the meaning of the instruction is "to inject 1000 cc over a period of 10 hours."

Requirements for information processing and decision making—The nurse is required to process information while making pharmacological calculations and to integrate information from different sources (physician's instruction, dilution in the bag, infusion set details, etc.). The nurse must make decisions regarding the results of calculations—whether they are logical or not. What action should be taken if the nurse is not confident of his/her knowledge? As mentioned in the previous section, the nurse must decide on the period the transfusion has to last, as this information is not always explicitly written down.

Memory requirements—The nurse must rely on short-term memory to remember the results of calculations and interim results; the nurse must also rely on long-term memory for control operations in order to recall information on the drug and dosages administered in the past.

Attentiveness, alertness, and concentration requirements—The nurse is required to maintain high levels of alertness and concentration, as she needs to make many calculations that demand mental effort. The nurse is also required to divide attention between various operations being performed within the framework of the process. Attentiveness demands are of the highest order where a dangerous drug is involved and incorrect dosages could have fatal consequences.

As shown above, administration of magnesium is a complex process with a myriad of demands made on the nurse carrying out his/her role. The medical calculations cannot be taken lightly and the stages and operations are numerous. Gaps between worker capability and role demands are liable to reveal themselves in this situation, gaps that could result in errors as shown in Figure 6.2.

DISCUSSION

Analysis of the *magnesium sulphate administration* process revealed main focal problems: (a) accuracy of dosage calculations, (b) the level of nurse's knowledge of the drug and its dosage calculations, and (c) the ability to obtain information from outside sources. As detailed below, these focal problems are influenced by the task, human and physical environment, as well as the general working conditions.

Accuracy of Drug Dosage Calculations

Administration of magnesium sulphate requires calculation of the amount of fluid to be withdrawn from the drug ampoule, the number of drops per minute, as well as conversion of units of measurement (if the physician's instructions are not in units that the nurse is used to). These operations require application of different formulas, that is, administration of a single unit drug requires many calculations that could give rise to confusion.

The ability to make accurate calculations is affected by the ability to divide one's attention, information processing ability, alertness, and concentration. The nurse prepares the drug at two locations: the nurse's station, where the patient's chart is checked, calculations are made, fluid extracted from the bottle containing magnesium, and filling the infusion bag; and at the patient's bedside, where the IV bag is attached to a regulator and the drip per minute rate is set. The nurses' station is at the center of the ward. Patient information and their files are kept at this location available for those seeking information (family members, patients, visitors). Usually, the location is busy, noisy, and a hive of activity and irritation. The patient's bedside is also problematic—there are usually other patients around as well as the patient's family. And the nurses, in addition to the task of attaching the infusion bag to the regulator, have to consider other patients at the same time.

It is clear from this description that the nurses on their shift work in a noisy and crowded environment, which alone creates a cognitive load, a heavy and pressurized load, while they are performing several tasks simultaneously and are often distracted. The nurse's working conditions can lead to calculation errors, leading to an erroneous result assumed to be correct, thus avoiding a repeat examination of the result or consultation with another nurse. The ability of nurses to maintain alertness and concentration over time under these conditions is not high and therefore the probability of calculation error by the nurse who is required to make tens of calculations daily, especially when it comes to multiple calculations for the administration of magnesium.

The case cited at the chapter outset is one example of miscalculation. We have not studied the case in depth and therefore must qualify our statements: under the working conditions of the nurse, when making one medical calculation of many, an error in calculation is not unlikely, simply moving the decimal point one place increases the dosage a 100-fold.

THE NURSE'S GENERAL KNOWLEDGE OF THE DRUG AND DOSAGE CALCULATIONS

Knowledge required to perform the task of administering magnesium sulphate is: familiarity with the drug, working knowledge of pharmacological calculations to be performed, its administration, minimum and maximum limits of dosage administration, accepted dosages, and normal and abnormal effects of the drug.

Medical calculations require identification of the required calculations and means of implementation. Pharmacological calculation knowledge is acquired at the nursing school stage and is not subsequently updated or refreshed. In actual role performance, the nurses do not apply all the calculations learned, as they specialize in specific departments, limiting their proficiency to calculations regularly required for their usual duties. During the study, discrepancies were found between departments in the frequency of administration of magnesium sulphate. In most hospital departments, the drug is administered quite rarely, and the nurses in these departments do not recall the calculations for this drug, consequently, a higher error probability. On the other hand, nurses engaged in administering magnesium sulphate on a more frequent basis (department of obstetrics, for example) reported, as expected, that they remember by heart the calculations, the quantities, and the conversion factors. Moreover, if the nurse needs help on this issue, dosage tables are available in the department (to obviate the need to calculate).

Analysis of the nurse's ability revealed a lack of knowledge among nurses who administer magnesium infrequently. We discovered that not all nurses are familiar with magnesium to the extent that ensures its safe administration. Not all nurses are aware of permissible dosage ranges, the accepted dosage levels, risks, and side effects. Therefore, those unfamiliar with the details of this information would not be familiar with the stages of implementation, dealing with situations of uncertainty or errors, and ways and means of error prevention. In this situation, such lack of knowledge makes it difficult to exercise control over the results of the calculations—the nurse does not know whether or not the calculated dosage is within the permitted range or whether or not a mistake has been made and an overdosage calculated. Moreover, it is difficult for the nurse to properly monitor the patient's condition after the drug has been administered, since he/she does not recognize the expected or abnormal reactions.

The knowledge necessary to perform the role (under the given conditions) is far greater than current knowledge in practice. Nurses administering the drug very rarely know enough about the drug and do not know how to calculate the dosage. This increases the error probability in dosage, errors that cannot be detected due to drug unfamiliarity.

THE ABILITY TO OBTAIN INFORMATION ON THE DRUG FROM OUTSIDE SOURCES

Familiarity with magnesium is imperative for safe administration, as described above. Nevertheless, in the departments where administration of magnesium is not commonplace, there is no database accessible to the nurse without him/her having to make unreasonable efforts. Where should information about magnesium be found?

There are different ways and means to obtain information on magnesium in the hospital: on the ampoule (but this information is limited and does not relate to administering the drug; see below); drug manuals (usually found at the nurses' workstation); from nurses/other physicians; or from pharmacists at the pharmacy. However, in most cases the nurses do not use these information channels, as obtaining information through these channels requires an unreasonable effort. Drug manuals demand investment of time in searching for and reading the information—and the time is not available. Other team members are usually busy and are also working under pressure and are unable to offer any assistance. Finding information about the drug demands from the nurse investment of unreasonable time and effort, and therefore the chances that the nurse will carry out the search are low. This state of affairs increases the probability of dosage errors that are not evident due to lack of familiarity with the drug.

Most of the nurses, if not all of them, try not to administer unfamiliar drugs. The complexities involved in searching for information forces most of the nurses to search for information at the most accessible place—the ampoule or the original package of the drug. Magnesium sulphate comes in alternative packaging formats—varying in volume, solution concentration, and labels. Two companies are marketing the drug to hospitals in Israel. It is likely that one could find a different bottle of magnesium in each hospital department. Original bottles containing the drug may be found in a few wards (as received from the drug company), while in other wards one may find bottles of solution prepared by the hospital pharmacy.

LABELS ON DRUG PACKAGES

Although the drug ampoule is the nurse's main source of information on the drug they need, the information needed to safely administer the drug is not included. The inscription on the label is confusing and can lead to errors. Thus, the error probability of dosage in the process of drug administration increases significantly. The findings with respect to the labels on the medicine bottle were as follows:

> Each type of ampoule has its own label. Every type of an ampoule contains a different concentration (from 25% to 50%). There is even no uniformity of information presented on the various types of packages: the pharmaceutical company or the pharmacy decides for itself what should be written on the label. For example, vital information such as allowable magnesium dosage ranges is missing, that is, no information on the maximum amount of material that may be administered and no warning that higher dosages could have fatal consequences.
> The listing on the labels supplied by the pharmaceutical companies shows:
> 4.06 mEq Mg^{++}/ml or 4 mEq Mg^{++}/ml, that is, for every milliliter there are 4.06 mEq of magnesium sulphate (see Photo 6.1). The nurse is almost never required to use this unit of measurement, as the instruction for administration is usually listed in grams. A nurse needing to calculate the drug dosage looks for the amount of substance in grams per milliliter of solution. What is written on the label is confusing. A nurse without sufficient knowledge, a nurse who is inexperienced in administering

Magnesium Sulphate Dosage

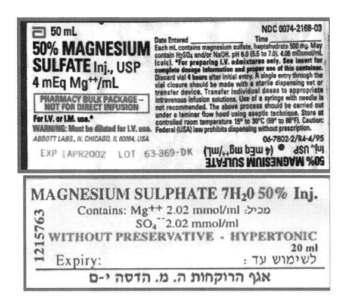

PHOTO 6.1 (See color insert.) Magnesium sulphate bottle—front and back labels.

the substance, a nurse under time pressure, or a nurse who is distracted by an additional task to be performed is liable to presume that each milliliter contains 4 mg of magnesium and not 4 milliequivalents (mEq) of magnesium, and hence a mistaken dosage is a certainty. A bottle containing 50 ml at a concentration of 50% has 500 mg (0.5 g) of magnesium in each milliliter of solution. A nurse can easily inject 125 times (500/4) the amount to be injected.

There is no clear and emphasized written reference that each milliliter of solution contains 500 mg of magnesium (50% bottle concentrations). A clearly written statement would exempt the nurse from pondering the significance of the solution concentration. This would seem to be a simple solution under optimal conditions (realization that 50% signifies that each milliliter contains 500 mg or 0.5 g) but not at all simple under the nurse's working conditions—noise, crowdedness, simultaneous multitasking, and rarely needing to make this calculation.

The magnesium element is notated as: Mg^{++}. This notation is very similar to the notation for milligram measure—mg. A nurse, who does not know the volume of substance per milliliter of solution and is searching for this information, could easily confuse mg (milligram) with Mg^{++} (magnesium). Because specific information is anticipated (volume in milligrams, as per the physician's instruction), information similar to the anticipated information could cause confusion. The probability of such confusion increases when the nurse's concentration and attention are not focused exclusively on searching for the information (for example, when under pressure or performing a number of tasks simultaneously). The mistake described would result in dosage error.

As if that were not enough, a spelling mistake was noted on the label of the bottle supplied by one of the companies: the magnesium element was notated as mg^{++} instead of Mg^{++}. It goes without saying that this is a critical mistake.

Magnesium Ampoule Volume

Another item of external information available to the nurse is the volume in the ampoule. The volume of a magnesium bottle may be 50 ml at 50% concentration. In other words, each bottle contains 25 grams of active substance. This amount of magnesium is much greater than the lethal amount (permissible adult dosage—4 mg per day). As the solution does not contain a preservative (as written on the label) and since the Mg package must be discarded after a number of hours, the size of the bottle is misleading when considering the accepted amount for injection purposes. In fact, almost always only a small percentage of the solution is extracted from the ampoule, we can use a flacon instead of bottle after which the flacon is discarded. Therefore, when a nurse obtains, for example, a result indicating that 0.16 ml should be extracted from the flacon, the calculation may seem logical, as the bottle contains 50 ml of solution, an amount 300 times greater than the calculated amount. This situation is misleading as the nurse is well aware that the bottle is intended for one time use.

CONCLUSIONS, THE EFFECTS ON DRUG ADMINISTRATION PROCESSES, AND RECOMMENDED SOLUTIONS

The three focal points where gaps were found between task requirements and the nurse's capabilities are intertwined and together constitute a high risk for dosage errors. The difficulty in carrying out complex calculations is compounded when the nurse is not familiar with the drug, the required calculations, and accepted drug dosages. The high error probability is added to by the limited capability of the nurse to obtain sufficient information about the drug without an effort.

This study did not undertake investigation of errors that occurred but rather a proactive study of the process of magnesium sulphate administration. No attempt was made to identify the chain of events in any specific case but rather to identify the central problems inherent in the process of drug administration. The solution to these problems will prevent future mistakes, in other words rectify the process.

A case analysis of the process of administration of magnesium sulphate exposes the nature of the problems that characterize drug administration in general. The process carried out by the nurse is similar for all drugs administered in the same manner. The workload, noise, overcrowding, and time pressure remain constant. The difference is the drug to be administered and related calculations, which vary according to units of measure, physician's instruction, and so on.

It is recommended to provide the nurse with appropriate conditions and assistance in carrying out his/her role. The expectation of a nurse to know all the characteristics of a drug that is administered only once a year is unrealistic. The work environment should provide this information in as accessible, fast, and reasonably effortless manner as possible. This will reduce the probability of errors stemming

Magnesium Sulphate Dosage

from unfamiliarity with the drug and required calculations. Moreover, the knowledge will allow the nurse to better control dose calculations and detect abnormalities. The difficulty in carrying out many calculations during the day is likely to be resolved in the proper planning of the work environment.

Naturally, the above does not in any way diminish the importance of coping with errors stemming from written instructions (an issue not covered by the case analysis).

Practical recommendations for improvements designed to reduce the risk of errors in drug administration are given below. The solutions are proposed for different levels of complexity, some at a high cost and some at a low cost. The solutions are subdivided into three groups: engineering solutions, procedural–organizational solutions, and training–learning solutions.

ENGINEERING SOLUTIONS

1. Help for calculations:
   1.1 *Exclusive preparation of drugs in the pharmacy*—The nurse is not bothered by having to perform calculations and is available for other tasks.
   1.2 *Calculation tool*—A smart pharmacological calculator enables all calculations for all existing drugs. This calculator requires the nurse to input the relevant variables and provides final results of the calculation. The calculator is small enough to be carried in any pocket. The calculation program is preinstalled in the pocket calculator; however, the majority of nurses do not have on their person these devices. One such device in each department would suffice for those cases where nurses are required to administer drugs on an infrequent basis.
   1.3 *Calculation program*—A macro, which can be incorporated in any application (such as Excel). The program requires the nurse to input the calculation's relevant variables before providing final results. A macro was prepared for this program for installation in departments with a computer (Figure 6.2).
   1.4 *Manual spreadsheet*—A structured framed sheet with empty cells in which the nurse can record the appropriate numbers. The sheet is designed to assist the nurse in converting the physician's instruction to an amount to be extracted from the bottle and to calculate the number of drops per minute without "brain racking." The sheet indicates the values required for the calculation and their location, as shown below (Figures 6.3 and 6.4):
2. Access to information about the drug and its availability:
   2.1 *Drug information bank*—Preparation of a database on drugs, in simple Hebrew and fixed and logical structure. The information database will encompass general information on the drug, permissible dosage ranges, acceptable dose ranges, risks, possible responses, calculations, and so forth. The information database must be available and accessible to every nurse, for swift and simple operation, preferably installed on a computer deployed in each department where database searches can be made according to different indices.

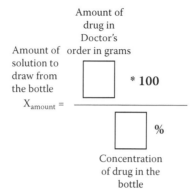

FIGURE 6.3 Calculation of required solution for recommended dosage.

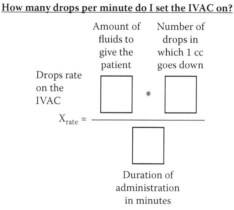

FIGURE 6.4 Calculation of magnesium drops per minute.

3. Changing bottle labels:
    3.1 *Adding the maximum dosage*—The nurse will have control over the physician's instruction and the values obtained from the calculation, reducing the probability of overdosage.
    3.2 Guideline with respect to the expiry period of the drug before discarding.
    3.3 *The text*—Label characters must be visible and legible, able to be read quickly and without mistakes. Based on research studies, standard character sizes, width–height ratio, color, contrast, and so forth, were determined. Further information can be found in Mark S. Sanders and Ernest J. McCormick's book *Human Factors in Engineering and Design*. A few recommendations from the book:

3.3.1 The recommended ratio between the character line width and character height is 1:8–1:10 for black characters on a white background.

3.3.2 The recommended ratio between the character line width and character height is 1:6–1:10 for white characters on a black background.

3.3.3 For reading close up (up to 35.5 cm), the recommended character size is 9–11 points (a point = 0.35 mm). At present, the label character size is 7 points.

3.3.4 In general, highlighting critical information (such as dosage ranges) in a different color is recommended. In the case of the magnesium label, the use of red characters on a white background accentuates the text relative to the remaining text and should be used to highlight data ranges.

4. Determination of bottle volume according to usage requirements:

    4.1 Those drug bottles where the contents that are not used must be quickly disposed of, the bottle volume should be determined according to the usage requirements. In other words, the bottle volume shall not exceed the maximum dosage. This volume will serve as a clue or reference point to the nurse as to the accepted amount or maximum permissible amount to be administered to the patient.

Procedural–Organizational Solutions

5. *Uniform registration of instructions for administering drugs by the responsible physicians*—Use of grams only and not mEq or milligrams, to prevent the nurse making unnecessary and confusing calculations.

6. *Control of administration*—At the start of each shift, a check on the amount of substance administered in the previous shift, *after* an independent calculation.

7. *Notification of a rare drug instruction*—The physician prescribing a drug that is not administered on a frequent basis in the department must make a special point of informing the nurses of the manner of drug administration, reading the instruction, relevant calculations, risks and side effects. In any event, the nurse must take full responsibility for completing all the necessary information for safe drug administration.

Training Improvements

8. *Post establishment drug database training*—Benefits and usage.

9. *Representation of errors and near misses and discussion*—In order to learn, to draw conclusions, and attempts to improve, and also to raise awareness of failures, problems, and limitations that all team members are subject to daily.

REFERENCES

1. Bates, D.W., Spell, N., Cullen, D.J., et al. (1997). The costs of adverse drug events in hospitalized patients. *JAMA*, 277:307–311.
2. Embrey. (1986). Approaches to aiding and training operator's diagnosis in abnormal situations. *Chemistry and Industry*, 7, 454–459.
3. Kohn, L.T., Corrigan, J., and Donaldson, M.S., eds. (2000). *To Err Is Human: Building a Safer Health System*. Washington, DC: Institute of Medicine, National Academy Press.
4. Leape, L.L. and Brennan, T.A. (1991). Adverse events, negligence in hospitalized patients: Results from the Harvard medical practice study. Perspective Health Risk Manage. *Perspectives in Healthcare Risk Management*, Spring; 11(2): 2–8.
5. Office of Drug Safety Annual Report FY (2004). Research Food and Drug Administration, Center for Drug Evaluation. An FDA report, U.S. Government.
6. Phillips, D.P., Christenfeld, N., and Glynn, L.M. (1998). Increase in U.S. medication-error deaths between 1983 and 1993. *The Lancet*, 351:643–644.
7. Rasmussen, J. (1983). Skills, rules and knowledge: Signals, signs and symbols, and other distinctions in human performance models. *IEEE Transactions on Systems, Men and Cybernetics*, SMC-13, 257–266.
8. Rasmussen, J. (1986). *Information Processing and Human Machine Interaction: An Approach to Cognitive Engineering*. New York: North Holland.
9. Roth, E.M. and Woods, D.D. (1988). Aiding human performance: I. Cognitive analysis. *Le Travail Humain*, 51, 39–64.
10. Swain, A.D. and Guttmann, H.E. (1983). *A Handbook of Human Reliability Analysis with Emphasis on Nuclear Power Plant Applications*. Nureg/CR–1278. Washington, DC: USNRC.
11. Oliven, A., Michalake, I., Zalman, D., Dorman, E., Yeshurunb, D., and Odeha, M. (2005). Prevention of prescription errors by computerized, on-line surveillance of drug order entry. *International Journal of Medical Informatics*, 74, 377–386.

7 Human Engineering and Safety Aspects in Neonatal Care Units
Analysis and Appraisal

Yael Auerbach-Shpak, Efrat Kedmi Shahar, and Sivan Kramer[*]

CONTENTS

Method 99
   Research Sample 99
   Mapping the Unit 100
   Interviews 100
   Observations 100
   Questionnaires 100
Findings and Discussion 101
   Workload 101
   Number of Operations per Unit Time 101
   Cognitive Load 102
   Environmental Conditions 102
   Common Load Characteristics 103
      Number of Operations per Unit Time 103
      Cognitive Load 105
      Environmental Conditions 106
   Unique Load Characteristics for Each of the Three Units 107
   Number of Operations per Unit Time—Large Neonatal Intensive Care
   Unit (NICU) 107
   The Additional Cognitive Load of Drug Preparation—Midsize NICU 108
   The Influence of Environmental Conditions—Small NICU 109
   Summary 110

---

[*] Yael Auerbach-Shpak, Efrat Kedmi Shahar, Sivan Kramer, and Daniel Gopher. "Aspects of Human Engineering and Safety in Neonatal Intensive Care Units—Analysis and Assessment." Center for Work Safety and Human Engineering, Technion, with the participation and assistance of Ilan Arad, Dorit Ben-Nun, Lea Sirota, Naomi Znandler, Daniel Reich, and Bruria Korlaro.

Communication and Information Transfer... 110
    Shift Transfer of Nursing Staff (Handover) 112
    Physicians' Rounds 112
    Summary 113
Proposals for Improvement... 114
Large NICU... 115
Midsize NICU 115
Small NICU 116
References... 116

What was the fate of infants born prematurely with low birth weight in previous generations, before the development of the incubator? There is no complete answer to this question, in the absence of precise records, and the majority of births took place at home. But, as infant mortality in general was high, including those born with normal weight, there is reason to assume that prematurely born infants died within a short time of their birth. Already, by the second half of the 18th century, it was recognized that proper heating arrangements increased the premature newborn chances of survival. In and around 1880, the French government decided to stimulate the population to increase the birthrate, fearing a shortage of "cannon fodder." Thus, the French began to treat both premature and newborn infants, whose chances, up to that time to survive, were almost zero. In 1880, the pediatrician Stephane Tarnier published the amazing results of treatment of prematurely born infants whose body weight was less than 2000 g: mortality decreased by 40%, thanks to the use of an appliance that the mother could take home and once there, properly supervise the newborn infant's welfare.

    The facility was gradually perfected—making it possible to maintain a constant humidity and ensuring ventilation of the capsule; a transparent cover was fitted so that the tiny patient could be observed, as well as other improvements and enhancements.

    A Chicago maternity hospital established a unit with four incubators where nurses operated the devices, and for the first time, a significant fall in the infant mortality rate was recorded. The big breakthrough as far as the general public and the medical establishment were concerned occurred after these installations went on exhibition and were hailed as a technical medical achievement. Already, in 1896, "living premature babies" were presented in an incubator in Berlin. In 1901, in Buffalo, there were nine incubators, far more sophisticated than their predecessors! Later the enthusiasm abated and it was unclear who should take care of these tiny babies—a pediatrician or obstetrician? Either way, the common belief was useless.

    Many years passed before the profession we know as *neonatology* (newborn medicine) saw the light of day: neonatologists are those receiving the premature infant in the delivery room (today it is possible to anticipate premature birth and make the necessary preparations for such an event). Neonatologists are trained to inject intravenously into a tiny vein that is not obvious to the untrained eye. They also supervise the progress of premature babies until they are in a self-supporting state and can be returned to their mother's care.

* * *

Human Engineering and Safety Aspects in Neonatal Care Units

Israel has about 150,000 births per year. Ten percent of newborns will spend their first days on earth, the critical period, in intensive care units for premature babies, in sophisticated and advanced incubators, a far cry from the facilities available for premature babies in France 200 years ago and more. The incubator is well lit and ventilated. Within the incubator, the newborn baby can be weighed, put on a respirator, and connected to alarm and monitoring systems.

Over the years, physicians have acquired a vast amount of knowledge on the physiology of the tiny newborn, premature newborn, and understanding of diseased states. Today, it is possible to prepare for a low birth weight newborn baby and even perform intrauterine repair of anomalies in order to receive the patient in an optimal state. But when the premature baby is admitted to the intensive care unit, he/she becomes part of a support system, which is nevertheless fraught with risks. The heavy workload and manpower shortage prevent provision of care and treatment at the required levels. As this *patient* cannot be bedded in the ward corridor, when another premature baby is born, another incubator is added (without additional manpower to care for the babies) and other premature babies are transferred to an intermediate ward, even if they are not mature enough, so that the new premature baby, in distress, can be admitted. If not for the dedication and hard work of the premature baby care teams, it would not be possible to maintain this system.

The unit for premature baby care, neonatal intensive care unit (NICU), appears to have been designed by an engineer who has never worked in this hostile working environment.

The Association of Neonatologists in the United States has recommended allocation of a working area around the incubator allowing convenient access for a nurse and a suitable location for placing auxiliary equipment, as well as a chair for the baby's mother so that she may keep an eye on and be able to physically touch her baby. To achieve all this, the required area for each premature baby is about 12 square meters and the distance between incubators should be about 2.5 meters (in a nonpartitioned room). The Association also recommends at least 20 power outlets for each bed (these standards are keyboard accessible for all hospital planners and directors). Storage space for instruments is also needed, as well as a special room for quarantining infected babies and space for parents (also part of the premature baby care team!) and of course an appropriate ratio between the number of premature newborns and the number of nurses responsible for their care. These conclusions, unanimously agreed upon at a conference in Orlando, Florida in early 2009, are not much different than the conclusions reached by Israeli committees who also diagnosed the distressful problem and proposed a manpower standard of 1:1.15, that is, one nurse "plus" to each bed. Namely, a unit with four intensive care stations should be attended by 12 nurses daily. Each additional bed would require additional standards.

A tour of Israel's largest NICU reveals that all these units, even the most up-to-date, are greatly at odds with the recommendations for U.S. units: most suffer from a shortage of skilled medical and nursing staff, as well as an onerous workload on both nurses and physicians.

This chapter addresses human factors engineering aspects of neonatal units with an emphasis on the intensive care units caring for premature infants and newborns in the more sensitive medical status.

* * *

The NICU admits premature newborns or newborn babies who have suffered some distress during labor and that in the absence of being admitted to the unit will not survive. The fact is that most of their body systems have not matured and therefore they need to be maintained on life support systems, requiring sophisticated equipment whose accurate operation is mandatory, as there is no room for error—any error whatsoever can cause irreparable damage to the tiny patient.

The Central Bureau of Statistics in Israel reports that 8–9% of babies born each year (about 12,600 births) are underweight, that is, less than 2500 g.[1] The State Comptroller reports an increase in the number of premature births while at the same time special care units are not increasing in number and the labor force complement remains static, hence, the units are unable to meet the strict and complex requirements of premature baby treatment.[2] This report addressing medical and nursing manpower shortages, the gap between hospital bed requirements and the number of actual beds (above standard), the shortage of new and customized equipment, and hospitalization grants, indicated the extent of the problem and presented a picture of overloaded departments operating with inadequate budgets and equipment, struggling to provide the necessary treatment required for the proper care of premature infants hospitalized in the departments.

Routine work in the NICU is dynamic and high tension, attributable to the premature babies' changing medical pattern and the need for staff to act quickly and treat many patients who cannot complain of pain or breathlessness. Therefore, the babies are continuously monitored around the clock and at each occupied incubator detailed information is being gathered, demanding immediate attention, and information is generated by the team (physicians, consultants, technicians). All this detailed information is collected and stored in different places.

Units feature three treatment level protocols and formats based on the physiological condition of the premature baby:

Intensive care—(Level III): For premature babies, needing close and constant supervision and monitoring.
Intermediate care—(Level II): For babies beyond the critical period stage but still in need of supervision, oxygen-enriched air, and appropriate nutrition.
Mature baby—Treatment for babies about to transfer to regular wards or released to return home.

Due to the multiplicity of devices surrounding each baby treatment station (about 10 different devices), the treatment station is crowded to capacity, as well as producing considerable noise. The premature newborn may have to remain in the unit from a few days to several months. The longer the baby remains in the NICU, the greater the cumulative information.

The medical and nursing staffs differ in their respective levels of training and experience. At the time of this study (2005), the nursing staff comprised practical nurses, registered nurses, and registered nurses having undergone intensive care training (premature infants, newborns, babies, and children). The different training levels

Human Engineering and Safety Aspects in Neonatal Care Units

allow for unified levels of treatment operations. With a higher level of nursing training, the nurses are allowed to carry out many more tasks. There are varying degrees of expertise and performance ability even among the medical staff, from specialized physicians who are required to have experience in treating premature babies as part of their training in pediatrics, to senior physicians (who also differ one from the other in the level of knowledge, specialization, and experience). The differences described affect the ability of each team member to perform his/her tasks. In each of the departments we visited, the team complement was uneven and generally dependent upon the departmental norms. For example, maybe less than half of the nursing staff had designated training and there may also be medical teams where the number of experts is insufficient. Apart from the differences between the departments we visited, the team composition, medical and nursing, is different for each shift, differences that affect the continuity and the quality of the treatment process. These characteristics make it more difficult to carry out routine operations in these important units.

In this chapter we have chosen to concentrate on two key work aspects where errors may occur, increasing the probability of adverse events: (a) workload of nursing staff and (b) communication and information transfer between teams and within teams.

These two aspects, despite their differences, are interdependent and influence each other: the workload may affect the rate of transmission of information, while the rate and modes of communication between team members may impact workload. These aspects and their interrelation remain almost untouched by systematic research. Topics which received the most attention were the physical layout and design of the individual unit and the ward (e.g., monitors alarms, lighting, etc.). Other topics have been family involvement in the treatment process and job design of the clinical staff.

METHOD

RESEARCH SAMPLE

The study was conducted in three neonatal intensive care units, differing in several characteristics (Table 7.1).

TABLE 7.1
The Composition of the Three Units

Large	Medium	Small	Category
40	22	16	No. of patients
62	33	38	No. of nurses
7 seniors	4 seniors	5 seniors	No. of physicians
2 fellows	2–3 residents	1 fellows	
2–4 residents		2–4 residents	
In a huge department plus additional space for storage	3 rooms plus additional space for storage and feeding	2 rooms No other additional space	Physical space

Mapping the Unit

Familiarity with the unit was reached by studying the physical structure and workstations, review of environmental conditions (climate control, alarms), work procedures, work schedules and team presence, documentation and transfer of information, and training of medical and nursing teams.

The analysis was guided and driven by the perspectives and methods of human factors engineering and cognitive psychology. The objective was to explicate and describe the cognitive conceptual framework and performance protocols of physicians and nurses to enable improvement of work format and protocols and contribute to the planning and design of information technology (IT) support systems for ICU decision making.[3] We examined the degree to which each of the three NICU units and their working environment was compatible with the requirements of the nursing and medical team assigned tasks.

Interviews

Interviews were conducted with team members—experienced physicians and nurses with different training backgrounds (senior physicians and interns, nurses with different qualification levels). To better understand unit complexities and locate problems, interviews were also held with representatives of external support for units (e.g., pharmacy, dairy kitchen, maintenance). The partially structured interviews were conducted by the human factors engineering team. At each interview, the interviewee's demographic variables, role description, communication routines, subjective rating of work difficulties, procedures, working with functionaries, physical structure, and equipment were documented.

Observations

Nurses were observed in the course of their work, from the moment of taking over the shift to handing over to the subsequent shift. Hourly formal records were made throughout the 24 hours, three shifts (morning, evening, and night). Every action carried out by the nurse was recorded with the time of execution (e.g., garbage disposal, interacting with a monitor or device, bathing, etc.). Further targeted and closer observations were made of more complex processes that were identified in the preliminary interviews, such as the preparation of drugs and admission of new babies to the unit. Shift handover process was also documented. Information was also collected on the use of and recording in forms.

Observations made a special emphasis on the transfer of transferring information, orally or in writing. All observations were made by human engineering personnel who had previously conducted the interviews. Observations were recorded in the Observation Form.

Questionnaires

Important issues regarding the transfer of information, physical space, equipment, patterns, and work procedures, were further clarified through questionnaires, which were completed by different team members. The questionnaires were adapted to the types of respondents (medical/nursing) who responded to the subject matter relevant to their specific roles.

The clinical team was also asked to complete another questionnaire that set out to evaluate the wear and tear on team members: they were asked to read 10 sentences, describing feelings about their work, and to rate the level which corresponds to their feelings about their work.

FINDINGS AND DISCUSSION

The nursing team's workload and problems in communication and information transfer between and within teams are the two central problems evident in NICUs. However, each problem had different manifestations in each of the three departments being studied, corresponding to the unique characteristics of each department—manpower, the number of patients, work procedures—these had different modes of influence on workload and the ways in which information is transferred in the unit.

Workload

Workload is a multidimensional concept which is affected by multiple determinants which may not be directly measureable.[4,6] The study considered workload as the product of three integrated components: the number of operations performed per unit of time, the cognitive demands (difficulty) of these operations, and the environmental conditions under which tasks are performed. A low workload does not present a handicap to dealing with the tasks at hand, while a high workload is conductive to errors and mishaps. For example, when a nurse concentrates on calculating the dosage of a drug and there is a sudden phone call, or when she needs to attend to the sounding alarm of a monitoring device while she is busy with drug administration, workload and attention interference are immediately magnified. Under such conditions, the likelihood of mishaps and errors are considerably increased.[5]

Within the framework of the NICU study we sought to locate the main determinants of workload in the unit and examine ways to reduce it in order to create modes of work that take into account the requirements the individual team member has in terms of his/her work and abilities. First, we briefly describe each of the workload factors before detailing and illustrating how each component affects the workings of the NICU.

Number of Operations per Unit Time

The first workload component is the number of operations or activities per unit time. If this number per unit time increases, the workload will also increase. In the course of the NICU study, this proportion was calculated based on the total number of activities performed by a nurse on his/her shift. When analyzing the data, the number of operations was summarized and divided by the number of hours on duty of the entire shift, subtracting the nurse's rest break periods (for example, a nurse performs 659 operations during his/her shift). Net work time was 7.54 hours. For this example, the calculated average activity ratio for this nurse was 87.36 operations per hour [Wl = A/Wt].

Cognitive Load

The second workload component is cognitive load (mental), the demand of the mental actions needed when processing the required information for performing a specific operation. Different tasks impose different cognitive loading, according to the task properties and mutual interaction between the person and the task.[6] A complex task, involving searching for and processing information, requires a mental effort that could lead to a sharp dip in performance. Mental stress can also derive, for example, from simultaneous performance of a number of tasks (driving, eating, telephone call). Simultaneous performance of several tasks is made possible by allocation of mental resources between different tasks or by ranking tasks according to degree of importance. But deciding on an order of priorities implies rejection of certain actions or implementation of those actions at an inferior level. There have been many attempts in the literature to define *mental stress*, but it is difficult to define with its many facets.[6] The mental load is influenced by the type of task and the ratio between the mental resources available to the team member and the resources needed for the task.[4] It may be concluded that a human being's information processing system is limited by his/her resources.[6] Mental stress is a function of task requirements, that is, the knowledge needed to carry out the task, level of difficulty, and task complexity; and also a function of the capabilities of the operator, that is, the level of training, skills, the degree of dedication, and physiological factors such as fatigue.

Measurement of mental workload is complex and is based primarily on task analysis. In order to assess the mental load incurred by carrying out tasks in the present study, the cognitive demands for each task were analyzed, as well as the best conditions for the requirements to be met. For example, an operation that demands concentration and focused attention should take place in a quiet environment in which there is no excessive distraction from the task at hand. In the absence of favorable conditions for operations, the consequent mental stress is a combination of the burden of carrying out the task itself and attempts to overcome the prevailing poor conditions. Cases in point were observed where discrepancies arose between the cognitive demands of the task and the ability of the team member to fulfill these demands. This type of situation calls for improvisation to compensate for the difficulty and paucity of optimal conditions.

Environmental Conditions

The third workload component is the environmental conditions—everything connected with task performance apart from the operator and the task itself, that is, the physical work environment: working area organization, temperature, ventilation, noise; and associated service environment and characteristics: service providers and assistants to the team member in fulfilling his/her role. For example, in a hot and noisy room, where any noise and distraction has different effect, performing any task that

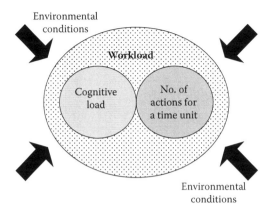

FIGURE 7.1 Determinants of workload.

requires close attention, even a simple task, will differ from performance of exactly the same task in a quiet and comfortably heated room. The environmental conditions of the NICU were assessed by mapping the physical environment of the department by measuring the distances between the different treatment locations and therapy rooms and specialist rooms, as well as organization of workstations (Figure 7.1).

Workload is the individual's interaction with his/her task.[6] The workload increases with the growing number of operations per unit time and widening gap between task demands and individual abilities. Workload intensifies under environmental conditions that make the individual's routine operations more difficult. The NICU study objective was to uncover workload sources in the department and examine ways and means to lighten the workload in order to arrive at a system that takes into account team member requirements and abilities while performing his/her task. Later in this chapter we will detail and demonstrate how each of the components listed above affect the workings of the NICU.

All the NICUs visited suffered heavy workloads, with respect to all the components discussed above, but a closer examination of the load factors revealed that each department had a unique profile. The load factors common to the three departments and unique factors for each department are described in this section. The focus shall be on the role of the nurse in the NICU, as the nurse's role stood out in the physical and cognitive workload, as described below.

COMMON LOAD CHARACTERISTICS

To recapitulate, the model described above encompasses three workload components, common to all departments, each one a component of the model.

Number of Operations per Unit Time

The number of operations per unit time is influenced by the number of tasks that the nurses have to perform and by the ratio between the number of nurses and the number of premature infants under their care.

A summary of the data of the three departments is found in Table 7.2.

TABLE 7.2
Departmental Data

Unit / Category	Small	Medium	Large
Mean number of activities per hour	57 (41–81)	63 (45–78)	74 (55–99)
Nurse to newborn mean ratio	NICU 1:2–3 Intermediate 1:4	NICU 1:3 Intermediate 1:5–6	NICU 1:3 Intermediate 1:5–6

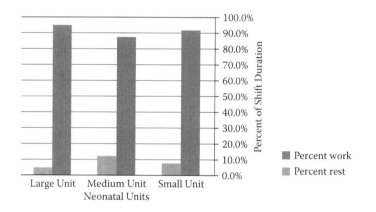

FIGURE 7.2 Percentage of work and rest times in the three units.

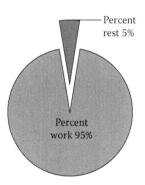

FIGURE 7.3 Percent of work and rest times.

The average number of operations per hour in all units is very high: 57 to 74 (Figure 7.2, see also Figure 7.3). That is to say, a nurse performs more than one operation per minute! This index is also influenced by the ratio between the nurses and the premature babies—as the number of babies under the responsibility of the nurse increases, so the number of operations to be carried out by the nurse will also increase. This workload, resulting from the number of operations, also affects the duration of work and rest during the shift. It was found that the amount of rest during the shift was very low (5–12.5% of the total shift duration), independent of the size of the NICU or the shift (morning, evening, or night). This situation stems from the nature of the treatment of premature infants, requiring constant monitoring within a fixed hourly cycle of 1 to 4 hours, according to the medical condition of the premature infant. For the sake of comparison, under the current labor laws, 8-hour work in three shifts calls for a minimal 16% rest time of the total work time.

Cognitive Load

The NICU nurse task demands concentration, alertness, and close attention to the situation of the premature infants throughout his/her shift. The nurse must divide his/her attention between infants in his/her care (and at times, other premature infants), in order to function efficiently during routine treatment and to act promptly in an emergency. Close attention to the premature babies is mandatory throughout the shift. At all times and at the same time, the nurse carries out many other operations that also demand concentration and attention.

The nurse is required to organize his/her tasks in order of priority—urgent tasks at the top of the list, to be dealt with initially. The act of prioritization compounds the workload, since it entails making decisions and demands on time and concentration. Moreover, the nurse is required to demonstrate flexible thinking, allowing changes to be made in the order of precedence as a result of unexpected changes (the infant's condition, the number of infants in the department, etc.).

The departmental load and very nature of the work in the NICU decree that there be collaboration and constant mutual assistance by the team members. This work pattern requires that the nurses divide their attention between their tasks and the tasks of other nurses, for whom they are providing assistance.

During any shift, hundreds of items of information accumulate around each infant, from the various monitors as well as observations made by team members. These are vital to the formation of an up-to-date medical situation map of the premature infant, essential for immediate treatment for any decision that has to be made on the manner of treatment. As the length of duration of stay in the unit is prolonged, the total volume of information increases exponentially and so does the cognitive load of monitoring and processing this information.

Another factor that increases the workload on the nursing team levels is associated with nurses' qualification and accreditation. As already indicated, the nursing team consists of nurses with varying levels of qualification and only a small number of nurses are qualified to perform certain procedures. In all three units, the number of neonatal qualified nurses is approximately 50%. Quite often, only one neonatal qualified nurse will be working on a shift, a situation, but not the only one, that increases the overall load on the nursing team, one in which an uncertified nurse is confronted with a task that she is not formally authorized to carry out and therefore must look for the senior nurse to carry out the required operation. Even if an authorized nurse was found, time can pass before the nurse is available, which would mean unwanted delay. Furthermore, the load on the authorized nurse will have increased, since she must perform these operations for a large number of premature babies during the shift and carry out more complex operations that require considerable concentration and attention, amounting to a greater cognitive load.

Environmental Conditions

The physical layout of the unit also affects the workload of team members. The effect is most pronounced at the team's workstation around the incubator, in organization of and layout of items in rooms with units, and in the overall organizational structure of the NICU.

Spatial organization of the unit is an important variable that may have a positive or negative effect on the workload and the quality of care for premature babies. From the observations made in the three units it was evident that nurses acting around the incubator devote a significant amount of time to organizing the working space, a factor that is dependent on different features of the room structure and available space. In general, the units were overcrowded due to the myriad of incubators, auxiliary equipment, and expendable equipment in routine use. It was observed that nurses find it difficult to maneuver within and between workstations.

The NICU central workstation is the incubator and its auxiliary equipment. Each incubator is surrounded by a considerable number of diverse equipment: monitoring devices, an infusion stand, drip counter, automated syringe pumps, oxygen blender, respirators and breathing aids, equipment tray, medical gas supply system, and electrical outlets. This, therefore, represents a highly loaded workstation. However, there is no standard, uniformity, or consistency in the order of equipment at the stations. In all the units visited, whenever a team member needed to take care of a premature baby, it was also necessary to research and readjust to the configuration of equipment in the incubator area. The problem gets worse for nurses who have to supervise several infants. For example, the manual respirator is not placed in an agreed location and the nurse who is required to operate the respirator is obliged to expend energy to find it. Each incubator is surrounded by interweaving power cables feeding the device (the teams have named the cable cluster *the spaghetti*). If a device has to be disconnected or relocated, it is difficult to find the appropriate cable and the nurse is forced to adopt uncomfortable body postures.

The typical nursing team member is responsible for a range of routine tasks to be carried out, as well as additional tasks resulting from the premature infant's dynamic situation. These situations are communicated via an audible alarm. Most devices in the NICU emit audible alarms. When a device exceeds its permissible limits, an audible alarm is triggered that can be heard. This situation raises a number of problems. First, a high percentage of alarms are false alarms that nevertheless require testing and validation.[8] The alarms were found to give rise to stress symptoms both in the team and in the babies.[9] Another problem is the ability to relate the alarm signals with the devices, exacerbated when devices of different generations are involved, each having a different alarm signal. In this case, team members encounter difficulties in forming a mental model of all the alarms and quickly relating a specific alarm to the relevant device and problem, this is especially so in the case of new personnel. These problems and the premature infant's dynamic situation intensify the team's consciousness of their burden.[8]

Human Engineering and Safety Aspects in Neonatal Care Units

UNIQUE LOAD CHARACTERISTICS FOR EACH OF THE THREE UNITS

In addition to load factors common to all NICUs, each NICU is characterized by a different load pattern, stemming from the dominant load component of the unit's activities.

The unique load component in each of the NICUs:

Number of Operations per Unit Time	Cognitive Load	Environmental Conditions
Large unit	Midsize unit	Small unit

NUMBER OF OPERATIONS PER UNIT TIME—LARGE NEONATAL INTENSIVE CARE UNIT (NICU)

The major load factor in the large NICU visited was the number of per unit time, attributable to the large number of neonates (~40), the spaced layout of the unit, encompassing many designated rooms (such as a dispensary, laundry room, nursing room, etc.), and large treatment rooms.

One of the most prominent factors affecting the workload and activity load is the ratio between the number of nurses and the number of premature infants in the care of each nurse. In the United States, the recommended ratio is 1/1, one nurse to each premature baby. In Israel, because of several factors, the load ratio is considerably higher. In the present unit in the intensive care, the morning shift nurse takes care of 2–3 premature infants requiring supervision and intensive care. A similar situation exists in other treatment rooms (intermediated and prerelease) where an even larger number of premature infants are in the care of a solitary nurse. This ratio between the number of nurses and the number of premature infants differs from day to night shifts—the number of nurses on the night shifts is smaller and therefore one premature baby on average is added to the workload. When one nurse must take care of so many premature infants, the number of operations that the nurse performs in each time unit increases, currently standing at 74 operations on average per hour. This is the highest number of operations observed in the three NICUs.

Another factor affecting the number of operations is the size of the NICU within which they are conducted. A large and spaced structure forces team members to travel long distances from room to room. The engagement in a specific function designated room also implies prolonged absences from the neonate incubator room. For example, it was observed that a nurse, in order to carry out various nursing operations, such as the preparation of drugs, was away from the incubator room for about 54 cumulative minutes during his/her shift. This activity consumes a considerable part of the nurses' time and adds to their workload and that of their colleagues. When a nurse is absent from the room, the remaining nurse must take over the supervision of the premature infants who are in the direct care of the absentee nurse.

THE ADDITIONAL COGNITIVE LOAD OF DRUG PREPARATION—MIDSIZE NICU

In the midsize NICU, it was found that a substantial increase in load and interference to direct care work of nurses resulted from the requirement that they themselves prepare the drugs to be administered to the neonates. Unlike other departments, where drugs are prepared in the pharmacy (unit dose), in this unit the nurses themselves prepared the drugs, several times on each shift.

By means of cognitive task analysis, backed up by observations and interviews, the drug preparation process was described along several types of demands and at each type of cognitive demand, its implication on the cognitive load of the nurse was examined. It is important to note that tasks were performed in a noisy environment with many possible cooccurring events. There was not a secluded location where the nurse could prepare the medications quietly and without pressure of time. Environmental conditions were affected by constant bleeping sounds from different devices, interference from other team members, and other disturbances, which interfere with the preparation of the drug, which in itself is complex and demanding.

Task analysis showed that the drug preparation task can be conceived to comprise four major types of cognitive demands:

Processing and interpreting written material—The drug preparation process begins with reading the information from the physician's Instructions Form. At times, the physician's handwriting is illegible and the nurse is obliged to make an extra effort to decipher the instructions.[10]

Attention demands[7]—The drug preparation process requires considerable attention and focus from the nurse and consideration of many specific details. For example, in calculating the dosage, the nurse is required to focus his/her attention on mental and manual calculations while other potentially distracting occurrences are going on around the nurse—noise in the treatment room, other tasks to be carried out at the same time and he/she is constantly in attendance to every other occurrence happening in the room.

Search for and location of information—Calculation and preparation requires multiple search, detection, and fetching activities from the medications room and its surroundings.

Decision making—Decisions and selections have to be made at each preparation. At each step, the nurse must decide which action should be performed from a spectrum of possible actions. At times, these decisions are complex in nature, as all the possibilities must be taken into account and given due consideration.[7] For example, the nurses are required to label the drug, providing comprehensive drug information, including the number of hours and frequency of drug administration. There are several types of labels, each differ with respect to several parameters. The nurse has to decide on the type of label on which to record these details. The wrong choice of label can lead to errors in scheduling and frequency of administration.

Taken together, the requirement to prepare medication by them was a single factor that was clearly significant in increasing the cognitive load on the nurse and influencing her work efficiency.

Human Engineering and Safety Aspects in Neonatal Care Units

THE INFLUENCE OF ENVIRONMENTAL CONDITIONS—SMALL NICU

Environmental conditions, physical and human, constitute the dominant contributory factor to the workload in a small NICU. The small NICU suffered from small and overcrowded rooms. Eight to ten (8–19) premature babies were located in close proximity one to the other and the narrow gap between the incubators was not sufficient to allow standing space for more than two people at a time. The situation became even more problematic during doctors' rounds or a visit from a specialist consultant bearing additional large and cumbersome instrumentation (such as an X-ray machine or echocardiography, which may be up to 1 meter in width). Hence, a nurse who has to perform her functions encounters difficulties in approaching the station and all the paraphernalia needs for the task at hand. The nurse must maneuver between various functionaries in the unit (other nurses, physicians, technicians, etc.) and between the various devices positioned around the incubator. All this maneuvering in itself is difficult enough and increases the workload and complicates nursing care.

The American Standard recommends 2.4 meter spacing between adjacent incubators. The average distance noted in this study was found to be 99 cm, resulting in severe overcrowding in the treatment room. According to the infection prevention teams in the hospitals, overcrowding contributes to the spread of infections. From interviews with departmental teams and completed questionnaires, it was clear that treatment room overcrowding made their work difficult and made them more conscious of the load (Photo 7.1).

This unit had no designated floor space, making it difficult for staff to work, forcing the nurse to somehow improvise floor space and waste precious time in arranging the necessary equipment and adaptation of the makeshift platform to his/her needs.

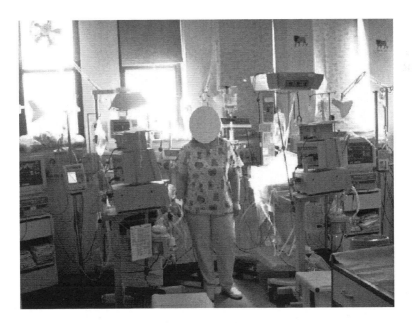

PHOTO 7.1 Physical separation and space between incubators.

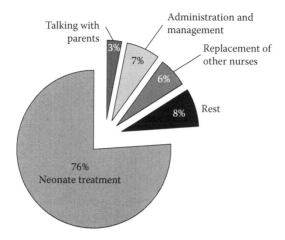

FIGURE 7.4 Average profile of nurse activities during a shift.

Moreover, due to the lack of designated floor space, a number of actions are carried out simultaneously at the same location, a state of affairs that is ripe for error.

Service given by external units and in unit auxiliary personnel may be another environmental work condition, which may be a factor affecting the departmental load, in particular, working arrangements with the departmental service units (auxiliary manpower). This working arrangement was far from satisfactory and there was no clear division of responsibility between the teams. Therefore, the burden on nursing staff became greater, as nurses performed tasks that are part of the auxiliary staff responsibility. On average, for 7% of shift time, the nurse performs operations that are within the framework of the NICU service unit responsibilities (medical secretary, auxiliary manpower, and subservice): administration, shift organization, answering telephone calls, searching for equipment and drugs, arranging equipment and other items in the room, and so on. These actions place a burden on the nurse, making it difficult to perform tasks at hand and thus increase her overall workload (Figure 7.4).

SUMMARY

The workload, with its three components, is heavy in all three NICUs although the load factors differ. The unique contribution of the present NICU study was the tracing of the individual departmental load profiles. The three workload factors were evident in all departments, but in each department a different factor stood out that made it possible to *tailor* proposed solutions to each NICU according to its dimensions. The proposed solutions were adjusted to each specific department.

COMMUNICATION AND INFORMATION TRANSFER

The term *communication and information transfer* refers to the extent to which medical and nursing staffs share the baby's recent medical status information. Communication problems and information transfer are the main causes for mishaps and adverse events in medical environments.[12] Each work task is dependent upon the information input and processing of that information. Improved communication and information transfer depend on protocols, procedures, and positive cooperation between members of the unit. Despite the obvious importance of good communication, the pattern is still the exception rather than the rule in the medical domain. Only 3% of the observed NICU shift activities involved both physicians and nurses, 2% of the activities involved verbal communication between physicians and nurses, while 37% of the errors observed in the unit were derived from these 2%.[13]

The NICU is characterized by an enormous volume of information, which is added to and changed frequently. The information is received from a host of professional workers (senior physicians, specialists, nurses, technicians, and others) as well as different instruments. This information is recorded or filed in a number of locations (command forms, nursing records, patient medical folder, etc.). To make the treatment decision the team draws upon information from all available sources (baby's observations, information from other team members, forms, and records) and only rarely is reliance placed upon a single source of information.[3] The team is required to invest considerable resources and time in integrating the required information into an overall updated picture, thus creation of a combined, unified situation map presents a difficult task, though essential for proper treatment. A study of adverse events in the NICU revealed that 14% of adverse events stem from problems of communication and information transfer.[14] Moreover, the study of near miss incidents showed that communication failures are at the core of 37% of them.[15]

The workload and communication and information transfer processes are interdependent and affect each other in a vicious circle: a person burdened by myriad tasks and pressure tends to rely on the information he believes to be the most important; he does not use all the existing information but focuses on readily available items. In other words, the workload limits the information available to team members at any given moment. Difficulties in transfer of information among team members increase the load and prompt the investment of significantly greater resources in collecting information. Also, uncertainty arising from lack of information contributes significantly to the mental burden.

The present study examined the patterns of communication in the three units— among the nursing staff, the medical staff, and between the two teams.

A number of information nodes in the daily work routines were identified where the information about the infant was passed between team members and first and foremost:

1. Nurse shift change (two or three times a day, depending on shift length).
2. Medical team information transfer, several times per day, depending on the size of the department and its needs.
3. Physicians' rounds, several times a day, varying complements.

Transfer of information between shifts at shift changeover is important in order that each team member has all the information needed. Transfer of information between functionaries (physicians and nurses) is important for the creation of a complete medical status map of the premature baby.

Transfer of information differed slightly in each of the three departments, according to workload, number of staff, the physical structure, and the requirements that the department seeks to fulfill through information transfer. Details of shift transfer of nursing staff and physicians' rounds in the different departments are given below.

SHIFT TRANSFER OF NURSING STAFF (HANDOVER)

Shift change is a potential failure point, as the incoming shift must digest a considerable amount of information. At the shift changeover, authority, responsibility, and information (formal and informal) is handed over. One could expect that shift changeover be structured and similar in all departments and between all shifts, but several factors stand in the way of such uniformity: the nursing team treating each neonate is not fixed between shifts and may also sometimes change within shift. Hence, existing and incoming nurses may differ in their acquaintance with the baby. This leads to differential gaps of formal, informal, and history information that need to be communicated and complemented. The difficulty is compounded when the nurse treating the premature baby in any particular shift will not necessarily be treating the same premature baby in subsequent shifts. Under these circumstances, a proper conduct and well-structured handover protocol is of special importance.

In the three observed units, handover lasted about half an hour, during which essential information about the infants was reviewed and passed on, encompassing general and specific information pertaining to the outgoing shift. Nurses from all units complained that information transfer was far from satisfactory and they are less than confident that the incoming shift is getting all the essential information pertaining to the baby. It was apparent from the interviews that it was not clear as to the precise information to be exchanged during handover. There were also considerable differences in the global conventions of handover.

For example, during the routine handovers in the midsize NICU, all nurses move together across all incubators. Only then is each nurse informed which neonates she will personally be responsible for and gets further briefings from the departing nurse. This routine was designed to enable all incoming team members to mutually assist one another when necessary, in each of the unit's rooms. This strategy is practiced many times over throughout the shifts and is important in the context of departmental work. Thus, during the 30-minute handover, the information about all 22 babies hospitalized in the unit is handed over. The time is divided between the premature babies according to their respective condition of severity and the short time remaining is dedicated to the transfer of information. Team members are not assigned to babies until all the information has been handed over. Therefore, they do not and cannot pay attention to all the details handed over. This process creates further limits on the team member's familiarity with the patient, which can lead to errors.

PHYSICIANS' ROUNDS

Physicians' rounds are one of the most important and fundamental junctions for exchanging, communicating, and passing on information to the unit staff and generation of an updated medical status map of the baby under treatment. During the physician's round, decisions are made that delineate the plan of treatment for the baby for the next 24 hours. In each of the units visited, the physician's round was conducted in a different style, some of which was determined by the physical size and spatial layout of the unit and in other cases by the number of neonates compared to on shift physicians. During the physician's round, efficient cooperation between team members and nursing and medical staff enhances decision making and quality of patient care.[16] Proper care should be based on cooperation and comprehensive information exchange between nurses and physicians.[17] This topic has been extensively researched and accepted as an important and direct positive influence on the quality of care.[18,19] Despite the obvious advantages inherent in conjunction with this, it is still unusual to find this well-covered and well-structured protocol of rounds in the majority of hospital wards.[18] In the present NICUs, it was found that in many cases the physician's round was conducted without the participation of any of the specific nurse staff engaged in treatment of the visited baby, even though she is likely to be the one with the most up-to-date knowledge of the patient's state of health and recent history. She is also the one that remains at the incubator when the round moves on. The nurse's nonparticipation in the physician's round also mars the generation of the nurse's situation map, since he/she does not receive firsthand all the information about the state of the infant and does not participate in any discussion and planning or treatment protocol.

One example of this problem was the very informal and unstructured physicians' rounds format, which was found in the small NICU. Physicians' rounds were conducted without a formal structure, and depended on the senior physician in charge of the round. The nursing staff rarely participated in the round and was therefore unable to contribute or gain from the review and discussion. Finally, the information was usually presented by one of the interns. There was no advance preparation of information and there was concern that important information could go missing or errors occur during information transfer.

Another example of the difficulty in transferring medical information was found in the large NICU, a unit characterized by its heavy workload on physicians. The ratio between the number of premature infants and physicians is 40:5–6, in other words, each physician is in charge, on average, of eight babies. Each physician makes detailed visits, alone, to the babies in his/her charge and subsequently makes decisions without joint discussion with colleagues, due to time constraints. Assuming that each visit lasts about 15 minutes (evident from observations), if visits were made together with all the physicians present, visits would last for about 10 hours! In this single physician round, the physicians lack in-depth discussion and consultation, which is an inherent part of hospital care quality routine. In this case, high workloads have a direct impact on decision making and quality of care (Photo 7.2).

Summary

Taken together, we have seen multiple factors that affect the quality and completeness of communication and information transfer. Transfer of information is essential

PHOTO 7.2 Doctors' rounds.

for reaching an appropriate medical status map for each individual member of the team and the joined team. Treatment of patients in an intensive medical environment where critically ill patients are in a dynamic state, requires constant supervision, necessitates information transfer processes, cooperation, and coordination between different team personnel, locations, and time frames. Coordination is not a trivial matter but can be difficult and challenging. Despite the many programs drawn up for patient treatment, plans for continuous treatment are subject to frequent and unexpected changes that disrupt the continuity of information transfer and continuity of treatment. These gaps arise out of the dynamics of medical treatment (such as unexpected fluctuations in the patient's condition and manpower changes) and nonspecific medical work processes (such as shift changeover or physicians' rounds)[20,21] that may cause communication failures between team members. Communication failures, especially if they stem from inadequate knowledge transfer between team members, are one of the main causal factors for abnormal environmental events.[13]

Difficulties in transfer of information emerged—among the nursing staff, the medical staff, and between the two teams.

PROPOSALS FOR IMPROVEMENT

Based on the analysis of the workload profiles, communication, and information transfer problems in the three neonatal units, specific recommendations were made in each unit to reduce workload, improve communication, and reduce errors.

The specific problems in each unit and possible solutions were described. Problems emerged over a range of issues that were examined. The solutions proposed would not involve a change in manpower, medical or nursing, since the solutions for many of the problems are not dependent on adding more nurses or doctors, on the contrary, the problem could worsen.

LARGE NICU

The two main problems that were addressed in the large NICU were overall, schedule of work and easy documenting and accessing of information.

Examples of the main solution and subsidiary solution proposed for these two problems were as follows:

1. Development of a revised scheduling of work schedules and team compositions in shifts.

    A revised strategy of planning work schedules and shift composition. A major problem of the large system has been the continuous and variable inflow of neonates and the dynamic changes in care requirements. This type of changing requirements called for a major change in the approach to the planning of work schedules and shift composition, moving from separate work schedule planning for individuals, to an emphasis on integral and moduler teams as the elementary unit for which work schedule is constructed. This shift from individuals to team increases the flexibility and degrees of freedom, while maintaining coherence when coping with changing situation demands.

2. Introduction and incorporation of advanced information technologies to assist recording, searching, and transferring of information.

    Introduction of advances in information technology. Contemporary information technology provides many options for recording, integration, search, and transfer of information. It can give solutions to problems of information transfer, as well as reduce the considerable workload of recording and searching for information. Adoption of a computerized information system tailored to the NICU would assist substantially in solving the difficulties of transferring information and workload, especially in a unit suffering from chronic congestion and shortage of manpower, as evident in the large NICU. The system will improve the updating and transfer of information and thus raise the quality of care and safety, reduce the load on each employee, assist him/her in obtaining information, documentation, and formulation of a situation map. It may also improve the communication between senior and junior staff, thereby enhancing the quality of staff functionality.

MIDSIZE NICU

The main contributory factor to the midsize NICU workload was the nurses' cognitive load resulting from the necessity to prepare drugs in small dosages. In order to solve this problem it was recommended to move and relocate the drug preparation to the pharmacy. This has greatly reduced the nurses' cognitive load, as well as the number of activities performed by the nurses during the shift.

Another important recommendation was a change of the handover rounds and review process. The proposed solution attempts to solve the two problems—to provide assignment of neonates to a nurse prior to and not following the general review of all unit neonates.

SMALL NICU

The main problems observed in the small unit stemmed from the environmental conditions in which the staff operates and problems with information transfer due to a deficient process pertaining to the writing of instructions. Therefore, the main recommendations were for a physical reorganization and standardization of the unit space and incubator workstation.

To improve communication and information transfer, a formal protocol was proposed for physicians' rounds. This protocol included joint attendance of physicians and nurses, structured review of the patient by each of the care personnel, and a structured report and notation form.

REFERENCES

1. Special chapter within the Annual Report of Israel State Comptroller, on Care Processes in Neonatal Units 2004.
2. Israel Bureau of Statistics Yearly Report 2004.
3. Alberdi, E., Gilhooly, K., Hunter, J., Logie, R., Lyon, A., McIntosh, N., and Reiss, J. (2000). Computerisation and decision making in neonatal intensive care: A cognitive engineering investigation. *Journal of Clinical Monitoring and Computing*, 16, 85–94.
4. Wickens, C.D., Gordon, S.E., and Liu, Y. (1997). *An Introduction to Human Factor Engineering*. New York: Addison-Wesley.
5. Gopher, D. and Donchin, E. (1986). Workload: An examination of the concept. In: K.R. Boff, L. Kaufman and J.P. Thomas (eds.), *Handbook of Perception and Human Performance*, Vol. 2. New York: Wiley, 1–49.
6. Lefrak, L. (2002). Moving toward safer practice: Reducing medication errors in neonatal care. *Journal of Perinatal and Neonatal Nursing*, 16(2), 73–84.
7. Emsley, D. (2003). Multiple goals and managers' job-related tension and performance. *Journal of Managerial Psychology,* 18(4), 345–356.
8. Bitan,Y., Meyer, J., Shinar, D., and Zmora, E. (2004). Nurses' reactions to alarms in a neonatal intensive care unit. *Cognitive Technology Work*, 6, 239–246.
9. Sabar, R. and Zmora, E. (1997). Nurses' response to alarms from monitoring systems in NICU. *Pediatrics Research*, 41,1027.
10. Gupta, A.K., Cooper, E.A., Feldman, S.R., Fleischer, A.B., and Balkrishnan, R. (2003). Analysis of factors associated with increased prescription illegibility: Results from the national 164 ambulatory medical care survey, 1990–1998. *The American Journal of Managed Care*, 9, 548–552.
11. Reed, S.K. (1996). *Cognition: Theory and Application*. Pacific Grove: Thomson Publishing.
12. Bates, D.W. and Gawande, A.A. (2003). Improving safety information technology. *The New England Journal of Medicine*, 348, 2526–2534.
13. Donchin, Y., Gopher, D., Olin, M., Badichi, Y., Biesky, M., Sprung, C.L., et al. (2003). A look into the nature and cause of human errors in the intensive care unit. *Quality and Safety in Health Care*, 12, 143–147.

14. Frey, B., Kehrer, B., Losa, M., Braun, H., Berweger, L., Micallef, L., Ebenberger, L. (2000). Comprehensive critical incident monitoring in a neonatal-pediatric intensive care unit: Experience with the system approach. *Intensive Care Medicine*, 26(1), 69–74.
15. Tourgeman-Bashkin, O., Shinar, D., and Zmora, E. (2005). The causes and severity of human errors in intensive care units. *HEPS 2005, The International Conference*, Florence, Italy. 85–89.
16. Manias, E. and Street, A. (2001). Nurse-doctor interactions during critical care ward rounds. *Journal of Clinical Nursing,* 10(4), 442–450.
17. Sweet, S.J. and Norman, I.J. (1995). The nurse-doctor relationship: A selective literature review. *Journal of Advance Nursing*, 22(1), 165–170.
18. Chaboyer, P.W. and Patterson, E. (2001). Australian hospital generalist and critical care nurses' perceptions of doctors-nurse collaboration. *Nursing and Health Sciences*, 3, 73–79.
19. Krogstad, U., Hofoss, D., and Hjortdahl, P. (2004). Doctor and nurse perception of inter-professional co-operation in hospitals. *International Journal of Quality in Health Care*, 16(6), 491–497.
20. Xiao, Y., Seagull, F., Faraj, S., and Mackenzi, C.F. (2003). Coordination practices for patient safety: Knowledge, cultural and supporting artifact requirements. *Proceedings of International Ergonomic Association*.
21. Cook, I.R., Render, M., and Woods, D.D. (2000). Gaps in the continuity of care and progress on patient safety. *British Medical Journal*, 320, 791–794.

8 Applying the Principles of Human–Computer Interaction to Improve the Efficiency of the Emergency Medicine Unit

Nirit Gavish

CONTENTS

Introduction 119
Task Analysis 122
The Mental Model of the System 123
Conceptual Model 124
    The Principles of the Conceptual Model 124
    Format of the Screen and a Description of the Interactive Screen Areas 125
    Why Did We Choose to Present So Much Information on One Screen? 126
    The Navigation Format 127
    Design Details 128
Summary 129
References 130

INTRODUCTION

The department of emergency medicine, commonly referred to as the *ER*, is a hospital's showcase on the one hand and on the other it functions as a *first line of defense*. In other words, it serves to provide extremely rapid and lifesaving primary care, and avoids placing strain on other hospital departments by treating patients whose condition does not require hospitalization.

Since it is impossible to predict the rate at which people arrive and the severity of their problems, there is sometimes a heavy workload, which may result in an insufficient amount of professional personnel to respond to the arrivals or a shortage of beds for them to occupy. Hospital administrators place great importance on the satisfactory and efficient running of emergency rooms. This is expressed mainly by lengths of stay that are as short as possible for the arriving patients and their sense of satisfaction from the treatment they received.

The importance of an ER that operates well stems from the assumption that the ER is a mirror of the entire hospital. Satisfaction with the quality of care not only increases the prestige of the hospital but also may bring in "potential customers." That is, patients will prefer it to other hospitals and will come there for treatment or surgery (which will serve to bring in impressive income for the hospital).

For example, a large hospital in the United States set a goal that those who came to their ER should be seen by a doctor no more than 24 minutes after their arrival time. This would provide patients with the feeling that someone is looking after them and that the wait for treatment, even if it is prolonged, can pass without the added stress of feeling forgotten and neglected. After this change was effected, there was an 8% increase in the number of patients that went there for elective treatment (nonacute).

Hospitals are constantly faced with the challenge of continuously improving their emergency departments. Occasionally, following patient complaints or an abnormally heavy flow of patients, staff meetings are held and, based on the workers' experience, changes are put into effect. For example, in a hospital in central Israel, the head of the ER noticed that the hours between 10:00 A.M. to 12:00 P.M. are more congested. To lighten the load she decided to change the way the shift nurses should be assigned: instead of three shifts, as was customary, it was determined that one of the nurses should begin his/her morning shift 2 hours after the regular starting time and finish it 2 hours after the regular end of shift and that another nurse should begin his/her shift an hour later and complete the shift 3 hours after the end of the regular shift. By introducing this change, the nurse in charge could modulate the workload and coordinate a reinforced staff presence during the busy hours. The perceptions of the decision makers in the field are very important, but there needs to be a tool to help them reach the decisions in a more intelligent and well-founded manner.

The ER is a dynamic and complex system. Many factors are involved in determining how it should be operated and it comprises many complex relationships.

When the length of the orthopedic patients' stay in the ER is shortened it may result in a lengthier stay for the surgical patients. Similarly, when X-ray machines are added to shorten the waiting times for the results the bottleneck may pass to a different area. For example, it may result in a longer queue for essential laboratory test results.

To examine the methods through which the operation of the ER could be improved analytically, a simulation tool was developed at the Faculty of Industrial Engineering and Management of the Technion—the Israel Institute of Technology.

Simulation is a technique from the field of performance research used to model a system in an analytical manner. This technique is essential for modeling complex systems whose operation cannot be formulated through simple equations. This technique enables the simulation of realistic situations and processes over time, thereby making it possible to answer questions such as: How does the average length of stay of ER patients change when another nurse is added to the staff? How much does the doctors' workload increase if more patients come to the ER? What is the average number of people waiting in line for X-rays at 10 o'clock? The information obtained in this way allows administrators to choose between different alternatives for redesigning the ER, improving its performance and planning new emergency departments.

Human–Computer Interaction

Faculty, students, and medical professionals invested thousands of hours spanning a period of 5 years to develop this simulation tool. First, observations and interviews were conducted in order to characterize the process that a patient goes through from his/her entry to the ER until his/her treatment decision is made. Schedules were also examined to determine the period of time required for each part of the treatment: a doctor's exam upon admission, blood tests, a repeat exam before discharge, and so on. Based on this data, mathematical models that simulate activities have been developed, such as the walking distance covered by the medical staff within the ER. An organized database was prepared to store and retrieve the findings and the models, and finally these were organized into a software package that enables the examination of various components of the ER's operation and determines how they change as a result of modifications in the various parameters. The completed simulation tool has powerful analytical capabilities that can be used to assist the hospital staff improve the functioning of the emergency room.

Surprisingly, however, the use of simulation tools by staff members to make needed decisions regarding the operation of the ER does not depend on the nature of the tool and its application capabilities. It turns out that, although there is widespread use of medical simulation to help decision makers and improve the efficiency of systems, the simulation results are seldom implemented. Wilson[2] found that only 16 of the 200 simulation project results were implemented successfully. Ward et al.[3] argues that a simulation program project is only the beginning of the project, not the end, and most often the difficulty in changing existing behavior and political considerations make it extremely difficult to apply the results of a simulation. Gavish and Sinreich[4] have argued for a *black box* problem that prevents people from using the simulation. Yuviler-Gavish and Gopher[5] have demonstrated the general phenomenon of under-reliance on decision support systems for the long term.

How is it possible to encourage the potential users to take advantage of the simulation tool and to seriously consider its findings? The answer to this question is provided through the scientific–engineering–psychological field called Human–Computer Interaction (HCI), which deals with the interaction between sophisticated computerized systems and people. It examines the way that certain components of the computerized system's design affects its use, which may be expressed in a number of ways: the time required to perform activities, the possibility that errors do not occur, the staff's satisfaction with the system, the propensity for its continued future use, and the ease with which users can remember how to perform the activity even after the passage of time.

Other related tasks in this area include the task analysis of its component tasks, defining what it should accomplish, systems design, implementation, and the manner in which the systems are used. It affords a compilation of knowledge encompassing elements of engineering, such as the distance between two buttons that enable a quick and convenient transition between them; and psychological components, such as the optimal number of suggested items to embed in the menu choice. Human–computer interaction is uniquely capable of conducting research with practical significance, such as new, original suggestions for designers of computerized systems.

This chapter will describe the manner in which the user interface for the simulation tool was designed. The principles of human–computer interaction were used to encourage the team members to take advantage of the simulation tool that was designed for them. Our slogan was "Help people help themselves." The knowledge was collected, the tool exists, and its potential applications were developed. Now only one task remained—a small but significant task—to motivate the people to take advantage of all of this.

TASK ANALYSIS

Task analysis represents the basis for designing the user interface: characterizing the people who are to use the system, the tasks that need to be performed in the system, the work environment, and the frequency of its use.

It is of utmost importance to identify the primary users in the system. This decision will greatly influence the character of the system being planned. In our case, we needed to choose between two possible user populations who differed quite significantly from one another. The system is intended to assist in making strategic decisions related to the operational management of the ER: How many doctors and nurses will there be? In which shifts? How will the patients be assigned to the various rooms? And so on. That being the case, who will use the system? The answer is seemingly clear: those who are responsible for making these strategic decisions; ER and hospital administration staff members. These *users* are responsible for performing a wide range of tasks in various fields, so they cannot be expected to delve into the analysis of flowcharts, graphs, tables and statistical data, but would be more inclined to value concise data summaries that they can use to help them make quick decisions.

But there is another population of potential users, who work *behind the scenes*: engineers in the organization and methods (O&M) departments of hospitals, who are trained as industrial and management engineers. These users are charged with advising the administrators in the strategic planning of ER management and so they too benefit from using the system. O&M personnel are not only skilled in the use of charts, graphs, tables, data, and so forth—but these are the tools of their trade, as masks are for a doctor or scalpels for a surgeon. Therefore, it is likely they will want to invest the time and effort needed to analyze various forms of the ER measures.

The nature of the system will be determined by deciding who will be the primary users of the system; the administrators or the O&M personnel: Will it enable a simple and quick summarization of the data or provide a broader choice of alternatives for examining and analyzing the information? Following meetings held in the country's hospitals, we reached the conclusion that the main users of the system would be the O&M department personnel. Therefore, the decision was made to adapt the system to their specifications and concomitantly, enable a quick and simple course of action for the administrators as well. The system must provide two modes of operation: a workspace that guides the user to the fundamental information necessary and another that enables the in-depth study of the data, being able to view various data breakdowns, and enable different forms of presentation.

The second and most important decision on the nature of the system is defining the context of the system's use. Four tasks can be performed through the system. The first task is to examine the operation of the ER under the existing conditions—data such as average stay times, the load on the doctors and the nurses, the occupancy of the rooms, and the like. The second task is to find ways to improve the operation of the ER by interventions, such as by adding manpower or changing room allocations. The third task is to examine the effects of change, such as building an addition to the ER. The fourth task deviates from the central action flow of review > change > examine the effect of the change, which is continuously being executed, and deals with the possibility of designing new emergency rooms. The hospital can examine the possible effects of different system designs on the ER operation and, accordingly, decide on the character of the new ER. The system should support the flow of the primary activity; the actions performed most often in their expected order. Therefore, beyond the operational flexibility of the system and the unconstrained use of its application capabilities, a simple work mode should be built into the system, which will examine the condition of the existing ER, select methods of change as needed, and review the situation after the changes are made.

THE MENTAL MODEL OF THE SYSTEM

The proper design of the system requires a definition of the mental model according to which it will be developed. Mental models are psychological representations that the users have regarding the system they are using. According to Norman,[1] mental models are:

> Models that people have about themselves, about others, about the environment and about the systems that they interact with. Mental models are created based on experience, training and instruction. The mental model of the system is formed primarily through interpreting the perceived activities of the system and its visible structure.

The system's designer perceives the system from a specific mental model of his/her own, and shapes it on the basis of this mental model. The mental model of the system that the designer builds should approximate the mental model of the system's users. If a gap exists between the mental models of the designer and the user, it could result in system work errors and dissatisfaction with the system. For example, when a Web site is designed in a book format, that is, moving from a page to a previous page or to the one following it, whereas the user was expecting a network format (a mental model of a network), that is, having navigational options beyond that of going from one page to another, a navigational problem may arise. When building the system we tried to make our mental model as transparent and clear as possible for the user.

Our mental model for the system is a collection of files called *models*. When working with the system, only one model is active. Opening another model requires closing the model that had been active until then. Each model reflects the condition of the ER in a particular hospital during a particular period of time—past, present, future—or a hypothetical condition. The model includes values for the set of parameters that reflects the methodologies used in the ER, the number of patients arriving, the number of nurses and doctors available, shift assignments, X-ray methodology, and so on. If a simulation is run, the model also encompasses its findings on the operation of the ER: lengths of stay, workloads, waiting times, and so forth.

When the hospital gets the system, a basic model of the condition of the hospital on the day the system was constructed is embedded within it as well as models of other hospitals. From this basic model (hereafter referred to as *the database*), the compilation of models can be created. Based on the existing model, whether it is the basic model or another model within the collection of models, one can build a new model by opening the existing model, changing its parameters as needed, running the simulation, defining the desired breakdown of the results, and saving it with a new name. If the modified model is saved with the name of the existing model, it will delete the old model and save the new model in its place. This applies to all the models except for the basic ones, which cannot be deleted.

The modifications that the user incorporates in the model—such as in its parameters or the way the results are presented, are saved as long as the model is open. When the user closes the model, or opens another model in its place, he/she must save the changes or they will be lost. This model of saving is similar to that found in other systems that deal with files.

CONCEPTUAL MODEL

The conceptual model affords a comprehensive view of the system and the general definitions of the operational interface principles. It concerns the basic logic of the interface design, the structure of the screen, the navigational format, the principles of the user–system interaction, the degrees of freedom, and the flexibility. The conceptual model is the basic framework for the detailed planning of each and every screen.

THE PRINCIPLES OF THE CONCEPTUAL MODEL

The system is designed primarily for industrial engineers and management, thus providing them with detailed information in ways familiar to them, which allows them to analyze the findings. The system allows for an extensive use of graphs, tables, flowcharts, and statistical data. It enables the user to define how the data and its various breakdowns are presented, such as the doctors' workload during the day versus at night, a comparison of the patient's lengths of stay according to patient classification (internal, surgical, or orthopedic), and more.

The system supports the central activity sequence—an attempt to improve the operation of the ER by changing parameters, such as the number of nurses and doctors, the hours of a shift, sizes of the rooms, the percentage of patients referred for tests and more. The system guides the user step-by-step to achieve his/her objectives. It presents a list of possible objectives and the user selects his/her objective of choice, such as shortening the length of stay in the ER, reducing staff load, shortening waiting time, and so forth. Once the user selects the desired objective, the system indicates the parameters that should be changed to achieve it. Once the user modifies the parameters, he/she is guided to run the simulation so that the subsequent situation can be examined. The results of the simulation are the performance indicators of the ER's operation: workloads, lengths of stay, waiting times, and so forth. If the user wants to make additional modifications, he/she can return to the beginning of the loop.

Human–Computer Interaction

The system enables the user to select the parameters he/she wants to change, determine the values of the parameters at will and without the need for the system to intervene, and run the simulation at each stage to see the effect of modifications on the operational measures of the ER. This enables a combination of work guided by the system's suggestions and independent work by the user.

The system is also built to allow for flexibility and flow between its various subsystems; it contains a convenient mode of transitioning between the parameter views and its modifications and viewing the results according to various statistical breakdowns as defined by the user. This can be accomplished by using the system's guidance that provides recommendations for changing parameters, running the simulation, and viewing the results; or an unconstrained route of autonomous analyses of sensitivity: changing parameters and running the simulation to examine the subsequent results without the need for the system's guidance or an established order.

FORMAT OF THE SCREEN AND A DESCRIPTION OF THE INTERACTIVE SCREEN AREAS

The *interaction area* is a section of the screen used to frame a specific activity and specific information within the system. There are five interactive screen areas: menus, command buttons, model-building wizards, parameters, and results (Figures 8.1 and 8.2).

The model-building wizard and parameter areas are interchangeable, so the parameters can be changed either by the system's recommendations or independently.

The menus include actions such as opening an existing model, saving a model, running a simulation, and changing settings. The command buttons enable a majority of the most frequently used menu activities. The model-building wizard guides the user through the process of ER operation improvement through the use of an existing model and entering modifications. The parameters can be directly changed in the parameter pane or indirectly with the help of the model-building wizard. The results pane presents the findings of the active model and enables one to define

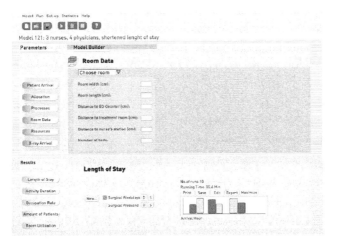

FIGURE 8.1 Interaction modes with enabled parameters set.

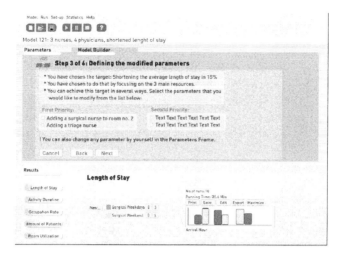

FIGURE 8.2 Interaction modes with active modeling wizard.

different breakdowns for view. The five interactive screen areas refer only to the open model—for example, "Save" will save the changes made in the open model. Some activities can be performed from the menu, such as presenting statistics and getting help.

WHY DID WE CHOOSE TO PRESENT SO MUCH INFORMATION ON ONE SCREEN?

A view of the screen graphics reveals an abundance of information. This resulted from a clear design decision, which affects the conceptual model: the display of a relatively large amount of information on one screen. The parameter and model-building wizard can be viewed alternatively through the inclusion of tabs at the top of the screen. At the bottom of the screen is the results pane. As a result, the parameters or the modeling wizard panes are always displayed simultaneously together with the results pane.

This design decision results in a screen that is more congested than conventional screens are. Most of the guidelines for determining the quantity of information on the screen advocate the presentation of as little information as possible—only that which is necessary for the user to perform the specific task during the interaction, and nothing else.[6,7] Tullis[8] calls for avoiding the tendency to display additional information just because it is available, since the additional load reduces the ability of users to discern the relevant information. Displaying large amounts of information on the screen is also contrary to the accepted rules of interface design,[9] according to which the capacity of the user should not be overloaded and that simplicity should be maintained as much as possible.

Why then did we disregard these recommendations? The answer stems from the job analysis we performed. From the job analysis it was decided that the main user population will be composed of industrial engineers and management, and the main objective of the system was to improve ER operation by changing the parameters.

Industrial and management engineers are accustomed to working with several sources of information simultaneously, processing and analyzing them through the integration of the information sources. It can be assumed that when making a decision on changing the parameters, the user population will seek to track the various outcome measures simultaneously. Changing the parameters is performed in the parameter pane or with the help of the modeling wizard, thus these two interactive areas should appear in conjunction with the results pane. The design decision supports the principles of the conceptual model of fluid and flexible operations. That is, since both the parameters and the results are visible to the user when performing the activities, he/she can switch between the panes seamlessly, without the need for opening and closing screens.

The successful design of the interface does not necessarily depend on a consistent implementation of principles and recommendations; sometimes the correct method actually entails disregarding accepted conventions and principles. Interface designers should be guided by adhering to the job analysis data and the conceptual model principles that they set up for themselves. The manner of realizing these elements is truly more complex than that recommended by conventional guidelines, but in some cases, as in this one, the optimal way is a unique, independent path.

THE NAVIGATION FORMAT

Navigation formatting style defines the manner that the user moves within the system: when will new windows be opened? How will the wizard be operated? How will the transition between the interaction areas be performed? And so on. The navigation format in an interface is based on free and flexible transitioning between the interactive panes. The parameters pane and the model-building wizard are located on two different tabs. The transition from tab to tab is done, as is customary, by clicking on the tab. The results panel is located in parallel at the bottom of the screen, so that going to it does not require any navigational activity.

Pop-up screens open from some of the menus and action buttons. For example, in the set-up menu, one of the options is the Operating Method and selecting it opens a screen in which the operational methodology of the emergency room is determined (Figure 8.3).

Navigating the model-building wizard between the different screens in which the improvement objective, the parameters to be changed, and so forth, are selected, is executed in a guided manner by the "Next," "Back," and "Cancel" buttons.

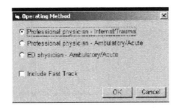

FIGURE 8.3 An example of the set-up screen when structuring navigation modes.

The parameters area enables one to view and update the different types of parameters, such as the number of active personnel, organization of the shifts, and sizes of the rooms. In addition, this pane enables a view of the results according to various measures, such as workloads and waiting times. The switch buttons make it possible to navigate between the types of parameters in the parameters pane and the types of measures in the results pane. Only one switch at a time can be operated and the rest are inactive.

Design Details

The design detailing stage focuses on the small details that make up the whole picture. At this stage the distribution of content between different screens, the manner in which each screen operates, and the precise appearance of the screens is decided on. In addition, this stage includes the determination of the controls to be used to determine the required functionality, the location and size of the controls, the text to appear on the screens—inclusive of font, as well as the decisions related to the pixel-level.

There are many principles that guide this stage and most of them can be learned—for example, those for determining the correct distribution of information between the screens to balance the load on each and reduce the need for scrolling to be able to see all the information provided, the principle guiding the selection of the appropriate controls; of grouping the screen content in a manner that is meaningful to the user, and how to align information vertically and horizontally to reduce visual clutter.

The following is an illustration of two screen sections that illustrate what is done during the design detailing stage (Figure 8.4).

This is a section of the screen from the parameters panel. It allows one to view and change the parameters related to the arrival of patients to the ER. Once the category of patients (surgical, orthopedic, etc.) is selected, a table appears which makes it possible to view the number of such patients who arrived at the ER during each

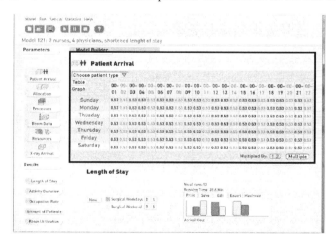

FIGURE 8.4 An illustrated design of a screen format for monitoring patients' arrivals.

Human–Computer Interaction

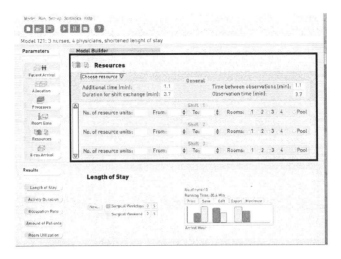

FIGURE 8.5 An illustrated design of a screen format focusing on clinical staff resources.

hour of a particular day. This information can also be viewed in a graphic format. It is possible to change the number of patients arriving by typing the number directly into the appropriate box, or by multiplying all the table numbers by a certain factor. One enters the factor directly into the text box on the right lower side of the pane and clicks on the "Multiply" button alongside it. The required information with respect to patient arrival is organized according to the size allocated for it on the screen, which does not require scrolling, and the arrival data are organized in a table that is easy to view and update (Figure 8.5).

This screen segment is also from the parameters pane. It deals with data relating to the ER personnel: nurses, doctors, X-ray technicians, and so forth. One selects the appropriate resource, fills in the general data, and integrates it according to shifts, in the appropriate rooms. The general data is kept separate from the shift data, and each shift (Shift 1, Shift 2…) is organized in its own area. The data are aligned vertically and horizontally, so that the visual load is reasonable despite the quantity of information on the screen.

SUMMARY

This chapter illustrates how to utilize the human–computer interaction in order to improve the operation of the ER. It reports on the deliberations that went into the creation of a user interface suitable for a simulation tool. However, the creation of sophisticated, multi-option simulation tools is not sufficient for improving the operation of the ER; in order that people apply this tool and take advantage of it as much as possible, an appropriate interface must be designed according to the data from the task analysis conducted for the use of the system. It must provide a mental model that is clear and easy to grasp. To achieve this, the conceptual and navigational principles to be formulated should consider the known guidelines and recommendations, but also diverge from them and adhere to the task analysis as a guide.

After the initial deliberations, the design should be implemented in detail, including the appearance, size, and function of each screen. This chapter describes the planning that was involved in only a few of the tens of screens in the system.

With the advancement of technology and the development of intelligent systems that aid in decision making, the importance of the field of human–computer interfaces increases.

In order for the person who you are creating the system for will advance and not lag behind in these technologies, the interface must be planned carefully. A successful and convenient interface, which connects the individual to the tool, is not just a charming feature to beautify the system. A good interface is essential for the proper utilization of man's new tools—sophisticated systems.

REFERENCES

1. Norman, D.A. (1988). *The Design of Everyday Things*. New York: Doubleday/Currency, p. 17.
2. Wilson, J.C.T. (1981). Implementation of computer simulation projects in health care. *Journal of Operational Research Society,* 32, 825–832.
3. Sanchez, S.M., Ferrin, D.M., Ogazon, T., Sepúlveda, J.A., and Ward, T.J. (2000). Emerging issues in healthcare simulation. *Proceedings of the 2000 Winter Simulation Conference.*
4. Gavish, N. and Sinreich, D. (2009). Adaptive interactivity: User interface design for simulation systems. *The Huntsville Simulation Conference (HSC 2009)*. Huntsville, Alabama.
5. Yuviler-Gavish, N. and Gopher, D. (2011). Effect of descriptive information and experience on automation reliance. *Human Factors, 53,* 230–244.
6. Smith, S.L. and Moiser, J.N. (1986). *Guidelines for Designing User Interface Software (Technical Report ESD-TR-86-278)*. Hanscom Air Force Base, MA: USAF Electronic System Division.
7. Galitz, W.O. (1993). *User-Interface Screen Design*. Wellesley, MA: QED Information Sciences.
8. Tullis, T.S. (1997). Screen design. In: Helander, M., Landauer, F., and Prabhu, P. (eds.), *Handbook of Human–Computer Interaction* (2nd ed.). Amsterdam, The Netherlands: Elsevier Science B.V.
9. Marcus, A. (1997). Graphical user interfaces. In: Helander, M., Landauer, F., and Prabhu, P. (eds.), *Handbook of Human–Computer Interaction* (2nd ed.). Amsterdam, The Netherlands: Elsevier Science B.V.

9 Human Factors Contributions to the Design of a Medication Room

Zvi Straucher[*]

CONTENTS

Discussion and Conclusion 137
References 137

Error prevention when dispensing medication is a topic that is a concern of medical service providers worldwide. In the United States, the number of deaths related to taking medication is estimated to be 7,000 a year[1,2] (a number that is statistically equivalent to 170 deaths a year in Israel). A factor analysis of the events reveals that the poor physical design of the location in the hospital where the medications are prepared is one of the causes for the errors.

The transition to a new surgical ward at the Haemek Medical Center in Afula was an opportunity to include human factors engineers in the team planning of the medication workstation in the new department. A joint team that encompassed the hospital staff, the department architectural design team, and staff from the Center for Work Safety and Human Engineering at the Technion planned and designed a new dispensary workstation. The impact of the new physical design for the medication workstation was tested by examining it in comparison to the condition of the older workstations at the hospital. The project encompassed four successive stages.

A. Analysis of the existing situation and a hierarchical presentation of the problems requiring a solution.
B. Formulation/display suggestions for finding solutions from the viewpoint of human factors engineering.

---

[*] "The Implementation of Human Factors Design in Planning a Hospital Medication Room" by Zvi Straucher, Orna Mart, Ilan Paltin, and Mary Azriel. Research Center for Work Safety and Human Engineering, Technion with the participation and assistance of Orna Blondheim, Ahuva Tal, and the hospital management of the Haemek Medical Center, Afula.

C. Applying the design principles into two newly built departments in the hospital.
D. A comparative examination of the preparation of drugs by the department nurses in the old and new workstations.

Stage A: Analysis of the existing situation and the formulation of the problems requiring solution in order of importance. Interviews were conducted with department head nurses and nurses that prepare the medications for this purpose. Observations of the task process were conducted in different departments, the various approaches to the preparation of medications were analyzed, and the process was also observed in another hospital. This step was summarized along with a detailed description of the preparation process in its various forms. The interviews and observations were conducted by professionals from the Technion accompanied by the departmental nursing administrators and with the backing of the hospital management. Below is a list of the problems found in the process of preparing medications, in descending order of importance:

The location of the medication workstation, behind the main work desk in the department, in an open and accessible space, which exposed the nurse preparing the medications to constant interference, including requests for assistance and environmental noise. The nurse had no opportunity to work uninterrupted and undisturbed (Photo 9.1).

The height of the work surfaces in the existing workstations was not appropriate relative to the average body proportions of the standing nurses who work there, forcing them to adopt unusual postures and endure muscular strain, causing them discomfort when performing the task (Photo 9.2).

The labels on the medication packages, as well as the names and dosages of the medications located on the drawers where they were stored, were not

PHOTO 9.1 The old medication preparation workstation was as part of the nurses' general reception station.

Human Factors Contributions to the Design of a Medication Room

PHOTO 9.2 Height of work platform is too low for an average nurse in standing work mode.

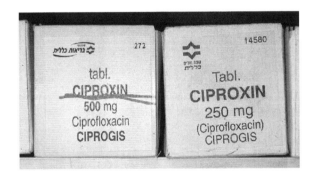

PHOTO 9.3 Nurses' manual with hand marks on existing medication labels to assist rapid discrimination.

optimal for the purpose of decoding the names and dosages in an error-free process (Photo 9.3).

The nurse preparing the medications needed to move about while collecting the medications. The nurses in the area were observed bumping into one another while preparing the medications.

The lighting and ventilation conditions in the nurses' work area were unsuitable for the demands of the process (understanding the prescription from the written instructions, visual search of the medication storage drawer, and collecting the correct medication in the required dosage).

The refrigerator doors and drug storage cupboard were sealed, which prevented the priming of the visual search process for the name of the medication requested.

The storage capacity of the medicine cabinet was not sufficient for all the drugs in the department.

Poor planning of the work area created a situation that prevented the preparation of medications by more than one nurse at any particular time.

Stage B: Suggestions for coping with the problems revealed in the existing drug preparation workstations.

All new wards will have a dedicated room set aside for preparing medications with controlled access, enabling a separate work area free of interference for the preparation of medications.

The work surfaces shall be designed for working while standing, suited to the 50th height percentile of the nurses who prepare the medications.

The lighting design and intensity will be designed according to the data regarding the reading and visual search tasks.[4]

The ventilation in the room will be designed to be suitable for an extended stay of two nurses, when the doors are closed.

The labels on the drawers in which the drugs are stored shall be designed to allow for accurate reading and interpretation of the name of the drug and the dosage required, minimizing the possibilities of errors occurring during the process.

The doors of the medication cabinet and the medication refrigerator in the dedicated room will be transparent and enable the identification of the required drug name without having to open them. The doors will allow for the priming of visual identification of the medication and the dosage required before the nurse extends his/her hand to collect the drug for distribution to patients.

The rooms will include planned *parking bays* for the drug carts, which will enable the carts to be *parked* while the drugs are being collected without interfering with the mutual work of two nurses, while maintaining a maximum amount of workspace and free access to all the functional areas in the room, with minimal nurse movement.

Stage C: Rooms in two hospital wards were designed according to the recommendations cited above.

Rooms dedicated for the preparation of drugs, with doors that cannot be opened except by using employee cards (Photo 9.4).

The furniture (medicine cabinets, refrigerator, and work surfaces) were constructed according to the principles described. Photo 9.5 shows the medicine cabinet and new medication refrigerator.

The drug storage drawers have been designed according to the recommendations of Ilan Palatin.[3] Photo 9.6 shows the new design of the labels.

The new labels, written with large letters and the dosage written in bold font, have been shown in a preliminary study to contribute to reducing errors.

The lighting and ventilation were installed as recommended by the Technion team.

Human Factors Contributions to the Design of a Medication Room   135

PHOTO 9.4 The door of the new medication room can be opened only with an employee ID card.

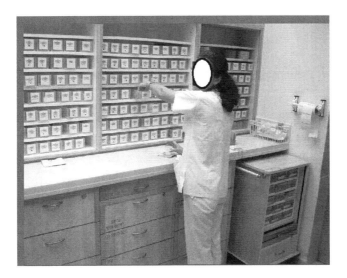

PHOTO 9.5 The new medication room with sliding work carts, two workstations, and redesigned cabinets and work platform.

Stage D: A comparative evaluation of the new medication room. Two months after the start of the work in the new departments, a test was conducted to examine the medication preparation in the new departments in comparison to the drug preparation in the older hospital departments. Two methods were used to perform the test:

a. A self-report questionnaire regarding satisfaction. 60 nurses (40 of whom continued to prepare the drugs under the previous conditions and 20 who now worked in the new departments) were asked about their satisfaction regarding various parameters of the working process. The satisfaction scores of the nurses in the dedicated rooms were significantly higher.

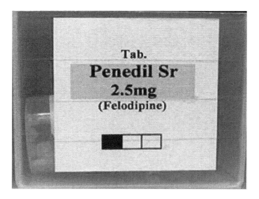

PHOTO 9.6 (See color insert.) The new medication labels, upper- and lowercase letters, background color of written name, and barographs of available number and dosage level of the medication.

b. A test simulating the drug preparation process was performed simultaneously with 20 nurses (10 nurses working under the old medication-preparing conditions and 10 working under the new conditions). They were asked to prepare 130 doses for dispensation. The Technion researchers observed the drug preparation and recorded:
- The number of distractions during the process.
- The number of times that the nurse reached into the wrong drawer for an incorrect dosage (recorded as an *almost mistake*).
- The number of errors regarding the type of drug or the dosage that was prepared.
- The time it took each nurse to prepare 10 doses of drugs.

Table 9.1 demonstrates that the differences in performance between the old and new medication room designs are statistically significant. Table 9.1 contains a summary of the nurses' performance in the new and older departments. (It is important to note that the nurses knew that this was being done for study purposes and the drugs prepared were not to be dispensed.)

TABLE 9.1
Illustrates the Findings of Observing 260 Prescriptions Being Prepared by 20 Nurses

New	Old	Measure
12	110	Distractions
3	17	"Near miss"
0	6	Errors
99 seconds	178 seconds	Average time for preparing three medications
SD = 45 s	SD = 84 s	by 10 nurses

DISCUSSION AND CONCLUSION

The above process has created a new work environment for nurses engaged in preparing medications in the surgical department in the Haemek Medical Center. The new workspace was designed to emphasize several critical factors in determining the probability of error in the preparation of drugs. The factors were: preventing distractions during the preparation of the medications; ensuring proper ergonomic conditions for performing the task (table height, lighting, ventilation, preliminary visual accessibility to perform the visual search), and redesigning the new label with the name of the drugs and their dosages—a design found to be optimal in a preliminary experiment.

Testing the process in the hospital 2 months after the work began in the new departments revealed a real and significant improvement, both according to the subjective opinions of the nurses, as well as in objective performance measures. The uniqueness of the process carried out in Haemek is that it is a pilot study, which examined in detail the work process that would be carried out at a new workstation that was to be newly designed according to parameters of needed improvement. There are no universal design principles for medication rooms, but rather the design must suit the work process and the composition of the population that it is being planned for—in our case, nurses who prepare the medications in Haemak hospital.

After the project, other departments in the hospital implemented this approach and the method used in marking the labels on the medication drawers was adopted by the management, Clalit Health Services, for medication storage drawers as well. Other Israeli hospitals have also implemented the design principles developed in this project. We hope they are applying the approach meticulously, ensuring that the design is suitable for the unique conditions and work processes of the departments and the populations for whom they are being planned.

Testing the process following its completion proved that correct and appropriate physical design of the work environment can demonstrate improved performance in medical environments, in quantitative features including performance time and number of errors.

REFERENCES

1. Anderson, D.J. and Webster, C.S. (2001). A systems approach to the reduction of medication error on the hospital ward. *Journal of Advanced Nursing* 35: 34–41.
2. Kohn, L.T., Corrigan, J.M., and Donaldson, M.S., eds. (2000). *To Err Is Human: Building a Safer Health System*. Washington, DC: Institute of Medicine, National Academy Press.
3. Palatin, I. (2004). Using Principles from the Research on Visual Search, Visual Grouping and Word Recognition for the Redesign of Medication Labels. (Master's thesis, Technion, Israel Institute of Technology.)
4. Van Cott, H.P. and Kinkade, R.G., eds. (1972). *Human Engineering Guide to Equipment Design*. Washington, DC: U.S. Department of Defense, Joint Services Steering Committee, p. 65.

10 The User-Centered Design of a Radiotherapy Chart

*Roni Sela and Yael Auerbach-Shpak**

CONTENTS

An Exemplary Account of an Event 140
Radiotherapy (Radiation Treatment) 141
The Radiotherapy Unit and the Radiotherapy Chart 141
The Course of Study 144
    A Detailed Cognitive Analysis of the Existing Radiotherapy Chart 145
    Characterization of the Content and Form Requirements of the New Chart 146
    Designing the New Chart 146
    The Introduction of the New Chart for Use in the Radiotherapy Unit 147
Cognitive Problems in Working with Forms and Solutions That Were
Integrated in the New Design 147
    Complete and Explicit Presentation of the Information 147
        The Lack of Designated Fields for Documenting Information Increases
        the Memory Load and Limits Accessibility of the Information 148
        Leaving Cells Blank Creates a Dangerous Ambiguity 149
    Legibility and Organization of the Information 149
        Lack of Consistency in Displaying Data Increases the Probability of
        Copying Mistakes and Reading Errors 150
        Lack of Human Factors Principles to Improve Information Legibility 150
    Supporting Quality Assurance and Means of Verification 151
        The Chart Does Not Provide a Format and Structure for Calculating
        and Monitoring Processes 151
        Problems in Identifying Changes Made during the Course of the
        Treatment and Verifying Them 152
    Modularity and Flexibility 152

---

* "A User-Centered Design of a Radiotherapy Chart" by Roni Sela, Yael Auerbach-Shpak, Zvi Straucher, Oded Klimer, Marina Rogachov, and Daniel Gopher. Center for Work Safety and Human Engineering, Technion, with the participation and assistance of Raquel Bar-Deroma, Chief Physicist; Riki Carmi, Chief Technologist; and Abraham Kuten, Director of the Oncology Division, Rambam Health Care Campus.

Questions and Challenges in Developing and Implementing the New Chart 153
Summary 160
References 161

Malignant neoplasms are not fought by surgical treatment and medication alone, but also by precise *bombing* of the core of the malignant tissue with ionizing radiation, a treatment performed in designated hospital centers. The physician ordering medication for a patient, whether through infusion or orally, does so as did many generations of physicians before him/her and by using explicit drug prescription procedures. The dosage and different methods used to deliver the drug are determined by the information a physician has regarding the patient's presenting condition: age, weight, and medications taken.

The calculation of the proper radiation dose in radiotherapy does not differ from prescribing a medication; however, it is based on an abundance of data from different sources and, in addition to the physician, important pieces of information must be supplied by physicists and technicians before operating the sophisticated machine. An unsuitable dose may be disastrous; for example, the patient may not receive the required dose of radiation rendering the treatment ineffective, or may receive too large a dose and the area exposed will be severely injured. Such mishaps occur more than once when implementing radiation devices.

AN EXEMPLARY ACCOUNT OF AN EVENT

> The usual radiation machine used for a patient had broken down, so we decided to treat him with another device. The units of radiation were recalculated and entered manually. Instead of going to lunch, I came to help. When I entered the number of radiation units, I typed 111 instead of 11. I calculated the total amount, the authentication system warned that there was a calculation error, but the dose appearing on the screen was still within the acceptable range. As a result, the patient received 100 excess RADs. It was a stressful day. We were short-staffed and I was working with three different machines that morning. Many patients were waiting at all the machines, and delays were caused at all of them, especially since the patients from the defective machine were brought in for treatment as well. It would have helped me if the sum of all radiation dosage units for the same field would have been recorded on the radiation chart as well.

(This is a report of an error in providing radiation therapy. Cited with the approval of the Radiation Oncology Safety Information System,[1] ROSIS, Web site—a public database that publishes reports received from European radiotherapy units in order to share their experiences of errors, adverse events, and the corrective actions taken to prevent them from recurring, while maintaining the confidentiality of those reporting the events and the patients.)

We will describe the project conducted by a joint team from the Technion—Center for Work Safety and Human Engineering—and the radiotherapy unit of the Rambam Medical Center in Haifa. The project was initiated by evaluating aspects of human factors engineering in the radiotherapy unit, after which the radiotherapy chart, which was observed to be the primary tool of communication and quality assurance in the unit, was evaluated in depth and redesigned. The new user-centered design was intended to develop information management, reduce human error in all stages of the treatment, and improve the communication between the interdisciplinary staff members.

This project was conducted through a proactive, systematic, and user-centered approach; that is, by viewing the individual as the focus of the system, which encompasses both the demands of the job (the task to be performed) and the environmental conditions. Our main concern was to examine the working conditions and the tasks carried out by the radiotherapy unit personnel, and to identify weak points that interfere with the cognitive processes of the service provider. Since, as is known, when cognitive processes such as memory, attention, calculation, or decision-making mechanisms are disrupted, it increases the probability of human errors, such as the mistake that occurred in the event described above.

RADIOTHERAPY (RADIATION TREATMENT)

Radiotherapy (the treatment of cancer through ionizing radiation) is the treatment of choice among approximately 60% of the patients with malignant disease. Sometimes radiotherapy is the only treatment and sometimes it is combined with pharmacological or surgical treatments. It is also used to prepare recipients of bone marrow transplants, since the radiation interferes with the production of various blood components. The radiation that reaches the body tissues produces electrically charged ions that damage the cellular DNA and prevents them from dividing and multiplying. Since the recovery capacity of cells that divide rapidly (such as cancerous cells, as well as skin and hair cells) is generally poorer than that of normal cells, a series of radiation treatments will damage the cancerous cells and inhibit the development of the tumor[2] without damaging the healthy body cells surrounding them. Hence, the importance of accuracy in dosage and location cannot be overemphasized. All radiation therapy involves the risk of damage and destruction of healthy tissues, and a small error in calculating the dose of radiation or inaccuracy of the radiation itself is sufficient to interfere with the healing rate and increase the negative side effects.[3,15] Moreover, the error rate in clinical radiotherapy is relatively lower than that of pharmacotherapy.[4,15] However, being that the secondary damage of the radiation may not appear until years after the treatment, it is usually difficult to link such damage directly to it, even if treatment errors were made.

THE RADIOTHERAPY UNIT AND THE RADIOTHERAPY CHART

Approximately 180 patients are treated daily in Haifa's Rambam Hospital radiotherapy unit, and about 2,000 new patients arrive each year from the entire northern area of the country. The unit employs four linear accelerators for treatment, some are older and some are newer and computerized. The unit operates in an outpatient format, and is an autonomous unit of the Oncology Institute, which also includes a clinic, a chemotherapy unit, and inpatient departments.

The radiotherapy process lasts about 15 weeks, and is divided into the following stages:

- Diagnosis and therapy protocol design
- Therapy planning and simulation
- The course of radiation—about 35 daily treatments
- Examination and follow-up

The staff members in this unit include oncologists, physicists, dosimetrists, radiation technicians, nurses, administrators, and welfare and maintenance personnel.

Each team member is responsible for a particular stage in the process and has a unique area of knowledge and expertise (with partial overlapping) for that stage of the treatment. Altogether, the team members represent a multidisciplinary unit that requires them to coordinate their activities to achieve a common goal: radiation treatment that is accurate and safe.[5]

Table 10.1 demonstrates the range of personnel and the activities performed during the radiotherapy treatment—about 30 staff members performing 7 different roles, carrying out at least 12 different activities related to transferring, receiving, and processing the necessary information. Note that the unit's workforce is much smaller than the required minimum according to international standards.[17]

From the moment that the patient enters the unit until his/her treatment is concluded, a vast amount of information is created, processed, and accumulated: treatment history, diagnosis, test results, treatment orders, calculations, reminders, target dates, planned measures, actual radiation doses given, and much more. A large portion of the information must be shared among all team members to ensure that the proper and accurate treatment protocol is followed. This information can be defined as a "unique database" that relates to the tasks and to the team itself.[6] The radiotherapy chart is a vital tool that should encompass and display all the details of this information.

The *Radiotherapy Chart* is a form used to record and document information in order to provide therapeutic protocol guidelines and monitor its implementation

TABLE 10.1
The Radiotherapy Unit Staff: Functions and Activities

Activities	Role
Therapeutic decisions, providing treatment orders, monitoring and documenting the patient's condition, consultation, responsible for the entire process.	Oncology physicians
Calculation of radiation measures, treatment planning, ensuring the quality of the process and the instrumentation.	Physicists
Treatment planning (imaging), executing treatment orders, documentation of the treatment.	Technicians
Calculating the radiation measures, planning the treatment.	Dosimetrists
Monitoring the examinations and documenting the patient's condition, providing advice.	Nurses
Schedule treatment appointments.	Medical secretary
Monitoring and documenting the patient's condition, providing specific advice.	Treatment collaborators (social worker, dietician)

The User-Centered Design of a Radiotherapy Chart

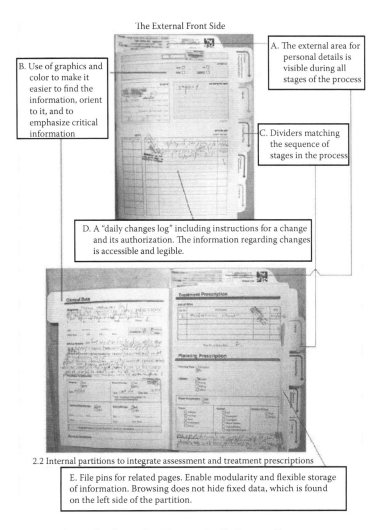

PHOTO 10.1 (See color insert.)   The new Radiotherapy Chart.

(Photo 10.1). In the unit we tested, the chart consisted of an A3 sheet of cardboard folded in half. The front was used to record patient data, care instructions, diagrams of the location to be targeted on the body, and how the patient was to be positioned on the device, as well as measures related to the method used to provide the radiation. The backside of the chart contained printed rows for recording the physician's diagnosis and details of the physician's or nurse's follow-up care. The inside of the chart contained data related to the measures of the RAD dosage and columns for documenting the measures of radiation actually provided. Inside the chart there was also a medium-sized envelope for storing various types of paperwork, results of tests, labels with the patient's personal information, and so forth. Photo 10.2b illustrates that this chart does not have sufficient space for all of the information required, and

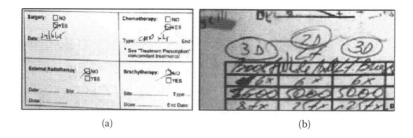

(a)   (b)

PHOTO 10.2 Important recorded information in the old chart without designed entry space (b) and its corresponding well-specified entry space in the newly designed chart (a).

that parts were recorded in a manner that was disorganized and difficult to work with. The problems inherent in this situation are discussed below.

In parallel to the manual chart, a computerized chart has been introduced to the radiotherapy unit that will someday replace it. Before continuing further, it should be noted that most of the usability problems to be presented later in this chapter may also apply to the computerized chart (although the physical solutions may be different).

Most of us consider forms such as the radiotherapy chart to be simple and straightforward tools. In fact, charts, and forms in general, are a significant component of work processes. In terms of human factors engineering, the radiotherapy chart is a cognitive artifact—a tool that represents actions regarding the information available to the user—storing and implementing it.[7,8] Cognitive artifacts are especially important for collaborative work in medical environments. They represent a link between the tangible physical nature of medical care delivery and the virtual world of computerized technologies.[9] The radiotherapy chart is the only consolidated accounting of all the information about a particular patient and the care he/she receives in the radiotherapy unit. Thus, the chart is the most important shared database for the work of the team throughout the entire process, as well as a tool that is essential for communication, monitoring, and quality assurance. The importance of the radiation chart can be derived from data reported on the ROSIS Web site dedicated to the subject. From the 709 adverse events in radiation therapy reported in the first 3 years of the site, about 40% were revealed by examining the radiotherapy chart (35% during the treatment itself and 25% at other monitored points during the course of the treatment).[1] Moreover, in a study at a large center for radiation therapy that examined errors and "near mistakes" that occurred during radiotherapy over a period of 10 years, it was found that many were associated with errors in documentation and most could be attributed to errors in data transmission or poor communication.[15] Therefore, the quality of the chart, in terms of its users' ability to benefit from it in performing the activities they are charged with, has a direct impact on the quality of care and patient safety and health.

THE COURSE OF STUDY

The project encompassed several stages: a comprehensive assessment of the radiotherapy unit, a detailed cognitive analysis of the radiotherapy chart, the characterization and design of a new chart, and the introduction of its use in the unit (Figure 10.1). We

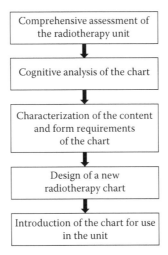

FIGURE 10.1 Stage of data collection analysis and redesign of the patient treatment charts. A comprehensive assessment of the radiotherapy unit.

will provide a detailed description of the method to illustrate the nature of working through a human factors approach as a structured and user-centered process.

To begin the assessment stage, the unit's general work processes were mapped through open interviews conducted with the heads of the subunits, representing the range of personnel, and by observing the process of treatment planning and of the therapy itself. Task analysis of the data was then performed, producing a flowchart depicting a typical treatment process, including the team members who participated in each stage. This analysis also identified and evaluated potential gaps between a person's ability to function in the system, the demands of the various tasks, and the physical environment. The conclusions were presented to the radiotherapy unit and hospital administrators. They indicated several potential areas requiring human factors engineering intervention, from which we chose to focus on the makeup of the radiation treatment chart and on redesigning it. This objective was selected since we determined that given the technological innovations of recent years and the overall computerized nature of treatment methods, new work processes have evolved that are dictated by new requirements for the storage and transfer of information. In light of all this, the radiotherapy chart was not adequate to fulfill the current needs. Improving the chart was intended to solve these and other problems, and transform it into a more effective and convenient tool.

A Detailed Cognitive Analysis of the Existing Radiotherapy Chart

The improvement process began with a focused analysis of the radiotherapy chart used in the unit in order to evaluate aspects of its cognitive[10] usability. The cognitive demands dictated by the properties of the chart were specifically examined, such as the availability of the required information, its absence, and how the information is presented. The assessment also examined the users' ability to cope with these demands. The main focus of this kind of assessment was to characterize the features of this valuable tool and examine how the design impacts the individual and the task.[7]

CHARACTERIZATION OF THE CONTENT AND FORM REQUIREMENTS OF THE NEW CHART

During the process of redesigning the radiotherapy chart, potential points of failure resulting from the old design were considered; problems that needed to be resolved in the new chart to reduce the likelihood of the potential failures and errors that could result, or to prevent them from occurring at all.

The concept for the redesign focused on two key aspects: its content (*what?*) and its form (*how*). The content aspect relates to the information that must be included on the chart and the form aspect relates to where the information is entered and its presentation, so it can be used conveniently and effectively.

To determine the information that must be included in the chart, we conducted structured interviews with staff members at all stages of treatment. By integrating the *information maps* received from the various users, a detailed list was created of all the items of information that were needed, grouped according to the functional stages of the treatment process. Some items of information were accompanied by specific requirements such as "a high degree of visibility is required—by all team members," "should be placed to enable comparison with other data," and so on. These specifications outlined the actual display of information during the physical designing of the chart in its new form. Radiotherapy charts from other radiotherapy units in Israel and abroad represented another source of information. These charts were also examined from the aspects of content and form, and the advantages and possible disadvantages of their structure. In addition to the specific requirements derived from characterizing the use of the radiation chart, the plan for designing the new chart was based on general principles of human factors engineering design. The chart should display the information (such as consistency, visibility, legibility, differentiation, compatibility) in a manner that would assist in perceiving the information, attention, and memory.[10,16]

DESIGNING THE NEW CHART

Based on the preparatory analysis, a prototype of a new radiotherapy chart was constructed—a hard copy model made of thin cardboard, which was as close as possible to the future real chart. Using such a prototype during the design process provides several advantages. These include providing support to the development team by enabling them to translate ideas into reality and enabling a medium for the communication of ideas, support for the product assessment, and in performing usability tests by using a concrete object and reacting to it.[10] Interestingly, the moment the new radiation chart was transformed from an idea to a tangible item, the team's as well as the management's responsiveness toward the research team increased, which positively influenced all those collaborating in the project's development. The prototype chart was brought to the radiotherapy unit, where it was piloted and examined by representatives of the various categories of users. As in all stages of developing the new chart's features and design, the stage in which the prototype was tested instigated a flurry of ideas and comments that continued till the joint determination of the nature of the finished product.

THE INTRODUCTION OF THE NEW CHART FOR USE IN THE RADIOTHERAPY UNIT

Once the final version of the new chart was determined and authorized, an initial number of copies were printed and the staff members were given instructions to familiarize themselves with the new chart and how to use it. Only then was it introduced for clinical use. After approximately 9 months (about 1,500 charts), the staff members comments regarding the convenience of working with the new chart and the structure of the information it contained were collected. Another assessment was conducted 9 months later (18 months from when it was begun to be used; about 3,000 charts). According to these comments, missing items of information were added and some that were not essential were deleted. No changes were made to the structure of the chart itself. Photo 10.3 illustrates the new chart. It is made from thin cardboard and contains several partitions in accordance with the stages of the work process (diagnosis and treatment orders, planning, documentation of its execution, nursing follow up, etc.). In the following section we will elaborate on the features of the new chart and how it presented solutions to the problems that characterized the old chart.

COGNITIVE PROBLEMS IN WORKING WITH FORMS AND SOLUTIONS THAT WERE INTEGRATED IN THE NEW DESIGN

Table 10.3 displays the abundance of problems revealed in the old chart, their implications, and how they were solved in the new design. The problems were grouped into four categories related to the possible pitfalls in working with forms:

A. A complete, comprehensive, and explicit presentation of all required information.
B. Legibility, readability, and organization of the information.
C. Support for quality assurance and methods of verification.
D. Modularity and flexibility.

Categories A and B are concerned with characteristics relating to the nature of the information and how it is presented, and categories C and D deal with the main objectives of the chart: to help monitor the processes and store information. Using examples we will illustrate the central cognitive problems in each category and the human factors engineering principles that contributed to their solutions.

COMPLETE AND EXPLICIT PRESENTATION OF THE INFORMATION

For all team members who use the chart, it is important to offer a complete and clear presentation of information in order that each staff member and all the users of the chart should have access to an updated and comprehensive account. Thus, the chart must contain fields for recording all the necessary information in a clear and unambiguous manner.

The Lack of Designated Fields for Documenting Information Increases the Memory Load and Limits Accessibility of the Information

When comparing between the list of items required and the actual number of fields present for recording data on the old chart, it was found that 50 out of 120 (42%) required information fields were not found in the old chart. Therefore, essential pieces of data (including sensitivity to the contrast media and concurrent oncological treatment) that lacked a designated field were recorded by the staff members in vacant areas of the chart (Photo 10.2b). The lack of sufficient fields is somewhat related to the changes in procedures and the state of the art medical equipment introduced to the system without these changes being reflected in the chart. The importance of having dedicated information fields in fixed locations stems from their serving as retrieval cues, which help the user instantly remember information that must be recorded in the chart. In addition, it helps the reader locate the information and use it when needed. Moreover, a familiar and fixed location on the chart increases the accessibility of the information and facilitates the detection of the needed material. If the information is placed arbitrarily or when fields for recording information are lacking, as in the previous chart, it increases the memory load and unnecessarily delays task performance. It is worthwhile noting that adding information to an area not designated for it, by taping on a note or jotting it down in the page margins, would not be possible on a computerized form as it is in handwritten forms. Details of information that are missing in a computerized form are much harder to locate and supplement. The solution implemented on the new chart was to create structured data fields according to the detailed list of all the required pieces of information (Photo 10.2a).

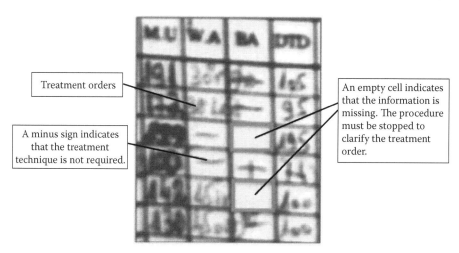

PHOTO 10.3 Agreed upon symbols to indicate *not required* in the old chart.

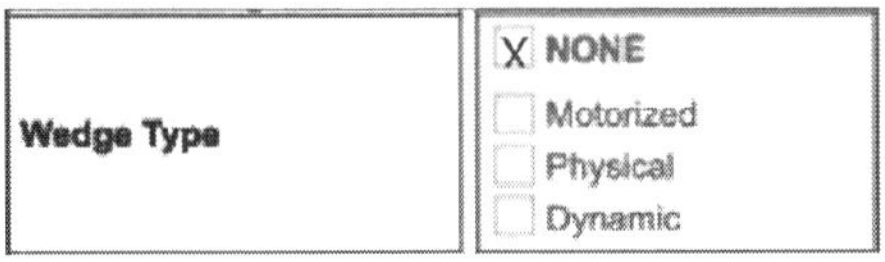

PHOTO 10.4 Designed entry for wedge level in the new chart. It was a blank field in the old chart.

Leaving Cells Blank Creates a Dangerous Ambiguity

A second example in this category is the ambiguity that could result from data fields left blank when filling in the chart. Radiation treatment processes differ from one another with respect to the equipment used, and the types of radiation and treatment aids. Not all methods are needed for every treatment. A wedge, for example, a device for radiation modulation, is required only for a few types of treatment. During the analysis of the work processes we realized that when a device was not necessary for a particular treatment, the relevant information field was left blank (nothing was written in it). This is a natural and spontaneous response, which nevertheless leads to a dangerously ambiguous situation: the person reading the instructions from the chart may assume, rightly, that the blank information field means that this treatment method is not required. However, there is another rare but dangerous possibility: that is, that the information field is empty because another staff member had forgotten to record the treatment orders. This type of ambiguity, which prevents a care provider from being able to distinguish between an aspect of treatment that was *forgotten* versus that it *is not required*, may lead to a critical error. In the wedge field, for example, this situation may lead to a dose of radiation that exceeds the dose that was to be given by 40%. There are two types of solutions to this problem. The immediate solution applied to the old chart was to use an agreed-upon symbol (a minus sign) to indicate a *not required* situation instead of leaving the cell empty. In other words, an information field must contain some sort of symbol after the chart is filled. When a treatment device is required, the suitable treatment order is listed, and when it is not required, a minus sign (–) was placed in the cell. If the information field remains blank, it is a warning: the required information is missing and the process must be stopped to clarify the treatment order (Photo 10.3).

Another solution to the problem of ambiguity resulting from cells that are left empty was implemented in the new chart: whenever the response categories were known in advance, an information field would be embedded in which there would be a checkbox for every possible answer, including a cell to mark when something is *not required* (Photo 10.4). This solution supports the control process of the chart and prevents a dangerous state of ambiguity.

LEGIBILITY AND ORGANIZATION OF THE INFORMATION

This category primarily encompasses aspects of the shape and structure of the information presented. Well-organized information enhances orientation in the radiotherapy

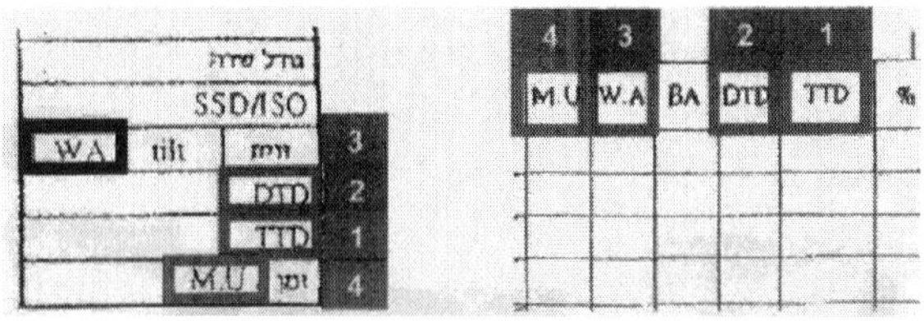

PHOTO 10.5 Lack of consistency of the sequence in which data is displayed in the old chart—on the outside of the chart (right) and on the inside (left).

chart and helps the user build a comprehensive, accurate image of the process. Some changes in this category have been established to assist the user in understanding the flow of the work process (functional workflow).

Lack of Consistency in Displaying Data Increases the Probability of Copying Mistakes and Reading Errors

A comparison between the process as represented on the chart and the process that actually takes place in the unit revealed some cases of mismatch and lack of consistency in the placement and sequence of the information fields (Photo 10.5). Information fields that were required at the same stage of the treatment were scattered in different areas of the chart, complicating and prolonging the search for the needed information. Some sequences of data on the chart did not match the sequence of the data in other documents or in the therapy devices from where the information was copied or in instruments in which this information had to be supplied. For example, reports printed in English from therapy devices had sequences of data listed from left to right but in the old chart where the data was copied, the sequence of the data was from right to left. As previously mentioned, this increased the probability of errors and mistakes in reading. Design solutions for these problems were, in part, to reduce the need to copy information by eliminating any unnecessary and redundant information fields, grouping related information fields next to one another, and matching the sequences of data to the flow of the manual and computerized procedures (Photo 10.1c).

Lack of Human Factors Principles to Improve Information Legibility

As was described, a table was sketched on the inside of the old chart containing a large number of crowded rows, impacting its legibility and tracking the sequence of data. Applying human factors principles to improve the legibility of information in the tables in the new chart included distinguishing long rows by using two alternating background shades to facilitate tracking the data along the length of each line (Photo 10.6).

Another means used to improve the readability and organization of the information was through the introduction of graphics and colors to visually emphasize a group of items and instantly distinguish it (Photo 10.1b,c). This was accomplished by including a color-coded arrangement, such that each partition representing a different stage of the process and the accompanying pages were given a unique color to emphasize the distinction, increase the user's orientation in the chart, and facilitate

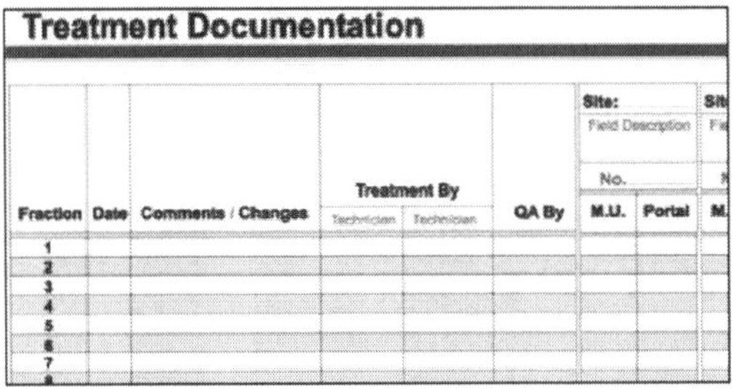

Treatment Documentation

Fraction	Date	Comments / Changes	Treatment By		QA By	Site: Field Description No.	
			Technician	Technician		M.U.	Portal
1							
2							
3							
4							
5							
6							
7							

PHOTO 10.6 Two alternating background shades to increase the table's legibility.

the association of related pages within each partition. Related fields of information were grouped together and colored with the same background color, which created visual clustering.

Supporting Quality Assurance and Means of Verification

An integral part of the work process is quality assurance and verification of the data, including systematic monitoring of the equipment's operation and of the precision of the calculations, measurements, and execution of the treatment orders. Here we present two examples that demonstrate the influence of proper design of the chart on these issues.

The Chart Does Not Provide a Format and Structure for Calculating and Monitoring Processes

The first example is related to the calculations: in an age in which machines perform most of our calculations, a person's ability to understand and verify the results is still very important, especially as simple arithmetic operations usually take place in one's head. In the area of health care, tasks involving calculations and their verification are very basic, common, and inherently prone to errors. But this potential for errors can be minimized relatively easily, by adding built-in means to support the process of calculation and examine the results, such as a formula or a table in which the calculation data were placed.

The design of the old radiation chart did not provide appropriate assistance for performing calculations and for easing the process of integrating the information. For example, when the general treatment guidelines are recorded, the total dose of radiation is divided into segments based on a partial and disorganized sequence of information fields. This complicates the process of entering the information for performing a relatively simple calculation and checking it, which may in turn lead to an error in the radiation dosage. The solution to this problem was to embed the data and sequence of the calculations to be performed within a table (Table 10.2).

In this manner an external representation of the calculation process serves as a memory aid and enables the information to be immediately comprehended.[11] This type of solution reduces memory load, organizes the calculation process, and enables a convenient means of checking it.

TABLE 10.2
Calculations Table

Description	
Prescription Volume	_____
Prescription Point	_____
Prescription Isodose	_____
Dose	_____
Dose per Fraction	_____
Total No. of Fractions	_____
No. of Fractions per Day	_____

Problems in Identifying Changes Made during the Course of the Treatment and Verifying Them

The treatment orders may be modified during the lengthy and complex process of treatment, for example, due to a change in the physical condition of the patient. It was difficult to emphasize these changes on the old chart. Changes in the instructions relate to the actions of different team members at different points in time. Since these modifications are generally entered in a number of different places on the chart, a staff member may not notice the required change because he/she would not think to look for such changes in the existing instructions. Failure to notice modifications in the information and in integrating them within the process may lead to errors that, in turn, will result in improper treatment.

The display of the revised information must be done in a way that draws the user's attention to it; the revised information must "present" itself to the user, instead of making the user search it out. One of the most important adjustments made in the new design was the initiation of a *daily changes log* on the front of the chart, which details changes in treatment orders and directs the user to their delineation within the chart, including a directive to perform the new instruction (Photo 10.1d). The change log draws attention to critical changes, increases their accessibility, and facilitates the monitoring of their actual performance.

MODULARITY AND FLEXIBILITY

Technology, designed to eventually enable paperless work, has paradoxically brought us to an intermediate stage in which the use of paper has increased in parallel to the increased use of computing.[12] Today, more than ever before, the information used in the process of radiation treatment is entered and retrieved both manually and on the computer, so that papers from external sources accumulate and must be attached to the radiation chart for documentation and review. The old chart was constructed as a two-sided simple form, and a medium-sized envelope was attached to it to store a limited number of related pages containing information that could not be recorded within the actual chart (such as a diagram of the field of radiation, labels with patient information, etc.).

Because of the rapid changes that occurred in therapeutic procedures and technology, the chart did not suffice for the abundance of data and accumulating associated pages. The envelope became so crammed with pages that it became increasingly difficult to find and retrieve them, various forms were stapled to the chart in a way that obscured important information written underneath them, pieces of important information (such as side effects of treatment) were not transmitted due to the lack of a convenient option for filing their pages, and in cases of prolonged or renewed treatments, new tables for reporting information had to be pasted in since all the columns for record keeping were full.

Therefore, the initial goal of the redesign was to increase the modularity and flexibility of the radiation chart: to ensure that the chart included information about the different types and lengths of treatments, and in fact, adapting the chart to integrate the transition between manual function to totally automated function. A practical design solution was to place filing pins within the various chart partitions to allow for the documentation of accompanying documents and adding tables for recording and documenting data. The filing pins in each partition were placed so that relevant information that had been previously recorded is not hidden even when browsing between the pages (Photo 10.1e) (Table 10.3).

QUESTIONS AND CHALLENGES IN DEVELOPING AND IMPLEMENTING THE NEW CHART

Why invest in a paper form in a computerized environment? As mentioned in the beginning of this chapter, one of the issues we deliberated over at the start of the project was whether to work with the existing paper form or to abandon it in favor of creating a computerized chart. On the one hand, a computerized chart is already being used for this purpose and the projected time until computerization is complete is about 5 years. So, if you are already investing, why not just invest in the future—the computerized form. One answer is that considering that 2,000 new patients are introduced to the unit each year, until the fully computerized form becomes the standard, about 10,000 people will be in treatment who will still require documentation in the manual chart. Another reason is that improving the manual form also directly bears on the improvement and proper nature of the informational requirements of the computerized form. The *information map*, based on the analysis of each of the team member's tasks, displays the information that each staff member needs to give and receive at every phase and during every activity in the treatment process. The correct execution of the process requires that they all possess the exact same information, whether it is delivered by computer or transmitted manually in a form. Moreover, although computerization denotes a technological change, the basic features of the interface between the individual and the information required to perform treatment tasks will not change. Both manual and computerized radiation charts require a comprehensive presentation of information that is accessible, clearly displayed, legible, and unambiguous. Thus, although our project focused on problems found in the paper forms, similar problems may become apparent in the work involving the computerized forms. As such, if these are not solved at this stage, the

TABLE 10.3
Summary of Problems in the Old Chart and Their Solutions in the New Chart

Category	Problems with the Old Chart	Detail of Their Impact	Design Solutions on the New Chart
A. Complete and Explicit Presentation of the Information			
Lack of information	Fifty of the 120 information fields needed (42%) did not exist and were therefore recorded informally in *agreed upon* locations on the chart.	Load on memory and difficulty locating information.	All the needed information fields are built into the chart.
Ambiguity of information	A single information field for all possible answers for a specific item. No option for indicating a procedure that is *not required*.	Load on memory, errors may occur due to an incorrect interpretation of an empty cell.	Introduction of fixed marking for the situation when a procedure is *not required*, and separate place for specifying each of the possible answers in every data field (Photo 10.4)
Excess information	Some of the information fields appear on the card more than once, and the information was copied unnecessarily.	Delayed the process, errors in copying information from one field to the other.	Elimination of repeated and redundant information fields.
	Presence of redundant information fields that are no longer being used.	Visual overload and congestion.	Delete the information fields that were not being used.
B. Legibility and Organization of the Information			
Functional workflow	2-Dimensional form without a clear separation between stages of the process.	Visual and conceptual overload, unnecessary information, confusion, delay of the process.	Cardboard binder with partitions separating the stages of the process (planning, treatment, monitoring, etc.) to organize the records.
	The internal structure of the arrangement of the chart does not follow the flow of the work process.	Difficulty in finding information, confusion, delay of the process.	Partitions organization and sequence profile information in accordance with the flow of work process.

The User-Centered Design of a Radiotherapy Chart

Category	Problem	Solution
	Mismatch between the direction in which the information is recorded on the chart (right to left) and the direction in the computerized and other printed documents (left to right).	Errors in entering and reading items of information.	Matching direction of the recordings on the chart to the sequence of information in the computerized and other printed documents.
	Some of the items of information that should be incorporated into a certain work stage are scattered in various places in the chart, including within the envelope (hidden or not accessible).		Grouping related data fields so that for each stage of work most of the relevant information for that stage will be presented.
Colors and shades	Chart only presented in two shades—black and white. Use of color was determined while writing a notation—emphasizing critical information (yellow) and changes in instructions (in red).	Although this does not harm cognitive processes of perception and understanding, this makes it difficult to view because the information is not prominent and important information does not get increased visual priority.	With separate partitioned card division, color coding was added to increase differentiation and orientation between the stages of the process and in order to facilitate association of related pages. Use of the color red to emphasize critical information (such as warnings about allergies). Visual grouping of related data fields by using the same background color.
	Rows in the table for recording data were long and crowded.	Difficulty in distinguishing between rows increases the probability of errors in copying: writing or reading data from the wrong row.	Creating a distinction between long rows in tables by using two alternating background shades.
Location and legibility	Small and crowded font.	Visual effort, resulting in delay and increased probability of a reading mistake.	Legibly sized font (12 point or larger).
	Headings are inconsistent with respect to letters and font sizes.	Visual overload. Difficulty in distinguishing the hierarchy of the information.	Create a hierarchy of headings by using consistent fonts, sizes, styles, and colors.
	The boxes for writing information are small and cramped.	Visual overload.	Increasing size of boxes for recording information and increased spaces between them.

(Continued)

TABLE 10.3 (Continued)
Summary of Problems in the Old Chart and Their Solutions in the New Chart

Category	Problems with the Old Chart	Detail of Their Impact	Design Solutions on the New Chart
	Not enough columns for writing instructions and documentation.	In many cases forced to add pages in a makeshift manner.	Add columns for recording instructions and documentation.
	Due to a lack of space, information updates are sometimes written over existing data, and important information is not visible.	Notable lack of information that was changed and a lack of visibility of the information necessary for treatment.	Different location and visual separation between the fields for frequently updated information and fields that contain fixed information to increase the prominence of the modified information and prevent the concealment of fixed information items.
	There are no built-in information fields for critical pieces of information such as warnings or allergies. The information recorded in *agreed upon* locations and emphasized by highlighting in yellow to attract attention.	Load on memory, information may not reach its destination, and therefore may cause errors in treatment.	Location of fields for critical pieces of information on the front of the binder to increase their visibility and prominence.
	Instructions for the placement of the patient on the device at different stages of the treatment or for different body parts are listed in the same place (different set-ups) and appear to the user at the same time.	Unnecessary visual overload, confusion between the different treatment instructions.	Create a separate page for each set-up, on pages stacked one on top of the other, so that only the relevant page appears to the user, and a minimum of information that is not relevant is presented. Pages are numbered to facilitate their orientation.
	Different terms used for fields that contain the same information.	Disrupts communication between different users.	Use of uniform terminology.
Language	Hebrew is the primary language. Some of the data field names listed in English. Writing in Hebrew and English.	Discrepancy between the computer and the manual system, different directions of writing and reading data.	English primary language, according to the computerized system. Hebrew is used only in some critical details (such as special warnings) and measures of monitoring treatment.

C. Support Quality Assurance and Means of Verification

No change log. Changes in the treatment instructions and in the treatment itself are recorded in red alongside the previous information to highlight the change.	Reduced prominence of changes that are made. Team members may not distinguish between treatment and changes and continue with the previous treatment, that is, incorrect treatment.	Initiate a log for changes on the front cover, referring to the changes recorded within the chart, including a structured place to note execution of the new instruction. Log increases the prominence of changes in the treatment instructions.
The area in which the names of the patients and their personal details can only be seen from the front of the chart. When the chart is opened or turned it is not known who it refers to.	Risk of a recording, reading, or attaching pages related to one patient to another patient's chart.	Set a prominent area at the top of the chart for personal information that can be seen during all stages of the treatment and allows for the clear identification of the patient as well as all the associated pages and information ascribed to him/her (Photo 10.1a).
No reminder that relates to critical pieces of information needed for therapeutic decisions (such as the date of a treatment or of a concurrent treatment).	Error due to lack of attention to information that influences the course of treatment.	Add reminders that refer the user to places where information critical to therapeutic decision making can be found (also helps in quality control after the decision is made).
Almost all accompanying paperwork (such as calculations, etc.) are placed in single, brown, medium-sized envelope attached to the chart and conceals other pieces of information written underneath it.	Reducing the possibility of attending to the concealed information and making it inaccessible.	Accompanying paperwork is filed using tabs and arranged so that the existing information is not hidden even when going from one page to another.
The number and size of cells is not sufficient for recording the details of the staff member that entered the information into the chart, executed the treatment instructions, or performed quality control.	Illegibility, confusing activities related to ensuring quality control and various inquiries related to the treatment instructions.	Structured data fields for recording the details of the staff member who entered information or performed quality control, next to the place where the information was recorded.

(Continued)

TABLE 10.3 (Continued)
Summary of Problems in the Old Chart and Their Solutions in the New Chart

Category	Problems with the Old Chart	Detail of Their Impact	Design Solutions on the New Chart
	Some of the table columns for documenting the treatment are not used because of changes in the number of data fields. The total overall dosage is far from where the relevant data is recorded.	Waste of valuable space, risk of calculation errors that interfere with reviewing the process.	Instead of the permanent headers used for the regular documentation columns, stickers serve to label the header for the column containing the total overall dosages. The sticker enables the total column to be placed according to the number of radiation fields, without defining the number of columns (radiation fields) ahead of time.
	Partial and disorganized sequence of information fields for the calculation of the distribution of the overall radiation dose into treatment segments. Some of the details of the information needed for the calculation are not displayed.	Causes the performance of the calculation and checking it to be difficult and complicated. Risk of miscalculation and providing an incorrect dose of radiation.	Table includes all the details of the information needed according to the sequence of the calculation. Facilitates the calculation itself and its regulation.

D. Modularity and Flexibility

One medium-sized envelope contains all the related paperwork, which is folded. The information contained is not prominently displayed or accessible. Important pieces of information (such as side effects that are discovered during follow-up sessions) do not get addressed because there is no suitable place for filing the papers on which they are documented.

Risk of errors in providing care due to not relating to the information that *disappeared* in the envelope.

Filing pins for filing documents and the accompanying papers in the relevant partitions. The information is visible and accessible to the user when it is required (Photo 10.1e).

The chart was developed before the introduction of some of the computerized processes and therefore is not suitable for everyone.

Cumbersome and causes delays of the work process.

The design concept of the chart is adapted to the computerized chart and for different levels of computerization in the treatment planning processes and in the actual treatment.

The chart is not suitable for whole body treatment and has limited space for recording different treatments, the number of treatment sites, and the number of radiation fields. The staff must add an area for recording in an arbitrary manner, by attaching papers to the chart.

Visual overload, lack of organization, and concealed information.

The chart is suited for the entire range of radiation treatments, for repeated treatments, and the varying numbers of treatment fields and sites.

cognitive failures inherent in the paper form design may be carried over to the computer age. These problems would continue to cause difficulties, in addition to new problems that might be revealed with the introduction of computing[4,13]; problems that are just as serious and may be camouflaged under different layers of software, screens, and menus.

The new chart is bigger and more expensive: Designing the new chart was not problem-free.

- A weighty compromise had to be made in transitioning from a simple double-sided form to a complex folder (the volume of the chart increased due to the additional space required for the missing items of information and for organizing it in a more useful manner).
- The introduction of separate partitions, even though they were logically suited to reflect the flow of the process, requires browsing through the pages and could reduce one's ability to comprehend the overall status in one glance.
- The unique structure and use of color coding complicated the production of the chart and raised production costs.

These and other points were an integral part of the process of designing the new chart so that it would be more helpful in addressing the needs and limitations of the previous one. We believe that the new chart has an advantage in terms of the cost benefit, but this will only be determined over time.

The implementation of the new radiation chart occurred concurrently with other changes in the radiation unit (changes of equipment, computing, structure, etc.). The new chart was integrated in the unit as it was used within the normal work processes, thus being able to test it in actual practice. The chart requires those who use it to learn new tasks and to perform some familiar tasks in a different manner.[7] Since the chart is compatible with the stages of the work being performed and was developed with the complete collaboration of its users, we expected that its integration within the system would be relatively easy, both in terms of learning how to use it and in terms of the staff members' willingness to accept and recognize its benefits.

SUMMARY

Ostensibly, a *simple* form in a technology-rich environment is not the concern of the caregiver. Nevertheless, the form is a medical tool and the problems illustrated here deal with the failure of the interface between the individual and the information, and between the individual and the tools he/she operates. Typically, such problems do not receive much attention. In part, this is because of the medical team's tendency to continue to work around the problems, because of the belief that these problems stem from inadequate training or indifference on the part of their administrators or due to overfocusing on technology at the expense of other aspects of the job, such as the organization of the work process.[13] Using the examples provided we tried to illustrate the nature of the problems and their severity. The design of a radiation chart ill suited to the task added unnecessary difficulties to the already high stress, heavily loaded daily regimen of work being performed, causing each of its users to become prone to making serious errors.

The new design should allow for comfortable and efficient work, which minimizes the risk of error. When the structure and information processing required of the radiation chart are integrated within the task and the structure of the information and suited to the person's information processing capabilities, the cognitive abilities are broadened and supported for the whole system, including the person, the work tools (the chart), and the task.[7] In this way, the cognitive resources needed by the staff members can be directed to the performance of the actual task. By providing solutions for the design problems, the entire system can benefit and operate better.

This project represents the advantages of applying basic principles of cognitive psychology in the early stages of designing medical instruments. We saw that avoiding the cognitive aspects of the chart's usability could lead to serious mistakes. Therefore, the design of this form must be based on the assessment of a user's ability to use it and communicate properly through its use. The consideration of the user's cognitive processes (such as memory limitations) from the early design stages can increase the functionality and usability of the chart, and may also contribute to reducing costs and improving quality.[14]

The condition of radiotherapy centers in Israel has appeared in the headlines on numerous occasions and even underwent a comprehensive investigation in 2005 by the State Comptroller's Office. The State Comptroller's 2005 Report,[18] published in May 2006, related among other things to the findings and recommendations of the systematic investigation conducted by the Technion staff at Rambam's radiotherapy unit, which included some of what was presented in this chapter. We chose to spotlight this part of the system and to illustrate the options available for improving it by introducing principles of human factors engineering. In this chapter, we focused on the use of the radiation therapy chart to demonstrate how a user-centered process to improve the chart may reduce the probability of human error and ameliorate the way in which information is communicated throughout the entire system.

REFERENCES

1. Radiation Oncology Safety Information System, www.ROSIS.info.
2. Robinson, A. (1990). Treatment of cancer. *Mada: A Publication of the Weizmann Institute for Publications in the Natural Sciences and Technology*, (4), 192–195.
3. Roach, M. (2004). Reducing the toxicity associated with the use of radiotherapy in men with localized prostate cancer. *Urologic Clinics of North America*, 31, 353–366.
4. Macklis, R.M., Meier, T., and Weinhous, M.S. (1998). Error rates in clinical radiotherapy. *Journal of Clinical Oncology*, 16, 551–556.
5. Salas, E., Dickenson, T.L., Converse, S.A., and Tannenbaum, S.I. (1992). Toward an understanding of team performance and training. In: R. Swezey and E. Salas (eds.), *Teams: Their Training and Performance*. Norwood, NJ: Ablex, 3–29.
6. Cooke, N.J., Salas, E., Cannon-Bowers, J.A., and Stout, R.J. (2000). Measuring team knowledge. *Human Factors*, 42(1), 151–173.
7. Norman, D.A. (1991). Cognitive Artifacts. In: J.M. Carroll (ed.), *Designing Interaction: Psychology of the Human–Computer Interface*. Cambridge, MA: Cambridge University Press.

8. Norman, D.A. (2002). *The Design of Everyday Things*. New York: Basic Books.
9. Xiao, Y. (2005). Artifacts and collaborative work in healthcare: Methodological, theoretical, and technological implications of the tangible. *Journal of Biomedical Informatics*, 38, 26–33.
10. Wickens, C.D., Gordon, S.E., and Liu, Y. (1998). *An Introduction to Human Factors Engineering*. New York: Addison-Wesley Educational Publishers Inc.
11. Zhang, J. and Norman, D.A. (1994). Representations in distributed cognitive tasks. *Cognitive Science*, 18, 87–122.
12. Sellen, A.J. and Harper, R.E.R. (2003). *The Myth of the Paperless Office*. Cambridge, MA: MIT Press.
13. Koppel, R.K., Metlay, J.P., Cohen, A., Abaluck, B., Localio, A.R., Kimmel, S.E., and Strom, B.L. (2005). Role of computerized physician order entry systems in facilitating medication errors. *Journal of the American Medical Association*, 293(10), 1197–1203.
14. Gillan, D.J. and Schvaneveldt, R.W. (1999). Applying cognitive psychology: Bridging the gulf between basic research and cognitive artifacts. In: F.T. Durso et al. (eds.), *Handbook of Applied Cognition 1999*. West Sussex, England: John Wiley & Sons, 3–31.
15. Yeung, T.K., Bortolotto, K., Cosby, S., Hoar, M., and Lederer, E. (2005). Quality assurance in radiotherapy: Evaluation of errors and incidents recorded over a 10-year period. *Radiotherapy and Oncology*, 74(3), 283–291.
16. Sanders, M.S. and McCormick, E.J. (1993). *Human Factors in Engineering and Design*. New York: McGraw-Hill.
17. (2001). *The Status of Radiotherapy in Israel in 2001 and Recommendations for Improvement*. Rambam Hospital Department of Oncology.
18. State Comptroller. (2006). *56th Annual Report on the 2005 Accounts and Fiscal Year 2004*. Jerusalem: Ministry of Health—Radiation Treatment Units, May 2006, 427–470.

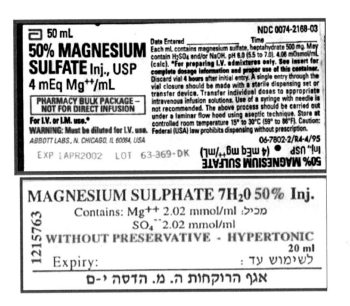

PHOTO 6.1 Magnesium sulphate bottle—front and back labels.

PHOTO 9.6 The new medication labels, upper- and lowercase letters, background color of written name, and barographs of available number and dosage level of the medication.

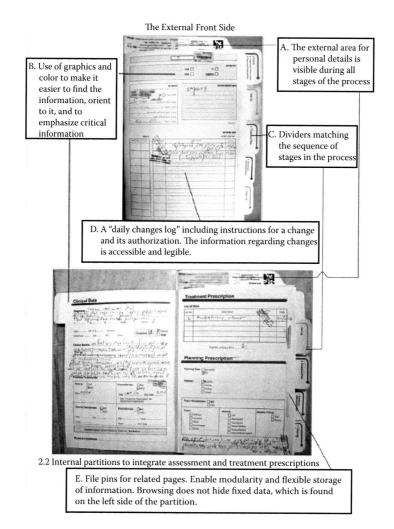

PHOTO 10.1 The new Radiotherapy Chart.

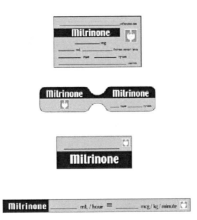

PHOTO A11.1 Samples of labels for IV line, syringe, and bottle. (From Porat N., Bitan Y., Sheffi D., et al., 2009, Use of Color-Coded Labels for Intravenous High-Risk Medications, *BMJ Quality and Safety*, 18, 6.[8])

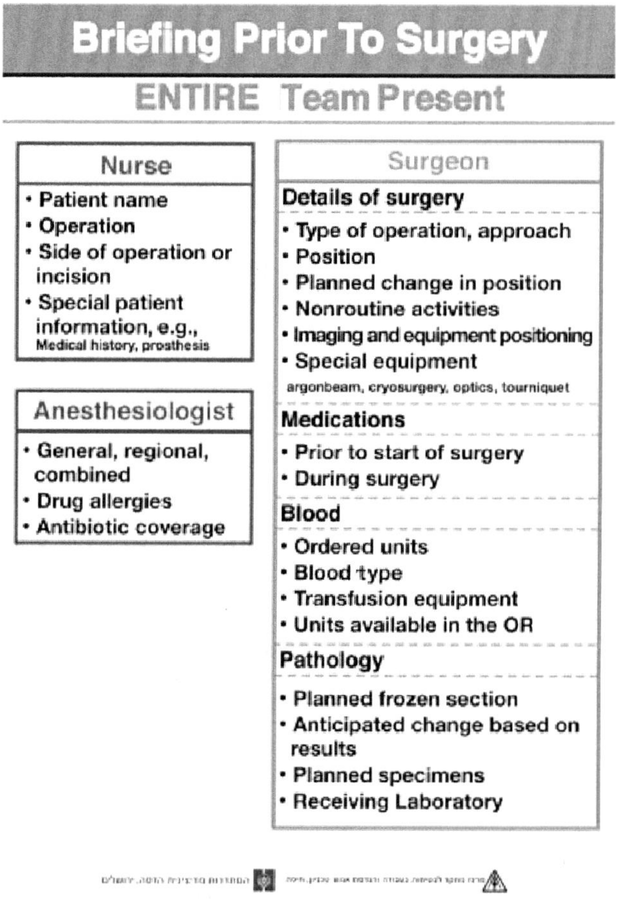

FIGURE 14.1 The presurgery briefing poster. (From Einav Y., Gopher D., Kara I., et al., 2010, Preoperative Briefing in the Operating Room: Shared Cognition, Team Work, and Patient Safety, *Chest*, 133, 443–449.[10])

11 Examining the Effectiveness of Using Designed Stickers for Labeling Drugs and Medical Tubing

Dorit Sheffi,[*] Yoel Donchin, Nurit Porat, and Yuval Bitan

CONTENTS

Introduction 163
Analysis of the Results 165
    Results 166
        Effect of Labeling Method 166
        Effect of Bed Order and the Interaction 167
        Questionnaire 169
Appendix A: Examples of the New Stickers 170
Appendix B: The Experimental Process 171
    Laboratory Simulation Designed to Test the Use of Labels Designed for
    Marking Drugs and Medical Tubing 171
References 174

INTRODUCTION

It is reported that one of the reasons for errors in dispensing medications is that different preparations have similar names and packaging.[1–3] For financial reasons, drug manufacturers do not take into account human factors engineering, and few studies exist that investigate how labels designed for the use of medical personnel should be marked.

In intensive care units, the slightest error can cost lives. The workstation around the patient's bed is crammed with a variety of plastic tubes used for monitoring

---

[*] Dorit Sheffi (MD thesis, Hadassah, Hebrew University Medical School, Jerusalem), Nurit Porat, Chana Rosenbaum, Yuval Bitan. This study was conducted at the Hadassah Medical Center. The stickers themselves can be viewed at the Center for Work Safety and Human Engineering Web site.

central arteries or veins, delivering drugs to the vein, ventilation, and collecting wastes. In addition, the medical tubing is surrounded by a variety of electrical cables. There is a reason why the medical staff calls this complex system *spaghetti*, and there is no doubt as to the importance of rapid and accurate identification of each of these tubes. A tangle of tubes can also be found in the *regular* departments; here, too, these different drug and infusion lines must be clearly marked.

When responding to an event that requires the rapid introduction of a drug through the appropriate line, most doctors and nurses deal with a heavy mental load in order to quickly identify the correct injection port. To identify the course of the plastic line easily and quickly, labels, often handwritten, are attached to identify the type of line (vein, central vein, and artery). Other unique and powerful systems are also found around the patient's bed enabling many different drugs to be dispensed, which are also labeled according to drug type and dosage.

In order to distinguish between one drug and another, it is not enough to properly mark the original drug packaging. The syringe into which the drug is drawn must also be properly tagged, as well as the tube that is inserted into the patient's vascular system. A clear label must take into account the variables that can affect the visibility of the marking (contrast), even under suboptimal visual conditions. These variables include the relationship between the font and the background colors (black on white is better than black on gray), contrast (black–white is better than the reverse), the font size (taking into consideration the distance from which the text must be identified), and letter spacing so that they can be easily distinguished.[4] Even gloss affects the label—it may cause reflections that make reading it more difficult. The ideal print uses a familiar and readable font that is not too narrow, inscriptions that are not cramped; the writing should be in sharp contrast since they are written in black on white, and make use of capital letters. In hospitals, over 150 different drugs are delivered via intravenous lines. The dose of these drugs varies according to the patient's weight and age, thus making it difficult to prepare labels for all of the drugs used. Moreover, the considerable number of marked labels that would be needed would require a designated storage area that would be easily accessible, and therein too lies potential for error. The alternative to premarked labels are drug label templates on which the drug manufacturer would list the name of the patient, the drug, the dosage and other identifying information.

It is difficult to test the efficiency of marked labels, although obviously they must be marked clearly. Assessing the efficiency of labeling has not been tested, except under laboratory conditions and in front of a computer monitor. The tests required health care workers to identify the drugs listed. The experimenters monitored the movements of their eyes as well as their ability to distinguish between the drugs.[5,6]

A medical team at Hadassah hospital, directed by Y. Donchin, developed and implemented a system for labeling drugs and infusion lines, which they developed through a number of different stages. They began by charting the names of drugs that could be misread because of the similarity in their names, or because various dosages are ordered. Drugs were divided into groups according to their mechanism of action—such as vasoactive drugs, muscle relaxants, and so on. The staff assigned a different background color to each group and determined the appropriate letter and font size. We conducted a simulation study to see whether the designed sticker was indeed more expedient and enabled more rapid and accurate identification of drugs and dosages. The research was conducted in the intensive care unit, and the study subjects were asked to perform tasks involving the identification of drugs, intravenous lines, and their use. Two simulated beds for critically ill patients were set up in the intensive care unit. A manikin with several infusion lines and a respirator was placed in each bed. On the right side we placed a central line to the jugular vein to measure central venous pressure (CVP) and two peripheral venous lines—one for intravenous delivery of fluids and the other for total parenteral nutrition (TPN). On the left side, four IV lines were inserted into the femoral vein, each of which was to provide a different drug through syringe pumps. An endotracheal tube was inserted into the trachea and attached to a respirator, a urinary catheter was placed in the bladder, and a monitor was positioned there as well.

One bed was used as a control, and the lines were labeled in the customary manner (blank white stickers and adhesive plasters on which the name of the drug was recorded in legible handwriting). The other bed was the intervention bed and the premarked labels were used. Sixty-three subjects participated in the study, of which 61 were nurses working in various intensive care units in the hospital with over 2½ years of experience in the ICU. Two other subjects were not included in the data analysis.

Each subject was assigned six tasks and was asked to describe his/her actions aloud. The time required to perform each task was measured with a stopwatch. Recordings were made of the subject's explanations during the performance of the task, and any errors that occurred were documented. Each subject performed the experimental tasks on both beds—32 subjects began with the control bed and then switched to the intervention bed, and 29 subjects began with the intervention bed and switched to the control bed.

The experimental tasks assigned were typical of those performed by the nursing staff in the ICU:

1. *Identification of a syringe*—Choose a syringe containing a certain drug in a specific dose from a table upon which several syringes were placed.
2. *Label a bag*—Label a bag containing a drug and its tubing and attach them to a specific vein in the patient's body.
3. *Identifying a pump*—Change the drug dosage in an active pump.
4. *Identification of a peripheral vein*—Rinse out an existing intravenous line.
5. *Description of all the lines and drugs*—Provide a description of each tube inserted into the patient's body and the drug being supplied through each one.
6. *Identifying an error*—Identification of a labeling error; contradictory labeling was established between the label found on an automatic syringe pump and that found on the tubing emerging from the syringe and into the patient's body.

ANALYSIS OF THE RESULTS

A quantitative analysis of the time taken to execute the tasks was conducted through the statistical analysis of the results using a crossover study design. The findings of such a study would provide an estimate of the effect of the type of labeling (adhesive plasters/stickers), the influence of the bed sequence (the control bed first or the intervention bed first), and the interaction between these variables. The statistical analysis was actually performed through a repeated measures analysis, in which the within-group effect referred to the time required to perform the task with adhesive tape versus stickers, the between-groups effect was represented by the order of the beds, and the interaction effect examined the relationship between the order of the beds and the method of labeling. A p-value of 5% or less was considered statistically significant.

RESULTS

Three main factors were taken into consideration in analyzing the results for each individual task and for the average performance of all of the tasks: the system of labeling, the order of the beds, and their interaction.

- *Labeling method*—Was the time taken to execute a specific task in the bed with the adhesive tape labels different than the time it took in the bed with the stickers, regardless of the order of the beds?
- *Order of the beds*—Was the time taken to execute a specific task in the two beds (adhesive, stickers) affected by the sequence of beds in which the tasks were performed, regardless of the method of labeling?
- *Interaction*—The nature of the relationship between the order of the beds and the labeling system. That is, if the delay between the time taken to execute a specific task in the adhesive tape bed and in the bed with the sticker labels depended on the sequencing of the beds in performing the tasks.

The average duration of each task and the standard deviation for both methods, as well as the statistical analysis of the crossover study results, are presented in Table 11.1.

During the experiment, on 15 occasions, a subject discovered the mistake in labeling even prior to being asked to do so. Thus, when the time came for him/her to perform the task of identifying the error, it only took him/her 0 seconds. This occurred both during the performance at the bed with adhesive plaster labels, at the bed with sticker labels, as well as in both circumstances.

As a result, we decided to examine the results of this task and the average performance of all the tasks, even after having eliminated these subjects from the analyses. These results are shown in Table 11.2.

Moreover, we examined the influence of demographic data on the experimental results. The demographic variable that had a significant effect was the subject's age, which was treated as a categorical variable (over/under age 40).

Effect of Labeling Method

The statistically significant advantage of the new labeling system using the designed stickers is evidenced by the mean time taken to perform all the tasks (Tables 11.1

TABLE 11.1
Results of Tasks Performed

Task	Overall Average Control Bed (SD) (Minute:Second)	Overall Average Intervention Bed (SD) (Minute:Second)	p Value Labeling Method Effect	p Value Bed Order Effect	p Value Labeling Method × Bed Order Interaction
Average time required to perform all tasks	00:28 (00:06)	00:25 (00:06)	<0.0001 (S)	0.33 (NS)	<0.0001 (S)
Syringe identification	00:06 (00:01)	00:07 (00:04)	0.029 (S)	0.54 (NS)	0.24 (NS)
Labeling of IV bag	01:20 (00:21)	01:07 (00:18)	<0.0001 (S)	0.25 (NS)	0.0002 (S)
Identification of syringe pump	00:05 (00:02)	00:05 (00:03)	0.93 (NS)	0.87 (NS)	0.062 (NS)
Identification of peripheral vein	00:10 (00:05)	00:11 (00:08)	0.73 (NS)	0.80 (NS)	0.0006 (S)
Description of all drugs and lines	00:50 (00:22)	00:45 (00:18)	0.04 (S)	0.69 (NS)	0.0001 (S)
Identification of an error in the treatment setting	00:17 (00:14)	00:15 (00:12)	0.4 (NS)	0.63 (NS)	0.03 (S)

Source: From Porat N., Bitan Y., Sheffi D., et al., 2009, Use of Color-Coded Labels for Intravenous High-Risk Medications, *BMJ Quality and Safety*, 18, 6.[8]

Note: NS, not significant; S, significant.

and 11.2) in the bag labeling task, the description of all the lines and drugs, and in the detection of an error (see [a] in Table 11.2), versus the system currently used (adhesive tape). The method of using the adhesive tape indicated a significant advantage in only one task; pump identification. For two of the tasks—peripheral vein identification and identification of the pump—neither method demonstrated a significant effect.

Effect of Bed Order and the Interaction

In eight of the nine sections of Tables 11.1 and 11.2, the bed order in the various tasks did not make a critical difference, but the interaction between the labeling method and bed order was significant in seven of the nine sections. Figure 11.1 depicts the

TABLE 11.2
Results[a] of Two Specific Tasks Performed

			p Value		
Task	Overall Average Control Bed (SD) (Minute:Second)	Overall Average Intervention Bed (SD) (Minute:Second)	Labeling Method Effect	Bed Order Effect	Labeling Method × Bed Order Interaction
Average time required to perform all tasks	00:27 (00:05)	00:24 (00:07)	0.0003 (S)	0.01 (S)	<0.0001 (S)
Identification of an error in the treatment setting	00:20 (00:13)	00:15 (00:12)	0.045 (S)	0.43 (NS)	0.0012 (S)

Source: Porat N., Bitan Y., Sheffi D., et al., 2009, Use of Color-Coded Labels for Intravenous High-Risk Medications, *BMJ Quality and Safety*, 18, 6.[8]

Note: NS, not significant; S, significant.

[a] Excluding the 15 cases where the participants found the labeling error before being asked.

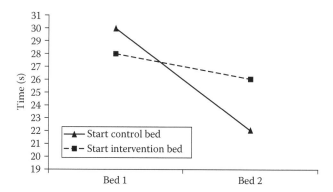

FIGURE 11.1 The overall averages of performing all tasks in beds with and without the new stickers. (From Porat N., Bitan Y., Sheffi D., et al., 2009, Use of Color-Coded Labels for Intravenous High-Risk Medications, *BMJ Quality and Safety*, 18, 6.[8])

mean results of the performance of all tasks. The standard deviation is presented as the standard error.

The average performance duration of all the tasks in the bed with the stickers was shorter than that of the beds with the adhesive plasters, independent of which bed the subjects began to perform the tasks.

Questionnaire

To summarize the comments provided in the questionnaires, 57 of the 61 subjects preferred the sticker method and only four subjects preferred the currently accepted method. In addition, comments were made regarding the specific design of the stickers (these suggestions will be taken into account when creating the final design of the stickers).

The design of the stickers was based on human engineering principles, and therefore the assumption was that human engineering-based labels would be preferred.

An analysis of the results and the questionnaire responses indicate that the new labeling method, based on stickers designed and prepared in advance, is preferable to the existing system. The new method provides a prominent label that is neater, making it easier to identify errors and facilitate orientation about the patient's bed in the intensive care unit. This method also allows for more convenient and rapid labeling of the medications and the medical tubing.

A statistically significant benefit was indicated by using the adhesive tape labeling method in the task of identifying the syringe. From the comments of the subjects during the task, we learned that this was not because of any difficulty in identifying the name of the drug on the syringes with the designed labels, but rather in locating the syringe with the correct dosage. According to the subjects, this was because this information was not positioned in the place where it would be currently found using the existing labeling method. Figure 11.1, presented in the results section, revealed the benefit of using the stickers, regardless of the order of the beds worked on during the experiment. That is, even if a learning process was involved for the tasks in moving from the first to the second bed, it did not affect the results statistically.

From the questionnaires we learned that the nursing staff prefers the method of predesigned stickers, but had specific comments regarding the design of a number of the labels. Some of the labels were redesigned according to the comments of the subjects and the hospital administration decided to introduce them into regular use.

The quantitative and qualitative results of the experiment supported our hypothesis that the use of designed labels enhances one's ability to distinguish and accurately identify the drugs and IV lines, and reduces the duration of the labeling process.

Simulation is a suitable method for resolving the question of whether a tag using designed labels is indeed an advantage. Newer methods of marking, instead of using predesigned labels, have been proposed by a team from the Department of Industrial Design at the Bezalel school. They suggested that a printer should be placed in the medication preparation room and each nurse should print the required labels, applying the developed software application. Through this method, there is no need to coordinate dozens of stickers (at considerable cost), and the nurse is given the opportunity to reread and rewrite the instructions (see Photo A11.1.).

The questionnaire and the simulation set-up can be found in the appendices at the end of this chapter. This will allow the reader to look for answers to other potential questions they may have related to hospital safety, ranging from the arrangement of the bed environment to the manner in which food is served to the patient.

APPENDIX A: EXAMPLES OF THE NEW STICKERS

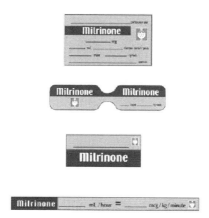

PHOTO A11.1 (See color insert.) Samples of labels for IV line, syringe, and bottle. (From Porat N., Bitan Y., Sheffi D., et al., 2009, Use of Color-Coded Labels for Intravenous High-Risk Medications, *BMJ Quality and Safety*, 18, 6.[8])

APPENDIX B: THE EXPERIMENTAL PROCESS

LABORATORY SIMULATION DESIGNED TO TEST THE USE OF LABELS
DESIGNED FOR MARKING DRUGS AND MEDICAL TUBING

Serial number_____
Date_____ Time_____

Background data:
Gender: M/F
Age: _____
Academic education: Registered nurse/Registered nurse-B.A./B.A./M.A./Ph.D.
Years of professional experience: _____ years
The department in which you work _____, years of experience in the department: _____ years
Job: {Work status $^2/_3$ / $^8/_9$ / Full time / Other: _____
The wording used to present the experiment to the subject:

Hello and thank you for agreeing to participate in this experiment.

The purpose of the experiment is to examine the labeling of drugs and IV lines in the intensive care unit. We are examining the difference between two types of labeling and not your abilities as a nurse. The experimental procedure will include a number of tasks. You will be asked to perform them as efficiently as possible, while verbally describing what you are doing and why. At the start of each task you are to stand next to the patient's chart (located next to the patient's bed). The experimental findings will be anonymous and your name will not be recorded anywhere in the results file. We would appreciate it if you would refrain from revealing the details of the experiment to the other subjects.

Thanks again and good luck.

Serial number_ Location during the course of the experiment: I/II

Bed in which adhesive tape is used:

Task	Duration (Min:Sec)	Identification (√/X)	Subject's Comments	Observer's Comments
The patient is experiencing pain, increase the rate of the FENTANYL infusion from 1 cc/min to 2 cc/min.	__ __ : __ __			
There is an error in how the drugs and IV lines in the system are labeled, identify the error. (The Fentanyl line is labeled Sandostatin.)	__ __ : __ __			

continued

Task	Duration (Min:Sec)	Identification (√/X)		Subject's Comments	Observer's Comments
This bag contains 150 cc of normal saline (N.S.) with 100 mg of lidocain. Mark the bag and the tubing and connect them to the central line in the femoral vein.	__ __ : __ __				
On the table in front of you are a number of syringes containing different drugs. Find the syringe containing 100 mg of PROCOR.	__ __ : __ __				
Rinse only the peripheral vein connected to the fluids (provide a syringe).	__ __ : __ __				
Please describe all the lines connected to the patient and which drugs he/she is receiving through them.	__ __ : __ __	C.L	CVP		
		P.L	Fluids		
		P.L	TPN		
		F.L	Dobutamine		
		F.L	Procor		
		F.L	Fentanyl		
		F.L	Sandostatin		

Serial number_____

Location during the course of the experiment: I/II

Bed in which stickers are used:

Task	Duration (Min:Sec)	Identification (√/X)	Subject's Comments	Observer's Comments
On the table in front of you are a number of syringes containing different drugs. Find the syringe containing 2 mg Fentanyl.	__ __ : __ __			
This bag contains 150 cc N.S. with 5000 units of Heparin. Label the bag and the tube and connect them to the central line in the femoral vein.	__ __ : __ __			

Task	Duration (Min:Sec)	Identification (√/X)		Subject's Comments	Observer's Comments
A high glucose level was indicated. Increase the insulin dose by 10 units.	__ __ : __ __				
Rinse the peripheral vein that is connected only to the fluids (provide a syringe).	__ __ : __ __				
Please describe all the lines connected to the patient and the medication he/she is receiving through them.	__ __ : __ __	C.L	CVP		
There is an error in how the drugs and lines in the system are marked, identify it. (The aminophylline tube is marked as insulin.)	__ __ : __ __	C.L	CVP		
		P.L	fluids		
		P.L	TPN		
		F.L	Morphine-Midazolam		
		F.L	Noradrenaline		
		F.L	Aminophylline		

Serial number_____

Questionnaire:

Thank you for participating in the experiment.

In comparing the two beds, what difference did you feel in doing your work?

_____

_____

Please answer the following questions regarding the new labeling displayed in front of you (on a separate page):

Circle the number that reflects your opinion best:

Convenience of using the stickers?	Very convenient 10 9 8 7 6 5 4 3 2 1 0 Not at all convenient.
How much does using the stickers contribute to orientation in the drug treatment of the patient?	Contributes a great deal 10 9 8 7 6 5 4 3 2 1 0 Does not contribute.
How much does the use of the stickers contribute to reducing the rate of errors?	Contributes a great deal 10 9 8 7 6 5 4 3 2 1 0 Does not contribute.
The sticker design?	Very clear 10 9 8 7 6 5 4 3 2 1 0 Not at all clear.

Circle the method that you would prefer to use:

Blank stickers and adhesive tape/stickers prepared ahead of time.

If you prefer the prepared stickers, which components of the system would you want to be marked in this way? _____

_____

Thoughts and comments:

_____
_____

Thank you for your cooperation.

REFERENCES

1. Quinn A., Bojko A., Gaddy C., and Israelski E. (2006). Better Drug Labeling for Pharmacists. *Ergonomic in Design*.
2. Bojko A., Gaddy C., Lew G., and Quinn A. (2005). Evaluation of Drug Label Design Using Eye Tracking. *Proceedings of the Human Factors and Ergonomics Society 49th Annual Meeting*, Orlando, FL.
3. Bojko A., Buffardi K., and Lew G. (2006). Eye Tracking Study on the Impact of the Manufacturer's Logo and Multilingual Description on Drug Selection Performance. *Proceedings of the Human Factors and Ergonomics Society 49th Annual Meeting*, Orlando, FL.
4. Wickens C.D., Gordon S.E., and Liu Y. (1997). *An Introduction to Human Factors Engineering*. New York: Longman.
5. Drug Safety Institute. Drug Labeling. www.drugsafetyinstitute.com
6. Drug Safety Institute. Medication Error. www.drugsafetyinstitute.com
7. Drug Safety Institute. Packing/Labeling Problems and Sound-a-Like/Look-a-Like Names. www.drugsafetyinstitute.com.
8. Porat N., Bitan Y., Sheffi D., et al. (2009). Use of color-coded labels for intravenous high-risk medications, *BMJ Quality and Safety*, 18, 6.

12 The Emperor's New Clothes—Design of Garments for the Operating Room Staff

*Anna Becker**

CONTENTS

Objectives of the Project	176
The Project	178
Environmental Conditions	186
Summary of the Questionnaire's Findings	188
Habits Regarding the Use of the Operating Room Clothing	188
Design Preferences for Operating Room Clothes	190
Significant Differences between Men and Women in Their Patterns of Use and Design Preferences for Operating Room Clothes	191
Significant Differences between the Different Professional Groups in Their Patterns of Operating Room Clothes Use and in Their Design Preferences	191
Summary	193
References	194

The image of the surgeon and of the surgical staff is familiar to anyone who watches television: blue or green scrubs, a cap to cover their hair, a mask over their nose and mouth, and shoe covers. That is how physicians appear when they emerge from surgery to talk to the family waiting outside, whereas in the room itself, they are also clad in a well-fastened gown from the neck down. Although work clothes should fit the worker and the working conditions especially, little attention has been paid to the proper attire of the operating room medical staff. In addition, regardless of the impressive changes occurring in the field of surgery, the dress of those who perform this work has not meaningfully changed in many years. Moreover, despite the different roles of the staff members, they all wear the same outfit—surgeons, anesthesiologists, nurses, technicians, medical engineers, and even the cleaning staff.

---

[*] Anna Becker: Final thesis [with honors], of a master's degree in the Department of Industrial Design, Technion. Supervised by Naomi Bitterman. The study was conducted primarily in the operating room at Hadassah University Medical Center, Ein Kerem, Jerusalem.

The outfit of the medical staff must be both comfortable and useful, and suit both the environment and the tasks of the staff members, while taking into account the need for cleanliness and sterility. The design should account for human engineering factors such as a person's dimensions and his/her type of job. The financial aspect should take into account quality processing and sewing to ensure the durability of the clothes under basic maintenance conditions and that they can be reused as much as possible.

An equally important aspect is the image presented: clothes reflect the characteristics of the wearer and testify as to his/her status and that of the hospital in general. Comfortable, aesthetic, and stylish clothing is said to help strengthen the relationship with the patient and provide the patient and the family with a sense of security.[1]

Detailed standards exist for work clothes in various professions, but no standards exist in Israel, or in the world, for operating room clothes. It is the hospital's responsibility to select the clothes of the medical staff (the administrative and housekeeping departments) and they are mainly concerned with economizing and preventing theft and damage of clothes within the hospital grounds.

The only relevant compulsory directive is directed toward the laundromats and other parties responsible for maintaining the clothes designed to provide protection against bacteria and viruses (surgical room clothing). They are required to test operating room clothing twice a year to determine if they are permeable to liquids and air. These tests are conducted at the Institute of Israeli Standards and at the textile lab of the School of Engineering and Design at Shenkar College in Ramat Gan. In this project, the clothing worn by the operating room personnel was reexamined for its suitability for the environmental conditions, the user's objectives, and their demands. The project was motivated by the numerous complaints made by the operating room staff and maintenance personnel, with respect to the design and quality of the operating room clothing.

OBJECTIVES OF THE PROJECT

The objectives were to assess the situation regarding the clothing (shirt and pants only) and the staff's needs, demands, and preferences, while considering the environmental conditions of the operating room suite and examining potential solutions for an improved outfit by merging the findings and the constraints in a way that was most compatible with the latest technological innovations in the textile industry: types of fibers, fabrics, forms, methods of processing, finishing, and sewing. Following this assessment, the medical staff in the operating room will be offered a prototype of an outfit that will be optimally suitable to the wearers (in terms of fabric, form, types of pockets, and their positions, etc.) and the buyer—the hospital (price, quality, and life-span of the outfits).

A work garment, which a person wears for most of the day, must provide protection against the risks that the employee faces yet be comfortable and not present a burden. The iron armor donned by medieval fighters is a good example of a *work outfit* that is uncomfortable and limits the movements of its wearer, in exchange for the protection it provides. Therefore, in designing the work outfit, one should consider the working conditions, environmental conditions, and personal characteristics of the workers. It is important to take into account the physical environmental factors, such as the temperature, humidity, ventilation, presence of ionizing radiation (X-rays, isotopes), and the presence of biological materials (possible contact with blood, perspiration, and other fluids). In addition, the type of physical efforts made by the employee and of course the circumstances in relation to the clothes' maintenance, storage, laundering, and security must be considered.

The environmental conditions of the hospital in general, and specifically of the operating room suite, defined as the *microclimate*—is where the work is done within a defined and fixed structure, in a fixed area, in a fixed location. Biological agents such as viruses, bacteria, other microorganisms, parasites and so on, may endanger the worker with infectious diseases. Throughout the design process, important topics such as the raw material and its characteristics were examined; but certainly the most important element is designing correctly from an ergonomic perspective, which should reduce the burden on the employee and increase the willingness to wear the clothes. It is preferable to design a garment of low weight and a comfortable form (without annoying corners, protrusions, and coarseness).

Within the operating room suite it is required to change from outdoor or personal clothes into operating room apparel—shirt and pants or coveralls. In a clothing system composed of shirts and pants, workers are expected to tuck their shirt into their pants; although shirts that hug the body can be left outside of the pants.

According to the U.S. Occupational Safety and Health Administration (OSHA) guidelines regarding personal protective equipment, staff members who do not have to scrub should wear a long-sleeved top fastened in such a way that allows for quick and easy opening of the garment, such as snaps.

In order to ensure the safety and comfort of operating room apparel within the operating room environment,[2] the following are a number of requirements regarding the fabrics from which they are sewn:

- They should not be woven or irritate the skin, and be sealed against environmental thermal energy exchange and perspiration secreted through the skin. The fabric should not be too stiff.
- They must be inflammable, free of toxic ingredients and pigments, devoid of harmful smells, and the colors should not be shiny and should minimize distortions from reflected light.[3]

The clothes must be laundered or cleaned after each use.[4] Work clothes contaminated with hazardous materials should not be stored or washed with the normal laundry and should be cleaned by implementing appropriate precautions. Operating room clothes should not be washed together with home laundry. Clothes that have become dirty or wet during the course of work must be changed.

Operating room garments should not be worn outside the operating room area. If this is unavoidable, an outer garment should be placed over the operating room clothes (preferably one that closes in the back). Upon returning to the intermediate area, the garment should be thrown into the designated laundry basket.[5,6]

THE PROJECT

The data collection for the task analysis and the characterization of the work environment was done through observations, followed by interviews and the gathering of relevant data from relevant literature (procedures and recommendations for work in the operating room, the structure and organization of the operating room). The observations were conducted in several large hospitals in Israel with the following objectives:

- The characterization of the operating room teams and their primary tasks.
- The characterization of the physical environment of the operating room suite.
- The characterization of the environmental factors in the operating room suite.
- The characterization of the operating room garments.
- The characterization of items commonly carried by the operating room staff.

The observations were conducted at the Hadassah Medical Center in Ein Kerem (Jerusalem), the Carmel Medical Center (Haifa), and the Sheba Medical Center (Tel Hashomer). The scheduling of the observations was independent of the operating schedule or any other factor, in order to get an impression that would be representative of all the operating room activities. The days that observations were conducted, no activity was performed that deviated from the routine operating room activities. All observations were conducted during the morning shift (7:30 A.M. to 3:00 P.M.).

The purpose of the observations was not concealed from the staff members. The observations were conducted in operating rooms, recovery rooms, and inpatient admission rooms, at points removed from the activity taking place in the room so as not to interfere with the smooth running of the surgery, the procedures performed before and after the surgery, and especially not to interfere with the movements of the staff members. Details and clarifications were provided by the staff members as much as possible, while taking care not to interfere with their activities. An attempt was made not to draw too much attention, so as not to cause the staff to act other than they normally would. From the start, it was made clear that the personal conversations between the operating room staff would not be recorded, that the observations were not for the purpose of recording the actual activities and their outcomes, and that the material recorded or photographed would not be used by any external third party.

Individual privacy was maintained throughout the data collection process. Patients were not photographed and personal details or details relating to their illnesses or of the treatment they received were not recorded or documented.

The observation was passive and entailed watching what was occurring in the operating room and the operating room suite by way of manual documentation and digital photography of important events (regarding the study) that occurred in the operating room (changes in posture, the use of objects, etc.).

To respect individual privacy, the manner in which the operating room clothes were put on and removed was based on personal experience alone. The observation in the locker room related to the arrangement of the furniture and the apparatus for issuing the clothes and returning them, messages to the employees, and how different pieces of clothing—hats, shoe covers, and so on—were laid out.

After the observations were concluded, personal interviews were conducted with the operating room personnel. Interviews with surgeons, anesthesiologists, and nurses helped in detecting common problems with the operating room attire, and an interview with a medical engineer helped characterize the environmental factors in the operating room suite.

To complement the observations in the hospital, tours of the central laundries designated for the washing of the medical equipment were arranged.

The members of the medical staff (surgeons, anesthesiologists, nurses, technicians, and the remaining operating room staff) were asked to complete a questionnaire to rate their satisfaction with their current clothes and choose a color they liked and a form that they felt was suitable. The final questionnaire was administered after some preliminary questionnaires were administered. It took about 15 minutes to complete the questionnaire. The questionnaire was anonymous and the only identifying details it contained were the name of the hospital, the respondent's area of expertise, and professional seniority. The contents of the questionnaire were based on the results of the operating room observations, the interviews with the laundry personnel and with a professional designer, and the results of a survey of examples of operating room garments from commercial sources and accessories from non-medical sources.

The questionnaire was composed of three parts. The first part (8 questions) addressed the patterns of operating room clothes use. The second part (13 questions) examined alternatives for the design of operating room clothes (assuming that there were no technological or financial constraints in constructing and marketing them) and was designed to reveal the respondents' expectations regarding their work clothes. Staff members were asked to choose solutions that they preferred regarding the operating room garments (forms, types of pockets, and their positioning, alternative solutions, etc.).

The third part of the questionnaire contained questions that addressed relevant personal details about the respondents, for the purpose of statistical analysis.

Technion-Israel Institute of Technology
Faculty of Architecture and Town Planning
Studies in Industrial Design

Hello,

Currently a project is being conducted at the Technion regarding the operating room staff's attire (shirt and pants only). As part of the project, the following questionnaire was designed to understand your expectations regarding your work clothes, and may help in improving them.

This series of questions has no right or wrong answers. Please answer in the manner that best suits you. For the success of the study, please try to pay close attention to the questions and answer honestly.

The survey is anonymous and will be used for research purposes only. It should take about 15 minutes to complete the questionnaire.

Thanks in advance for your cooperation.

Use of the Operating Room Clothes

1. Are the operating room clothes that you wear comfortable for you?
   ☐ Yes   ☐ No

2. Which of the following reasons cause you to change your operating room garment during the shift? You can select more than one answer.
   ☐ I don't customarily change my operating room garment during the shift.
   ☐ The garment is uncomfortable—the size is not suitable for me.
   ☐ The clothes were dirty and I did not notice.
   ☐ The clothes got dirty while I was working.
   ☐ The clothes are damaged (torn, for example).
   ☐ The clothes were wrinkled and I did not notice.
   ☐ The clothes got wet while I was working.
   ☐ Other reasons, please specify ...

3. Do you tend to wear an upper garment over the operating room clothes *when you are in* the operating room suite?
   ☐ Yes, always   ☐ Often   ☐ Rarely   ☐ Never

4. Why do you wear an upper garment over your operating room clothes? You can select more than one answer.
   ☐ The operating room is too cold
   ☐ It is too cold in the hallway
   ☐ I feel too exposed
   ☐ Other reasons, please specify ...

5. What upper garment do you usually wear over the operating room clothes?
   ☐ Another short-sleeved shirt
   ☐ A long-sleeved shirt
   ☐ A gown
   ☐ A fleece top
   ☐ Other...

6. Do you tend to wear an upper garment over the operating room clothes *when you leave* the operating room area?
   ☐ Yes, always
   ☐ Often
   ☐ Rarely
   ☐ Never

The Emperor's New Clothes

7. What objects do you carry with you while you are still dressed in the operating room clothes, and where do you tend to put them? Mark the appropriate boxes with an X.

Items	Shirt		Pants			Other Location	The Location Is Not Important
	Top Pocket	Side Pockets	Back Pocket	Side Pockets	Pants Belt		
Employee tag							
Medications							
Scissors							
Stethoscope							
Stamp							
Radiation monitoring device for X-rays							
Cell phone 1							
Cell phone 2							
Beeper							
Wallet							
Small change							
Pens							
Electronic diary							
Telephone notepad							
Glasses							
Glasses case							
Keys							
Watch							
Lipstick/balm							
Tissues							
Cigarettes							
Marking pen							
Tourniquet							
Candy/gum							
Other…							

Rank how frequently you do *each* of the following activities *while you are still dressed in the operating room clothes*. (Mark the appropriate box with an X.)

Very Little	Sometimes	Not Much	Frequently	Very Frequent	Type of Activity
					Walking
					Sitting
					Standing
					Running
					Bending
					Carrying objects
					Raising your arms
					Other....

Design of Operating Room Clothes

1. In your opinion, should the operating room clothes remain unisex?
   ☐ Yes   ☐ No

2. If not, how in your opinion can the separation be made?
   _____
   _____

3. Which colors would you choose for the operating room clothes? Mark an X on the hexagon with the colors you prefer from the two following scales (more than one answer can be selected).

4. Would you choose operating room clothes with color combinations?
   ☐ Yes   ☐ No

5. If you answered *yes*, which color combinations would you choose? Please specify. _____

6. Would you prefer printed designs on the clothing (stripes, dots, patterns, etc.)?
   ☐ Yes   ☐ No

7. If so, please specify which patterns you would choose. _____

8. There are a wide variety of collars and necklines for operating room shirts. Mark the option you prefer with a circle.

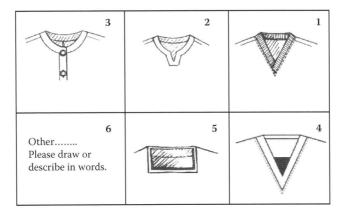

9. Which type of pant belt would you choose from among the following options? Mark the option you prefer with a circle.

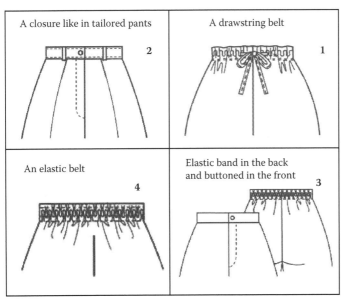

10. Mark on the sketches of the shirts below where you feel it is comfortable to store portable work equipment and personal effects (place an X in the checkboxes). You can select more than one preference.

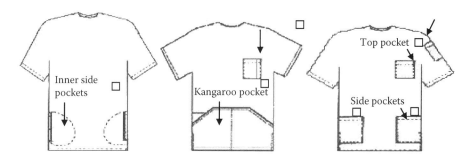

Other...Please draw or describe in words.

11. Mark on the sketches of pants below where you feel it would be comfortable to store portable work equipment and personal items (place an X in the checkbox). You can select more than one preference.

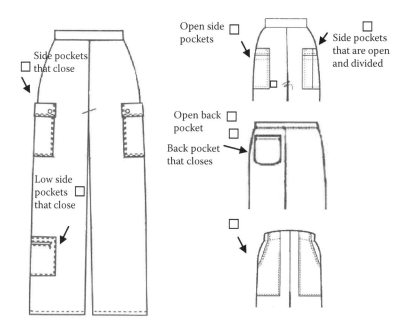

Other, please draw or describe in words_____

12. What would you be interested in changing in the operating room clothes, and how? Please specify on the rows below:
    A. Operating room shirt:
        ☐ I would not be interested in changing anything.
        ☐ The cut of the shirt (length, width, neckline)_____
        ☐ The cut of sleeves_____
        ☐ The location and number of pockets_____
        ☐ Type of fabric (touch, degree of ventilation, warmth)
        ☐ The notation of the garment's size_____
        ☐ Other_____
    B. Operating room pants
        ☐ I'm not interested in changing anything
        ☐ Shape of pants (length, width, zipper)_____
        ☐ Location and number of pockets_____
        ☐ Type of fabric (touch, degree of ventilation, warmth)
        ☐ The notation of the garment's size_____
        ☐ Other_____

13. Do the following solutions seem appropriate for mobile/portable? Work equipment and personal items in the operating room area? You can select more than one answer.
    ☐ Washable pouch
    ☐ A lightweight sweater that is easy to wash
    ☐ Overall
    ☐ Apron with pockets
    ☐ Set of pockets that can be moved around (attached with velcro)
    ☐ Other solutions, please specify_____
    ☐ The solutions presented do not seem to be appropriate

Personal Details

1. Gender:
   ☐ Male   ☐ Female
2. Age____
3. Occupation
   ☐ Surgeon   ☐ Anesthesiologist   ☐ Nurse   ☐ Technician
   ☐ Other_____
4. Professional seniority (years of work experience in the operating room)
5. Name of the hospital
6. Height (cm)
   ☐ 150–155   ☐ 156–160   ☐ 161–165   ☐ 166–170   ☐ 171–175
   ☐ 176–180   ☐ 181–185   ☐ 186–190   ☐ over 190 cm
7. Weight: (kg)
   ☐ 46–50   ☐ 51–55   ☐ 56–60   ☐ 61–65   ☐ 66–70
   ☐ 71–75   ☐ 76–80   ☐ 81–85   ☐ 86–90   ☐ over 90 kg

From our observation of the staff members at work, we saw that they each do different types of tasks. From the occupational analysis we performed, it appears that there are employees who must carry many accessories, whereas some of them must get rid of all the accessories on their person when the surgery begins. Currently, all operating room personnel have been dressed in the same garments.

ENVIRONMENTAL CONDITIONS

Temperature—The operating room suite is not influenced by external weather conditions. The temperature of the operating room is controlled, and its minimal value is 18°C. The temperature in the operating room hallway is 22°C and 21–24°C in the recovery room. The main reason for setting a relatively low temperature in the operating room is to prevent infection; high heat can cause perspiration to drip from the surgeon's brow into the surgical area. The needs of the staff working in short-sleeved clothing and those of the staff who wear warm sterile garments over their work clothes conflict.[7]

Humidity—The humidity in the operating room does not exceed 50% and its level is maintained by drying systems to prevent infection.

Ventilation—The operating room air undergoes absolute filtering (the air is filtered at a level of 99.999% or five-ninths. The air pressure in the operating room is positive and it is negative in the hallway, thus the air flows outwardly.[8]

Lighting—The entire operating room is lit with artificial lighting, that is, fluorescent lighting with a uniform filter distribution. The surgical lamps are halogen reflectors with a light intensity of 100–150 thousand lux. Special filters absorb approximately 99% of the heat they generate so that the reflectors do not produce heat.[9] In certain medical procedures (such as a laparoscopy and a hysteroscopy) the room is dark (so that the surgeons can clearly see what is shown on the monitor) and the surgical area lighting is intensely focused.

Clothing—The operating room clothes are uniform and are used by all staff who work in the operating room areas: surgeons, anesthesiologists, nurses, technicians, medical engineering staff, cleaning staff, and so on. The outfit worn includes the following items of clothing: a shirt, pants, a cap, and shoe covers or work shoes. As mentioned previously, the observations showed that the clothes are worn by the staff members during the entire shift and not only during their stay in the operating room. Generally, they wear them from the start of the workday and continue wearing them until the day ends—during the surgeries, in the recovery room, during department rounds, during daily academic and clinical activity, and so on.

The observations revealed that the operating room clothes fulfill various functions and undergo changes and improvisations according to the various ways they are used, such as for hanging keys, carrying a pager, a stethoscope, a mobile phone (or more than one), identification tags, markers and more (see Photos 12.1–12.4).

Most of the staff members wear different types of upper garments (personal or belonging to the department): flannel shirts, an additional operating room shirt, a long-sleeved shirt with knitted cuffs, an operating room gown, a fleece shirt and more. Generally, the upper garment is not sent to the laundry after each shift.

The Emperor's New Clothes

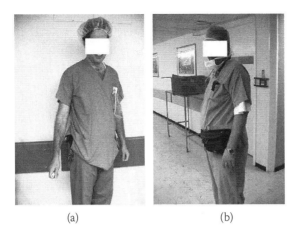

PHOTO 12.1 (a) An anesthesiologist at the entrance to the operating room, a typical example of how the identification tag and keys are hung. Typical appearances of pockets pulled downward. (b) A typical example of how the stethoscope is hung, how a pouch and a pager are hung, and a T-shirt under the operating room shirt.

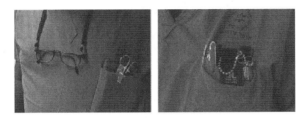

PHOTO 12.2 A typical example of the appearance of an upper pocket on the operating room shirt and the objects within it: pens, radiation monitoring devices for X-rays, scissors, an identification tag (in the pocket); and glasses hanging from the neck.

The staff members in the operating rooms carry personal items and all types of work tools with them. A list of the most common items appears below:

Identification tag—Medical workers must wear a visible tag.
Cellular phone—Internal cell phones (for work purposes in the hospital area only) and personal cell phones.
Pager (beeper)—A device that is vital during emergency situations (cellular phones are not always able to be used) and the administration requires that surgeons and anesthesiologists carry it.
Pens and markers, drugs, scissors, a stamp, a radiation hazard monitoring device for X-ray, a tourniquet, a diary or electronic diary, an address book, glasses and a glasses case, keys, a watch, lipstick, tissues, cigarettes, nose drops (Table 12.1).

The sample included 160 respondents from 11 hospitals in the country, including 61 surgeons, 41 anesthesiologists, 38 nurses, 10 technicians, 2 assistants, a secretary, an

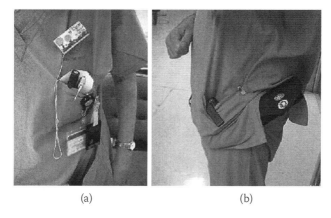

PHOTO 12.3 Operating room shirt pockets: (a) upper pocket. An example of how an identification tag, an employee tag, a radiation monitoring device for X-rays, and pens are carried. The pocket divider is composed of stiff paper. (b) Carrying a pager in the side pocket of the shirt, a glasses case hung on the pants belt.

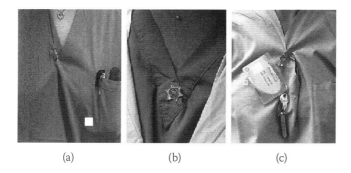

PHOTO 12.4 An example of a large neckline and solutions for closing it: (a) with a safety pin, (b) with a Star of David pin, and (c) with an identification tag and keys.

orderly, 2 girls from the national service, a dosimetrist, and a resident. Three of the respondents did not specify their occupation. The survey included 90 men and 67 women; three of the respondents did not specify their gender.

SUMMARY OF THE QUESTIONNAIRE'S FINDINGS

HABITS REGARDING THE USE OF THE OPERATING ROOM CLOTHING

- Characterization of the comfort of operating room clothes.
  - 81% of the respondents reported that the operating room clothes that they currently wear are generally comfortable for them.
  - The highest percentage of respondents reporting that the operating room clothing is generally not comfortable for them was among the nurses' group.
- Characterization of the main reasons for changing operating room clothes during a shift.

TABLE 12.1
The Weight of the Objects Most Commonly Carried in the Operating Room Clothes

Weight in Grams	Object
200	Wallet
150	Diary/Electronic diary
130	Pager
100–200	A set of keys
105	Cellular phone
85	Stethoscope
80	Address book
60	Stamp
40	Glasses case
40	Tourniquet
30	Pack of cigarettes
30	Glasses
30 (per box)	Medications (box)
30 (including the case)	Employee tag
25	Coins
25	Monitor of X-ray radiation hazard
20	Watch
20	Scissors
20	Lipstick
10	Pens/Markers
10	Candy/Gum

- 13% of the respondents do not tend to change their operation room clothes during a shift. The three most common reasons for changing the operating room clothes (indicated by about half the respondents) are: the clothes were dirty, got wet working, or the clothes were damaged).
- Characterization of the habits related to wearing an upper garment over the operating room clothes while in the operating room area and when leaving it.
  - A third of the respondents reported that they always or frequently tend to wear an upper garment over the operating room clothes while they were in the operating room area.
  - 80% of the respondents reported that the most common reason for wearing a garment over the operating room clothes was because the room was cold. About half of the respondents choose to wear a gown or a sweatshirt as an upper garment.
  - More than half of respondents reported that they always or frequently wear an upper garment over their operating room clothing when they leave the operating room area.
- Characterization of the objects carried in the operating room clothes and where they are carried.

- The objects carried in the operating room clothes (reported by over half the respondents) are: pens, a cell phone, keys, an employee tag, and a stamp.
  - Objects that are less commonly carried (reported by about one-third of the respondents) are: a wallet, a watch, a stethoscope, and scissors.
- The characterization of the physical activities performed by the operating room staff members.
- The common activities (as reported by over half of the respondents) are: standing, walking, sitting, and bending.

Design Preferences for Operating Room Clothes

- Characterization of the selected clothing style preferences.
  - 81% of all the respondents reported that they are interested in a unisex design. The respondents who reported the highest rate of interest in changing the design style was the group of anesthesiologists (30%).
- Characterization of the preferred colors, color combinations, and colored patterns selected for the clothing.
  - The three groups of colors preferred for operating room clothing (representing over half the respondents) are: shades of light blue, dark blue, and bright green.
  - A quarter of the respondents reported that they are interested in color combinations in the operating room clothes.
  - 14% of the respondents are interested in including colorful prints (stripes, dots, shapes, etc.) to the operating room clothes.
- Characterization of the preferences selected for the collars and necklines of the operating room shirt.
  - The preferred collar for the operating room shirt is a V-neck collar (selected by about a third of the respondents). Two collars that are less often preferred (selected by about one-quarter of the respondents) are a heart-shaped collar and a trapezoid collar with an outer V-neck.
- Characteristics of the preferred selection of an operating room pant belt.
  - The two types of pant belts most often preferred (as selected by half of the respondents) are: an elastic belt and one with a drawstring.
- Characteristics of the preferred selection of where to carry objects in the operating room shirt.
  - About half of the respondents indicated that they prefer to store objects in the side pockets and top pocket of their operating room shirt.
- Characteristics of the preferred places selected for carrying objects in the operating room shirt.
  - About a third of the respondents indicated that they prefer closed side pockets and side pockets with diagonal openings for storing objects in the operating room pants.
- Characterization of the changes desired in the operating room clothes.
  - Over a third of the respondents are interested in changing the placement and number of pockets in the operating room shirt, the type of fabric, and the cut of the shirt.

- Three-quarters of all respondents are interested in changing the placement and number of pockets in the operating room pants, and about a third would like to change the type of fabric and the cut of the pants.
- Characterization of alternative solutions for carrying portable work equipment and personal effects in the operating room area.
  - Three-quarters of the respondents are interested in trying alternative solutions.
- About a third of them are interested in trying a washable pouch or a system of detachable pockets as an alternative to carrying portable work equipment and personal items in the operating room, and a lightweight washable vest.

SIGNIFICANT DIFFERENCES BETWEEN MEN AND WOMEN IN THEIR PATTERNS OF USE AND DESIGN PREFERENCES FOR OPERATING ROOM CLOTHES

- Patterns of upper garment wear while in the area.
  - Men: Only 16% of the men regularly or frequently wear an upper garment over their operating room clothes.
  - Women: 60% of the women always or frequently wear an upper garment over their operating room clothes.
- Characterization of the types of items carried.
  - Men: 92% of the men carry a mobile phone, 88%—pens, 78%—an employee tag, 75%—keys, 66%—a stamp, 51%—a wallet.
  - Women: 92% of the women carry pens with them, 86%—keys, 79%—a cell phone, 77%—an employee tag, 58%—a stamp.
- Characterization of preferred types of collars and necklines selected for the operating room shirt.
  - Men: 30% of the men prefer a trapezoid collar with an outer V-neck in the operating room shirt.
  - Women: 25% of the women prefer a Chanel-type round collar.
- Characterization of alternative solutions for carrying portable work equipment and personal items in the operating room area.
  - Men: 51% of the men are interested in trying a washable pouch as an alternative solution for carrying portable work equipment and personal items in the surgical area.
  - Women: Women prefer a detachable pocket system (49%) and a lightweight, washable vest (43%).

SIGNIFICANT DIFFERENCES BETWEEN THE DIFFERENT PROFESSIONAL GROUPS IN THEIR PATTERNS OF OPERATING ROOM CLOTHES USE AND IN THEIR DESIGN PREFERENCES

- Characterization of the comfort of the operating room clothes.
  - Surgeons: 90% of the surgeons' group reported that the operating room clothes are generally comfortable for them.
  - Anesthesiologists: 73% of the anesthesiologists reported that they were satisfied with their operating room clothes.
  - Nurses: 71% of the nurses reported that the clothing they currently wear in the operating room is generally comfortable.

- Wear patterns related to upper garment while in the operating suite.
  - Surgeons: 13% of the surgeons' group regularly or frequently wear an upper garment over their operating room outfit.
  - 73% of them wear a gown as an upper garment.
  - Anesthesiologists: 37% of the anesthesiologists reported that they wear an upper garment over their operating room outfit when they are in the operating room. 53% wear a gown for this purpose.
  - Nurses: 58% of the nurses wear an upper garment. 66% of them generally wear jogging fleece.
- Characterization of the types of physical activities performed by the operating room staff.
  - The two main physical activities that most of the staff members reported doing were standing and walking.
  - Surgeons: In addition, the surgeons report that sitting was a physical activity performed (64%).
  - Anesthesiologists: The respondents among the anesthesiologists' group reported that a common activity is sitting (74%), 49% of the anesthetists raise their arms very frequently and bend during their shift.
  - Nurses: 78% of the nurses reported that they carried objects, 70% bend, and 49% raise their arms high up very frequently while performing their role. Only 11% of the nurses rated sitting as a frequent activity.
- Characterization of the selected preferences in the colors, color combinations, and colored patterns of the clothing.
  - Surgeons: About half the surgeons indicated a preference for the following groups of colors: dark shades of blue (79%), light shades of green (72%), light shades of blue (58%), light shades of purple (53%), dark shades of purple (49%). Only 7% of the surgeons' group are interested in surgical clothes with colorful prints.
  - Anesthesiologists: Over half of the anesthesiologists indicated a preference for the following groups of colors: light shades of green (80%), light shades of blue (68%), dark shades of blue (63%). 35% of the anesthesiologists are interested in operating room clothes with color combinations. 25% are interested in adding prints to their operating room clothes.
  - Nurses: 98% of the nurses indicated a preference for operating room clothes in the following color groups: light shades of blue (94%), dark shades of blue (53%), light shades of green, and 33%—shades of reddish brown. 34% of the nurses' group are interested in operating clothes with color combinations and 23% of them are interested in including colorful prints.

Based on these data a model was constructed that took all the factors into account. The ideal garment was to be disposable, adapted for gender, which has a special collar for the women, a special attachment for hanging the tag, and a clip for a telephone, and much more. But due to budget constraints and the uniqueness of the laundering,

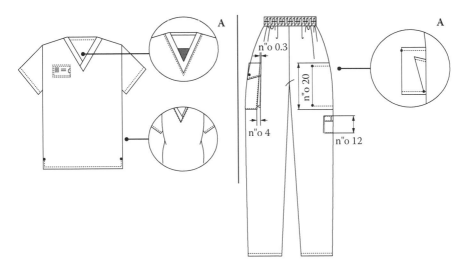

FIGURE 12.1 Prototype of the operating room shirt and pants. Front view (right). Back view (left).

FIGURE 12.2 Prototype of a pocket with a side opening for storing the employee's magnetic card.

this is not possible. Therefore, the following model was proposed: and the addition of a pocket for the (employee) tag (Figures 12.1 and 12.2).

SUMMARY

The current situation on the topic of operating room clothes was assessed for the first time, integrating relevant information from medical, commercial, and academic sources. The findings are a combination of the various users' needs for these clothes, as well as their requirements and preferences.

The project led to a series of innovative design recommendations for operating room clothes, without introducing any evolutionary changes. However, new and original changes were presented with respect to the composition of the material and the method for carrying objects.

This work proves that reliance on the opinions of the users alone may be misleading and that before reaching decisions an objective investigation of the environment, the users, and the object is necessary.

The applications stemming from this project may reduce the burden on the users who carry many objects on their person and free them of the need to find different solutions and improvisations to adapt the garment to their personal needs. In this way, it will be easier for the workers to fulfill their roles. The design and organizational changes will reduce the risk of occupational accidents and facilitate the maintenance of a high level of cleanliness in the operating room area.

This research may also increase awareness of the subject of extending the life of operating room clothing and draw the users' attention to the maintenance of their work clothes.

REFERENCES

1. (2003). Together to a century of quality. *Ростест. Москва*. Retrieved September, 23, 2004, from http://www.medstyleservice.ru/press/index.html.
2. (1998). Recommended practices for surgical attire. Association of Operating Room Nurses. AORN Technical Practices Coordinating Committee, *AORN Journal*, 68(6), 1048–1052.
3. (1986). Proposed recommended practices. Aseptic barrier materials for surgical gowns and drapes. *AORN Journal*, 44(4), 612–6.
4. Safety Regulations. (1997). Institute for Hygiene and Safety. *Body Protection*, 6, 58. (Hebrew).
5. (1998). Recommended practices for surgical attire. Association of Operating Room Nurses. AORN Technical Practices Coordinating Committee, *AORN Journal*, 68(6), 1048–1052.
6. Ducel, G., Fabry, J., and Nicolle, L., eds. (2002). Prevention of hospital-acquired infections. A practical guide, 2nd ed. *World Health Organization*, 32–40. Retrieved May 23, 2005, from http://www.who.int/csr/resources/publications/drugresist/en/whocdscsreph200212.pdf.
7. (2002). How to dress to dentists? Retrieved September, 24, 2004, from http://www.textile-press.ru/?id=1249.
8. Designing and Operating Heating, Ventilation and Air-Conditioning (HVAC) in Hospitals. (2002). *Deutsche Gesellschaft für Krankenhaushygiene*. Retrieved May, 23, 2005, from http://www.dgkh.de/cgi-local/byteserver.pl/pdfdata/leitlinien/rlt-anlagen_e_gelb.pdf/.
9. Lighting for operating rooms. Retrieved December, 29, 2005, from http://www.altaide.ru/www/chromophare.html.
10. Becker, A., Donchin, Y., and Bitterman, N.J. (2009). Operating room clothing: Design and ergonomic concepts. *Clin Anesth*. September 21(6), 459–61.

13 Thinking Patterns of Physicians and Nurses and the Communication between Them in the Intensive Care Unit

Yehuda Badihi* and Daniel Gopher

CONTENTS

Analysis of Physicians' and Nurses' Thinking Patterns 195
    Characteristics of the Data Processing—The Activities That Focus on
    Interpreting and Developing the Information... 198
    Organization of the Data 198
    Characteristics of Performance—Actions That Are Products of the Mental
    Processes and the Mental Model.. 199
Results.. 199
    Communication and Conveying Information between a Physician and a
    Nurse 202
    Character of Data Intake and Thinking Style...203
    The Quality of the Communication between the Physician and the
    Nurse—And the Thinking Style..203
Summary 204
References...205

ANALYSIS OF PHYSICIANS' AND NURSES' THINKING PATTERNS

A physician specialist can determine a diagnosis from a brief impression of the face of a patient entering the clinic. Nurses can recognize problems in the infusion while treating an unconscious patient. How does a professional arrive at the correct (or incorrect) conclusions? What is the mindset of the physician? And in what way does the nurse see and diagnose the condition of his/her patient? The choices made by medical staff members often literally involve life and death decisions. Thus, in

---

* Doctoral dissertation submitted to the Center for Work Safety and Human Engineering, Technion, Haifa. Under the supervision of Daniel Gopher and Yoel Donchin.

the study presented below we tried to understand the thought process occurring in the mind of the physician and the nurse when faced with a medical problem that requires such decisions.

How can the investigator "penetrate" into a subject's manner of thinking? Is it possible to trace the decision-making process of the physician in deciding that a patient is dehydrated and must immediately receive a 1000 ml infusion of a specific solution?

Currently, tools are available to investigate cognitive processes. One method used is to request subjects to describe (as well as possible) the manner in which they ascertained a given diagnosis. By asking them to "think out loud," their manner of thinking can be extracted by analyzing their replies. This type of report may enable the investigator to delineate the method by which the physician gathers the information he/she needs in order to arrive at medical decisions.

The condition of a patient in the intensive care unit often fluctuates—a sudden drop in blood pressure, a sudden worsening of his/her respiratory status, and so on. The responses of the physician and the nurse to such changes are based on what they know about the patient, predicated by their impression of the patient's condition, which they claim is somewhat akin to a *mental model* of the condition.

A mental model is a system through which knowledge is represented: a simplified representation of the real world as perceived by an individual that explains phenomena and thus predicts what may occur. With respect to a medical situation, such a model can help the professional intervene immediately to prevent deterioration in the patient's condition.

Johnson-Laird,[11] who developed and elaborated on the concept of the mental model, argues that it is not necessarily a *picture* or mental *image* of the world. Rather, a mental model can be distinguished by the *functional* components and structures it contains: a relationship between objects, attributes of objects, and the laws or rules of action. Thus, the nature of the mental model is expressed by its functionality and structure.

Several types of models are possible within a system, including:

- *Logical models*—Based on a description of the *logical structure* of a system as a method of understanding the system and how it can be used.
- *Analogical models*—Aid us in forming *parallels* and *points of similarity* between different systems.
- *Models of mapping*—Determining *relationships* between objects according to measures of distance, time, and/or function. These also include models of mapping the relationship between roles and actions: prioritizing objects according to their importance, the dependence of one object to another, and so forth.

Mental models, therefore, represent a *worldview*, which form the basis of an individual's cognitive and emotional attitudes.

The intensive care unit (ICU) was selected as the subject of this study because of the frequent changes and the need for *intensive* care, that is, to make quick decisions and execute them rapidly. The study was based in the intensive care unit at

Hadassah hospital in Ein Kerem, Jerusalem. The experiment was conducted with the authorization of the hospital's Helsinki committee and with the consent of all the participants. The ICU contains six beds and a few other beds in the recovery room of the hospital, to be used as needed. Two attending physicians closely supervise approximately six physicians, each at different stages of specialization. During the morning shifts one nurse is assigned to every two patients and some physicians are present among them as well. The night shift physicians transfer the information to the physicians and nurses during the morning rounds, which are designated exclusively for this purpose.

A. *The study population* was composed of the entire ICU medical and nursing staff—21 physicians and 17 nurses.

B. *The data collection* was based on the simulation of real cases that were documented in the months prior to the study. The physician or the nurse was provided with the patient's chart (an actual chart) and all the relevant data, such as diagnosis, age, allergies to medications, and the events that had occurred in the last 24 hours. In practice, the simulation reconstructed the ICU shift change procedure, performed next to the patient's bed. During the simulation, a microphone was attached to the subject and he/she was asked to "think aloud" and describe exactly what he/she was doing. The thought sequence could be interrupted in order to ask the subject (physician or nurse) what their thoughts were regarding the results of the various tests and about the information provided to them during the simulation. In the context of the simulation the physician was also asked to fill out the instruction sheet—generally done at the completion of the morning rounds—and explain how each instruction and order should be properly performed. The nurse was asked to perform the tests and treatments that are routinely performed once an hour and document it on the follow-up sheet. These activities, recorded both on tape and in the charts simultaneously, helped simulate the actual situation for the treating physician and nurse, and represented a necessary basis for the physician's and nurse's understanding of the patient's condition and the *thinking aloud* regarding the proper treatment.

The course of the simulation represented a reconstruction of about 20 minutes of the physician or nurse's activity in the unit. The thinking aloud was recorded on videotape and transcribed. Each word that was spoken during the simulation was then made accessible through a word processing program, enabling searches and semantic analysis. We used a verbal protocol analysis method based on the working theory of how thinking patterns and mental models could be derived from the syntactic structure of sentences and statements (using the SHAPA program developed by Penelope Sanderson to analyze these data).[2]

The subject's narrative was examined according to the approach developed by Pylyshyn, who proposed the concept of *functional architecture*. This concept connotes a similarity between the operations performed by a computer and the processes of human cognition. According to Pylyshyn, the operation of these two systems is based on the symbolic representation of behavioral principles. The levels of human thought are based on a continuous process of input—processing—output (= action). According to this approach, the mapping and classification of the processes (required by a physician or nurse to establish a mental model of the patient) can be grouped into three clusters. The mental model can be conceptualized through a computational/logical conceptual system.

During the study, 40 protocols were gathered, each composed of approximately 1,500 sentences.

Input data includes all the external clues—subjective or objective—identified by medical professionals in dealing with the transition from a given state to a desired state (judgment, decision, and problem solving). These clues may be passively or actively absorbed, through an initiated process, at *each of the stages* that precede making a decision.

We divided the tasks that focus on information intake into four groups:

A. *Passive information intake* (= data P)—The information provided by the physician/nurse finishing his/her shift as an integral part of the routine of transferring responsibility for the patient.
B. *Active information intake* (= data A)—Information that is obtained by the medical professionals who admit the patient.
C. *Results of the tests*—A physical examination of the patient, information from monitoring devices, results of the blood tests sent to the laboratory, and so on.
D. *Searching for information* (= search)—From various sources of information, especially after the departure of the night duty personnel (when the information must be found from the documentation and records).

CHARACTERISTICS OF THE DATA PROCESSING—THE ACTIVITIES THAT FOCUS ON INTERPRETING AND DEVELOPING THE INFORMATION

A. *Hypothesis*—A probable hypothesis regarding the overall condition of the patient, or the previous method of treatment, or records taken.
B. *Significance*—Logical inference of the patient's condition, or the previous method of treatment, or records taken.
C. *Factoring*—Factoring in the contribution of the processes or information in a given context in relation to time, importance, quantity, and so on.
D. *Status*—The determination of a piece of information by the physician or the nurse regarding the patient's condition in a given context.
E. *Calculation*—A mathematical calculation of numerical data during the activity.

ORGANIZATION OF THE DATA

The activities that focus on expanding and developing the information are:

A. *Planning*—The planning of how to construct an overall picture of the situation, through tests, treatments, or in documenting the information.
B. *Process*—A continuing component in a series of processes in the course of the physician's or nurse's activities, through tests, treatments, and documentation.
C. *Comparison*—Comparing pieces of information, processes, or conditions.

D. *Problem*—A problem that arises during the activity of the physician or the nurse and requires them to provide a suitable solution.

E. *Error*—An error that was previously made and is revealed during the simulation.

CHARACTERISTICS OF PERFORMANCE—ACTIONS THAT ARE PRODUCTS OF THE MENTAL PROCESSES AND THE MENTAL MODEL

Activities that focus on performance are:

A. *Order* (= order)—Giving an order and its recording by the physician.
B. *Treatment* (= treatment)—Providing the patient with treatment by the physician or the nurse.
C. *Documentation* (= documentation)—Documenting information through any method of documentation.

The analysis focuses on the structural aspects of the thought processes, rather than on their content, on the formal pattern of the thought process and deriving conclusions, not on the content or medical judgment.

RESULTS

The results indicate differences between physicians and nurses in terms of the thinking patterns, the number of steps undertaken before arriving at a decision, and the use each makes with the data components.

The physician primarily relies on previous knowledge, born of many years of study and vast clinical experience. The physician makes assumptions and verifies them. Alternatively, the nurse bases his/her decision on the clinical condition of the patient, responding and performing based on the fluctuating circumstances.

Furthermore, the physician and nurse were discovered to have different mental models for each patient, resulting in subsequent potential miscommunications and mishaps.

The method we have chosen allows us to observe not only each individual action involved in obtaining information, but to see the whole process, the thinking *path*, which begins with the receipt of the initial information and is completed with the implementation of any activities that result from the processes preceding it. This provides the answers regarding questions such as: what is the complexity of the processes (as noted from the initial stage to the final stage of implementation)? What occurs during each stage of the construction of a *worldview*? Where do the physician and nurse channel the process to? Gathering more information? Further processing the information? Implementing an action?

The process of constructing a mental picture of the patient's condition by a staff member begins when the patient is admitted to the unit, it continues with the change in shift, and is completed at the stage of implementing an order that concludes the process of constructing a worldview (Table 13.1).

The process begins when the physician/nurse finishing the shift provides information and transfers responsibility for the patient to the physician/nurse beginning

TABLE 13.1
Scope of the Various Activities in the Process of Thinking during a Shift—Synopsis of a Comparison between the Physician and the Nurse

Physician	Nurse
Initiated information (active)—50%	Initiated information (active)—20%
Presented information (passive)—20%	Presented information (passive)—30%
Examinations—10% of the total information collected	Examinations—30% of the total information collected
Information gathering process is slower during the construction of a mental model—30%	Information gathering process is faster during the construction of a mental model—50%
Overall, more information processing—50%	Overall, more organization of information—50%
Overall, more elaboration and enrichment of information	More routine examinations (unbiased)
Relatively many processes involving: processing—processing—processing…—execution	Relatively many processes involving: intake—processing—intake—processing—execution
More structured dependence on the thinking process (dependence on the previous statements made)	Relatively low structured dependence during the thinking process
Relatively many transitions between the various body parts = high structural dependence between the body parts—60% of the processes. More integrative information processing from different worlds of content.	Less structural dependence—35% of the processes of information gathering and processing. More *localized* information processing.

The Comparison of Thinking Patterns—Physicians and Nurses

Nurse	Physician
Passive intake of information—primarily through tests	Active and biased intake of information
Processing primarily to organize the information	More intensive information processing, primarily for the expansion and enrichment of the information
Shorter processes of execution encompassing many stages of information gathering (intake)	Longer processes of execution (including decisions), integrating intermediate stages of information processing
Pieces of information are the result of organization and additional input, not very dependent on previous information	Pieces of information are very dependent on one another (the later process occasionally results from the previous process)
Execution errors: negative correlation or no relationship with the body systems that incorporate more testing and planning (more common among nurses)	Execution errors: a significantly positive correlation with body systems that incorporate active information processing (more common among physicians)
The nurse's mental model is based on information that is organized and up to date and relies less on professional knowledge	The physician's mental model is based primarily on professional knowledge and working hypotheses derived from it, and relies less on information about the patient's clinical condition
The nurse mainly responds or accepts the guidelines and instructions and is to act on them	The physician is more dominant in his/her communications with the nurse: demands specific information, supervises or gives instructions

their shift. Therefore, 100% of the information is information that is received. This information is the first step in building "a picture of the situation," that is, the condition of the patient in bed number five, for example. This input data initiates other processes. For the physician—about 30% of the information provided makes it necessary to gather additional input information (proactive or passive), while 70% is used to process the information that was received.

Roughly halfway through the process, the nurse directs it toward obtaining more information, and only 40% is used for information processing. At this stage, the second stage, some of the information (16%) is directed toward the performance of activities (treatment or recording information), which is, in fact, the purpose of the process, and with this the process of constructing the thinking path is completed (Figures 13.1 and 13.2).

The same is also true regarding the other thinking paths, for the physician and for the nurse.

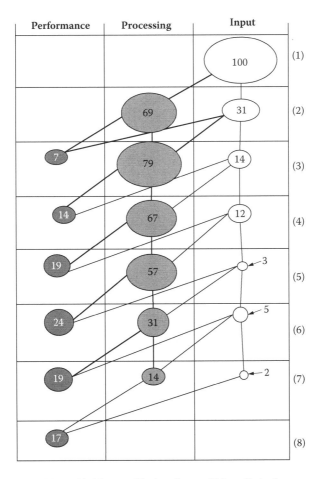

FIGURE 13.1 Physicians thinking profile (numbers within cells indicate step percent frequency in all profiles).

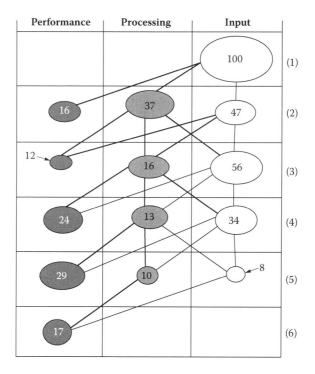

FIGURE 13.2 Nurses thinking profile (numbers within cells indicate step percent frequency in all profiles).

COMMUNICATION AND CONVEYING INFORMATION BETWEEN A PHYSICIAN AND A NURSE

The obvious difference between the physician's and nurse's thinking styles may explain the probability that it could impair the quality of their communication. Gopher, Donchin, and others[9] found that a current and accurate mental picture of the patient's condition is critical to one's ability to function in an environment where the data is constantly changing. In their opinion, a current mental picture of the condition requires that the various professionals—physicians, nurses, consultants, and technicians found at the patient's bedside throughout the day—communicate in a manner that allows for the continuous transfer of information over time. In the ICU, there is a system for presenting information centering on two formal means of communication: a monitoring sheet and the page on which the physician records the care instructions (*instruction sheet*). Two procedures are added to this system: the shift change and the physician's rounds.

It has been shown that one of the central problems is miscommunication and deficiencies in conveying information among all medical staff members (including between the physicians, and between the physicians and the nurses). This problem is reflected in all the above-mentioned channels: the formal system of presenting the information, during the shift change, and during the physicians' rounds. The data presented above indicate that the problem does not lie in the frequency of communications or how often it is utilized (even with respect to informal communications), but in the *nature* of the messages transmitted.

CHARACTER OF DATA INTAKE AND THINKING STYLE

The picture illustrated so far reflects a significant gap between physicians and nurses in three aspects of the thinking patterns that guide them in constructing the mental model of the patient.

A. *The dominance of the inner knowledge base over external information.* For physicians, this dominance leads them to ignore some of the relevant observable information to assist in the processes of information processing that he/she acts upon (referred to as "Do not confuse me with facts"). The physician needs, therefore, an external source of *unbiased* information, particularly at the beginning of the mental model construction process.

B. *The lack of a systematic manner of constructing a worldview.* Among other things, systematic refers to methodically integrating external information in the various processes used for analyses and organizing the foundational information as part of information processing. Both of these are more frequently characteristic of the nurse. Thus, the physician's need for more updated and detailed information is prominent in this respect as well.

C. *The level of complexity of the thinking structure.* The complexity of the thinking structure is greater for physicians than for nurses. At highly complex levels of the thinking process, validation or refutation of any piece of information may, as stated previously, influence a relatively large number of hypotheses/decisions. Such information may sometimes undermine the validity of the existing scheme in the emerging mental model. This leads us to conclude that what is required is the exposure to a maximal amount of relevant information, at the beginning stages of the patient's intake, so that the entire construction process will be well founded with as broad a base of information as possible.

THE QUALITY OF THE COMMUNICATION BETWEEN THE PHYSICIAN AND THE NURSE—AND THE THINKING STYLE

The situation described above highlights the discrepancy between the physician and the nurse in everything related to the scope of data collection and the methods used to receive it, especially at the initiation of a patient intake process. This incompatibility may result in the poor quality of the communication between the teams. Both the physician and the nurse share in the responsibility for this problem.

The physician is unaware, apparently, of the significance of a style of information gathering that is incomplete and biased, and of the need for him/her to somehow compensate for this. Since the physician is not proficient in the manner in which a nurse thinks, *he/she does not recognize that the nurse can compensate for this deficiency*, a compensation that is required by the physician at the critical junction of the process of constructing a mental model—during the patient's intake process.

The nurse is most suited to provide such compensation: The nurse's worldview is based first and foremost on gathering the information required by a physician. During this process, the nurse does not rely on biased input of information. The nurse is greatly assisted by the processes the doctor is lacking: preliminarily passive observations followed by active testing throughout the patient care process. Due to the extent of the different thinking styles in relation to the data collection processing, the nurse is unaware, apparently, of the need to update the physician regarding the important pieces of information required to validate/repudiate the physician's various theories regarding the patient. As previously stated, these matters are needed primarily in the initial stages of the process.

The formal information system, and the monitoring sheet in particular, contains abundant information, which should be able to fulfill the needs outlined above; in practice, however, it does not. As the data makes evident, even in this system the physician is exposed to the same biased processes. Moreover, the nurse's methodical perspective is not sufficiently reflected in this system of information. For example, it does not sufficiently emphasize anomalies in the various measures or interaction between them, or of the treatment history, all of which bear directly on the assumptions made by the physician at that time.

It is thus apparent that there is great significance to the quality of communication between the physician and the nurse, as is the need for vast improvement in this area. Better communication has great "added value" for the quality of information the physician is exposed to. As demonstrated above, these components are directly related to the quality of the physician's performance, as manifested by the error rates.

SUMMARY

Comparing the physician's patterns of thought with the nurse's allows us to hone the distinctions between their patterns of thinking as a possible tier in the causes for error and failures. The disparity between a physician and a nurse goes beyond a different functional architecture in their thinking style. The difference is located in their different roles, which facilitates a different orientation with regard to the thinking process. By virtue of his/her role the physician is required to make a diagnosis and order a specific treatment. Constructing a *world image* in order to diagnose and make medical decisions is a complex process; so it is natural that the tools used for thinking would also be characterized by a high level of complexity. The broad range of knowledge acquired during a physician's studies facilitates a high level of structural dependence among the various pieces of information, objects, and between the various body systems.

A high level of complexity, in itself, is not prone to errors. The knowledge base may be constructed systematically by anchoring it to the external world and using the observational data, especially at the beginning of the process. It is the interaction of the high complexity with a *lack of methodology*, accompanied by a narrow preliminary external knowledge base, which makes fertile ground for error. Unlike the other components, a preliminary information base that is narrow and biased, and an unsystematic construction of the mental model is not derived from the content or the specific role; it is apparently a behavioral norm acquired during the physician's training. The research data do not indicate a significant difference between those who are experts in these fields and those who are not. One conclusion is the need to take this drawback into consideration during the physician's clinical training.

In collecting data, the physician does not take enough advantage of the nurse's contributions nor of the information he/she collects. The physician may unconsciously compensate for his/her lack of firsthand unbiased information, with the help of the nurse and his/her emphasis on the clinical situation; to use her as an "anchor" and systematic complement enabling the physician to better organize and process the information. The responsibility for the discrepancy between the physician and the nurse lies with both parties. The background for this has been discussed above. Therefore, the solution lies in increasing the physician's awareness of the nurse's contribution with respect to the nurse's ongoing relationship with the patient and familiarity with the patient's status, and the importance of the objective data the nurse has collected, as compensation for the shortcomings of the physician's intuitive thinking process. On the other hand, the nurse should have a better understanding of the nature of information the physician requires, and what should be emphasized and pointed out to him/her. These are the much needed yet lacking anchor points in the thinking process that can support the expansion of the presented information during the physician's preliminary processing of the information. Such a solution may be realized, for example, in the context of the nurse's formal participation in the doctor's rounds. The rounds are in need of an established procedure for a preliminary and up-to-date report from the nurse. Participation in the physicians' rounds will help the nurse recognize which points in the information system he/she must locate and emphasize to the physician. This procedure would also provide the physician with a better foundation from which to process the additional information in the future. This is just one solution among many. Our intention is not to attempt to suggest solutions, but the situation described above exposes a possible source for execution errors: the disparate thinking processes of the physicians and the nurses, which impairs the quality of communication between them. This provides an opening from which it may be possible to identify suitable procedural solutions, develop more appropriate information systems, and possibly even make changes in the thinking processes and the physicians' and nurses' perception regarding each other's role, which should be clarified to both professionals during their initial training. Such solutions also require a careful examination both of the formal structures and of content in the unique environment of the intensive care unit.

REFERENCES

1. Benbassat, J., Margolis, C.Z., and Elstein, A.S. (1992). Intellectual, psychological and social roots of the incompatibility of analytic and intuitive clinical decision making. Unpublished paper, Israel: Ben-Gurion University.
2. Benbassat, J. and Bachar-Bassan, E. (1984). A comparison of initial diagnostic hypotheses of medical students and internists. *Journal of Medical Education*, 59, pp. 951–956.
3. Craik, K. (1943). *The Nature of Explanation*. Cambridge, MA: Cambridge Univ. Press.

4. Dreyfus, H.L. and Dreyfus, S.E. (1986). *Mind over Machine*. New York: The Free Press, pp. 16–51.
5. Evans, D.A. (1989). Issues of cognitive science medicine. In: Evans, D.A., Patel, V.L., (eds.), *Cognitive Science in Medicine. A Biomedical Modeling*. Cambridge, MA: MIT Press, pp. 1–19.
6. Gentner, D. (1983). Structure mapping: A theoretical framework. *Cognitive Science*, 7, 155–170.
7. Gentner, D. and Grudin, J. (1985). The evolution of mental metaphors, A 90-year retrospective. *American Psychologist*, 40 (2), 181–192.
8. Gopher, D., Badihi, Y., and Donchin, Y. (1994). Knowledge compilation of doctors and nurses on their patients and its relationship to errors. *Proceedings of the 12th Triennial Congress of the International Ergonomics Association*, Vol. V, 57–59.
9. Gopher, D., Donchin, Y., et al. (1990). *Characteristics of the Activity and the Analysis of Human Errors in the Intensive Care Unit*. Technion: Haifa Research Center for Safety at Work and Human Engineering.
10. Johnson-Laird, P.N. (1982). Mental models in the cognitive science. In: Norman, D.A. (ed.), *Perspective on Cognitive Science*. NJ: Albex Pub. Corp., pp. 147–191.
11. Johnson-Laird, P.N. (1983). *Mental Models*. Cambridge: Harvard Univ. Press, pp. 126–147, 396–447.
12. Johnson-Laird, P.N. (1988). How is meaning mentally represented? *International Social Science Journal*, 40, 45–62.
13. Kassirer, J.P. (1989). Diagnostic reasoning. *Annals of Internal Medicine*, 110, 893–900.
14. Kassirer, J.P. and Gorry, G.A. (1978). Clinical problem solving: A behavioral analysis. *Annals of Internal Medicine*, 89, 245–255.
15. Millward, R.B. (1985). Mind your (mental) models. *Psycholinguistics Research*, 14, 5, 427–446.
16. Moskowitz, A.J., Kuipers, B.J., and Kassirer, J.P. (1988). Dealing with uncertainty, risks and trade-offs: Clinical decisions, *Annals of Internal Medicine*, 108, 435–449.
17. Pylyshyn, Z.W. (1980). Computation and cognition, issues in the foundations of cognitive science. *Behavioral and Brain Sciences*, 3, 111–169.
18. Pylyshyn, Z.W. (1985). *Computation and Cognition*. Cambridge, MA: MIT Press, pp. 23–101, 193–285.
19. Pylyshyn, Z.W. (1980). Cognitive representation and the process-architecture distinction. *Behavioral and Brain Sciences*, 3(1), 154–169.
20. Reason, J. (1990). *Human Error*. New York: Cambridge University Press, pp. 1–53.
21. Smith, M., Morris, P.E., Levy, P., and Ellis, A.W. (1987). *Cognition in Action*. Hillsdale, NJ: Lawrence Erlbaum, pp. 47–69, 285–311.
22. Young, R.M. (1983). Surrogates on mapping: Two kinds of conceptual models for interactive devices. In: Gentner, D., Stevens, A.L. (eds.), *Mental Models*. Hillsdale, NJ: Lawrence Erlbaum, pp. 35–52.
23. Sanderson, P.M. (1995). *MacSHAPA Manual: Version 1.0.2*., University of Illinois at Urbana–Champaign, Urbana, IL: Department of Mechanical and Industrial Engineering.

14 The Operating Room Briefing

Yael Einav, Yoel Donchin, and Daniel Gopher*

CONTENTS

Why a Briefing? 208
The Development of the Briefing 210
The Briefing 210
Examining the Impact of the Briefing on the Number of Nonroutine Events
during Surgery 212
Summary 215
References 215

Anyone traveling on a passenger flight, even a frequent flyer, is required to watch a short movie on safety before takeoff, which includes instructions on how to tighten the seat belt and various emergency procedures. Before each flight, pilots, even those who have completed countless flights, will discuss the flight plan, routine or not, and will subsequently check the instructions in the procedure booklet and review the procedures they are to perform, step by step. Fighter pilots receive training before each and every mission and upon return analyze the incidents that occurred and report any system deficiencies. In more sophisticated systems, such as when launching a spacecraft, dozens of National Aeronautic and Space Administration (NASA) control stations operate essentially as a single unit and each station is aware of the information that each of the other stations have. Flight inspections are performed at several centers and the transfer of information between them is referred to as a *Hand Out* (literally: distribution).

The high frequency of errors that occur in the operating room (4.5 nonroutine events, on average; see Chapter 4), are partially as a result of the feeling of omnipotence among the people involved. Eventually, the high rates of errors became an epidemic prompting even the greatest skeptics to seek help from human factors engineers and experts in human behavior. It can be said that the progress in medical

---

* Yael Einav: doctoral dissertation submitted to the Center for Work Safety and Human Engineering, Technion, Haifa. Under the supervision of Daniel Gopher and Yoel Donchin. This observation was conducted in the operating rooms of the Hadassah Medical Center, Ein Kerem, Jerusalem, in cooperation with the nurses in the gynecology operating room and the orthopedics operating room (Orna Ben Yosef and Margaret Lun), and with Yitzchak Karah, the Director of Nursing Services of the Operating Rooms Division. Neri Laufer and Prof. Iri Leibergal also assisted. The posters were designed by Avital Zik.

technology does not take into account the limits of human ability. Our observations in the operating rooms (see Chapter 4) presented the vulnerabilities within the operating room system. The system was shown to be prone to many gaps, which are essentially "asking for error" and in this context even a small mistake can result in disaster.

In another study[1] researchers found that 50% of the information transmitted by staff members in the operating room is conveyed either too late or prematurely (relative to when the surgery is scheduled to take place), 36% of the information is incomplete or inaccurate, and in 21% of the times a key participant is absent. Thirty percent of the shared exchanges that occur in the operating room result in situations that place the patient at risk, increase the mental (cognitive) load, interfere with the routine, or increase the level of tension in the room.

One of the (cognitive) interpretations of the Technion researchers' findings is that each of the staff members in the operating room lack situational awareness and that the surgical team as a whole lacks a shared situational awareness. Situational awareness is a concept that refers to how well a given situation in a specific system is understood, and the ability to anticipate events that will occur in the system in the near future.[2] Studies have shown that *shared* situational awareness allows each staff member to be aware of what to expect from the other members and to coordinate their actions better.[3] In fact, a high level of situational awareness enables team members to utilize their expectations and existing models to propose solutions quickly and efficiently. Studies have shown that an accurate *understanding of the situation* is critical to making the right decision; the key determinant of the quality of the decision is a mental model of the situation—whether the model is correct or incorrect.

The surgical process involves a wealth of information, including crucial information that may influence the surgical procedure that staff members have to be cognizant of and take into consideration. Staff members are aware of the importance of the information, but its retrieval from various sources is very time-consuming.

In many cases, the Technion researchers observed that the staff began to work even in the absence of critical information, that is, without a complete, shared situational awareness. Such situations force staff members to make difficult, nearly impossible decisions, such as whether to cancel a surgery when the anesthetist notices that the blood test results of the patient who is already on the operating table, are missing. Or how to position the patient on the operating room table when the information can only be provided by the responsible surgeon, but he/she is not present in the room.

In an attempt to bridge the information gaps and improve the efficiency and safety of the operating room staff's work, we have developed a *staff briefing* (in collaboration with medical personnel who work in Hadassah Ein Kerem Hospital operating rooms), implemented it in 102 gynecological and orthopedic operations, and examined its effect on teamwork and patient safety.

WHY A BRIEFING?

A briefing is an activity aimed at providing information about a defined staff task. A briefing may be used to provide information, assign responsibilities, emphasize potential problems and risks, provide background information, clarify information,

The Operating Room Briefing

and provide guidance and instructions. A time-dependent briefing is transmitted immediately before the activity. Klein showed that a briefing by a supervisor prior to a team activity provides the members with the flexibility they need to adjust to changing circumstances during the course of the task.[4] Orasanu demonstrated that the ability of a team to improvise during the course of a mission depends on the information provided by the commander of the team *prior* to the mission.[5]

How does the briefing boost staff performance? A briefing creates expectations. People's expectations related to a certain situation directs their attention (which is always limited) to those elements in their environment that are expected to be the most useful for them. Contextual information that a person obtains before performing a task will affect how he/she copes with conflicting and unpredictable information during the course of the task.

Furthermore, the briefing provides a review of the information and enables it to be discussed, which strengthens the formation of associations between the various details. Human memory is constructed in such a way that all the pieces of information studied at any point throughout our lifetime is stored in our mind (long-term memory). We do not reflect on all the individual items of information we possess at every given moment; however, we can at any given moment, access immediate memory (working memory) and consider the details individually or simultaneously, more or less. What determines which details are in our memory and accessible to us? Our brain works such that every stimulus we perceive in the environment (see, hear, smell, etc.) forms associations in our memories with other (event) information that already exist, held in our memory and are thus *easier to retrieve*. That is, if someone reads the word *elephant*, it will raise memories of experiences, pictures, and memories related to an elephant in the reader's mind. All these details will surface in the reader's working memory and become more readily available. When we present certain pieces of information in a briefing, we not only make this information more readily available in the memories of those present at the briefing, but also any other associated information. The mind of each staff member contains hundreds of thousands of pieces of information that may be important during surgery. The briefing is like a beam of light that illuminates the part of a person's brain in which the most relevant information is apparently located. It directs the person to the place where the most relevant information can be most likely found. A number of studies have indicated that presenting task objectives and background data (even if not directly related to the task) prior to a task, contributes to a more accurate perception of stimuli in the environment.

Medical staff members may claim that the information is available to them even without a briefing and that they make sure to update themselves by reviewing the information in a patient's chart before entering the operating room. We will not discuss whether this is true or the percentage of cases in which this is true; suffice it to say that people who search for information tend to use strategies that are biased and irrational.[6–8] Some of these biases have been investigated in depth, such as *confirmation bias*: the tendency to prefer information that supports existing assumptions and beliefs; a *fallacy* in which a person feels he/she knows more than he/she actually does; *overconfidence effect*[9]; and biases based on the availability of information, and so forth. These biases cause a person to be unreliable in the search for information.

Therefore, there is no basis for assuming that each staff member searches for information in an efficient manner.

THE DEVELOPMENT OF THE BRIEFING

The study was conducted in two stages. In the first stage, the content and format of the briefing were developed. In the second stage, its effectiveness was examined by comparing what occurs during surgeries conducted without a prior briefing and those conducted after a briefing.

The first stage included interviews with staff members (nurses, anesthesiologists, and surgeons) and observations in the operating rooms. The purpose of the interviews was to become familiar with the surgical procedure and determine the information that is transmitted between the staff members and that is vital to the proper execution of the operation. At this point, the staff members were given questionnaires and asked to indicate the details that they felt would be important to include in the briefing, were it to be incorporated into common work practice. The problems characterizing the teamwork in the operating room were likewise investigated.

In the second stage of the study, observations were conducted in the operating room, as previously described in the study presented in Chapter 4. The purpose of the observations was to create a continuous recording of the events that take place in the operating room and detect failures and vulnerabilities. The observations formed the basis for identifying nonroutine events—the dependent variable in this study, which would affect the level of patient safety and efficiency of the work staff. These events were defined as those that reflect a compromise in work safety, inefficiencies in the process, or failures in communication or coordinated action. An irregular nonroutine event does not always result in a detectable accident or a mistake—it is just one of the many circumstances that may lead to error. In contrast to the study described in Chapter 4, this research did not examine procedural events, but focused only on events related to safety that could develop into a malfunction that would harm the patient if not corrected in time.

The observation started upon the patient's entry into the operating room and concluded with his/her transfer to the recovery room. The observers were directed to continuously record all the events in the operating room—conversations, activities, the entrance and exit of staff members, requests, and arguments (all the entries were anonymous); names of the team members and the surgeon were not recorded. The research team analyzed the raw observation narratives to generate a list of nonroutine events. The following is a comparison of the number of nonroutine events observed in the sample of surgeries performed without a briefing (130 surgeries) to the number of irregular events observed in the sample of surgeries that were conducted after a briefing (102 surgeries).

THE BRIEFING

The briefing that was developed encompasses three parts. Each part is to be verbally conveyed by the three central members of the operating team—nurse, anesthetist,

The Operating Room Briefing

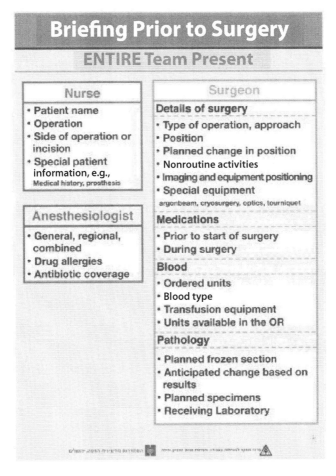

FIGURE 14.1 (See color insert.) The presurgery briefing poster. (From Einav Y., Gopher D., Kara I., et al., 2010, Preoperative Briefing in the Operating Room: Shared Cognition, Team Work, and Patient Safety, *Chest*, 133, 443–449.[10])

and surgeon—before surgery (see Figure 14.1). Each part of the briefing details information that a team member must explain to all team members involved in the operation. The information chosen to be included in the briefing must conform to at least one of two stipulations: it must be essential for the identification of the patient and of the surgery (the patient's name, type of surgery), or contribute to the formation of situational awareness and an accurate mental model. For example, the nurse should specify if the patient suffers from comorbid diseases and the surgeon should describe the surgical approach. In other words, this type of information refers to details that may not be perceived as contributing to the safety of the surgery or its efficacy, but may improve the individual or team work by influencing the formation of a shared situational awareness, thereby reducing the potential for errors. These pieces of information may help to delineate a clearer picture of the patient's clinical condition and of the activities that are expected to occur, and may therefore be useful in uncovering and processing information during the surgery, and increase the readiness of the staff to perform their required tasks.

The briefing poster. The format of the briefing is as follows: the participating team members all gather around the patient's operating table immediately upon being brought into the operating room or at least before the induction of anesthesia. During the briefing all other activities are stopped and the team members devote complete attention to the speakers. The briefing is designed to be similar to a large poster hanging in the operating room, and the staff members—the nurse, anesthetist, and surgeon—specify what is written on the poster and in the order determined by them (the order in which the information is presented is not dictated to those giving the briefing). The briefing is not documented, there is no need to record any of the items or hold a briefing form. The hanging poster is the only means applied for the briefing.

EXAMINING THE IMPACT OF THE BRIEFING ON THE NUMBER OF NONROUTINE EVENTS DURING SURGERY

Observations of 232 surgeries were conducted and 474 nonroutine events were documented. A significant 25% reduction in the number of nonroutine events occurred in the sample of surgeries conducted after a briefing, in contrast to the number of events in the sample of the surgeries conducted without a briefing (see Figure 14.2). In the sample of the surgeries conducted after a briefing, a 16% increase was found in the number of surgeries conducted without any events occurring at all, and a 5% and 11% decrease was found in the number of surgeries in which 1–2 nonroutine events and 3 or more nonroutine events, respectively, were documented.

The greatest decrease in nonroutine events was related to teamwork (a lack of coordination or of communication).

Another study finding was that the distribution of the various nonroutine events found in both departments was similar. Nonroutine events were classified

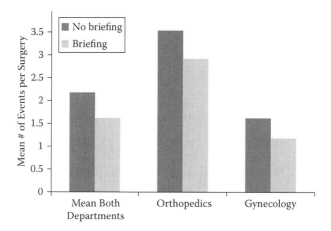

FIGURE 14.2 Average number of nonroutine events per surgery in the two operating theaters.

The Operating Room Briefing 213

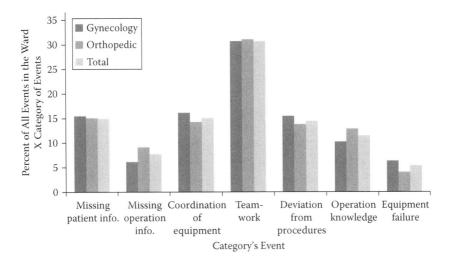

FIGURE 14.3 Frequency of nonroutine events in the different categories.

into seven categories: a lack of information about the patient, a lack of information about the operation, a problem in coordination of the equipment (preparation, operation), a problem involving teamwork, noncompliance with procedures, a problem in operational knowledge, and equipment failure. This classification was determined independently by three judges after reviewing all the nonroutine events from the observation recordings. Each judge compiled a list of categories after which all the judges came to an agreement with respect to a uniform system of classification that encompassed the seven categories (see Figure 14.3).

Figure 14.3 illustrates how the nonroutine events are distributed across the categories in both departments similarly. This finding strengthens the argument that the current problems in the surgical process are systemic, since the surgeries conducted in the gynecological and orthopedic departments are different—performed by different teams with different equipment and address medical problems that are quite different. Nevertheless, the distribution into categories is identical.

Figure 14.3 shows that the most frequent category of nonroutine events is that of *teamwork*. This finding is consistent with the findings of the researchers in Chapter 4 and the accepted findings worldwide in the study of teamwork in the operating room.

The study found high variability in the way in which the briefings were managed by the different teams. There were briefings that were conducted in a superficial manner merely in order to fulfill their obligation (for example, the surgeon conducted the briefing alone and concluded it by saying "the patient is healthy and young, we know what we're doing here, there is nothing to add"); and there were better briefings conducted according to protocol (the team reviewed all the relevant information and the briefing was presented by the three main staff members—a nurse, an anesthetist, and a surgeon).

To determine the effect of the quality of the briefings, a minimum list of details that must be included in the briefing was decided on. A briefing in which at least 40%

of the minimum information was presented was considered to be a "good" briefing and one that includes less than 40% of the information was considered to be inadequate. It was found that 65% of all the briefings conducted in the study were good briefings. The greatest decrease in the number of nonroutine events occurred in the gynecological department. In comparing the two study samples—those in which the surgeries were performed without a briefing at all and those performed after a good briefing—it was determined that 37% fewer nonroutine events were recorded in the sample that had included a good briefing.

No correlation was found between the duration of a surgical procedure and the number of nonroutine events that occurred during the procedure. However, an interesting trend was found: in the surgeries performed in the gynecology department, in which the majority of surgeries are relatively short (less than an hour), the greatest effect was found in the shorter procedures. In contrast, among the orthopedic surgeries, which are generally longer (more than 1 hour), results indicated that the longer the surgery, the more effective the briefing was. This trend indicates the possibility that the briefings with the greatest effect are those given prior to the more common surgeries.

The staff members who participated in the briefings were asked to express their opinions on the degree to which the briefing contributed to their individual work, teamwork, and patient safety. All the staff members (nurses, anesthesiologists, and surgeons) gave high ratings regarding all three of these criteria (4 or more on a scale of 1–5).

After the study was concluded, the teams began to conduct briefings as part of their daily routine, even without the presence of observers in the operating room, and as of today the briefing is an accepted routine (Photo 14.1).

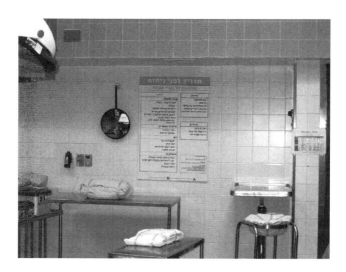

PHOTO 14.1 The poster on a visible wall of the operating room.

SUMMARY

This chapter describes a study that proposed a solution for a problem and examined its effectiveness. The solution presented is easy to implement. In light of its favorable properties it has since been implemented in the field. We have shown that the operating room medical team work under conditions that endanger both them and their patients. The approach taken was preventive rather than punitive, although we were careful to ensure that the administration of the briefing will not be recorded to prevent the staff from perceiving the study as an attempt to find the "guilty" parties. The preventative approach actually succeeded in securing the full cooperation of the medical staff and contributed to the implementation of the solution proposed, that is, the briefing. The medical staff assessed this new approach and found that it contributed to patient safety and to its work as a team. Most importantly, we have shown that medical teams, as opposed to the prevailing opinion, are willing to accept a change in behavior in order to try and improve their work.

This study also indicates that checklists and procedures that are not burdensome are appreciated by responsible medical teams.

REFERENCES

1. Lingard, L., Espin, S., Rubin, B., Whyte, S., Colmenares, M., Baker, G.R., et al. 2006. Getting teams to talk: Development and pilot implementation of a checklist to promote interprofessional communication in the OR. *Qual Saf Health Care*, 14(5): 340–346.
2. Endsley, M.R. 1995. Toward a theory of situation awareness in dynamic systems. *Human Factors,* 37(1): 32–64.
3. Salas, E., Prince, C., Baker, D.P., and Shrestha, L. 1995. Situation awareness in team performance: Implications for measurements and training. *Human Factors*, 37(1): 123–136.
4. Klein, G.A., Calderwood, R., and Clinton-Cirocco, A. 1986. Rapid decision making on the fireground, *Proceedings of Human Factors and Ergonomics Society 30th Annual Meeting*, Dayton, OH.
5. Orasanu, J.M. and Fischer, U. 1998. Information transfer and crew performance, *Proceedings of the 6th International Symposium on Aviation Psychology*,w Columbus, OH.
6. Bjork, R.A. 1996. Assessing our own competence: Heuristics and illusions. In: Gopher, D. and Koriat, A. (eds.), *Attention and Performance XVII*. Cambridge, MA: MIT Press.
7. Chen, J.Q. and Lee, S.M. 2003. An exploratory cognitive DSS for strategic decision making. *Decision Making Systems*, 36: 147–160.
8. Tversky, A. and Kahneman, D. 1974. Judgment under uncertainty: Heuristics and biases. *Science*, 185: 1124–1131.
9. Fichhoff, B., Slovic, P., and Lichtenstein, S. 1977. Knowing with certainty: The appropriateness of extreme confidence. *Journal of Experimental Psychology: Human Perception and Performance*, 3: 552–564.
10. Einav, Y., Gopher, D., Kara, I., Ben-Yosef, O., Lawn, M., Laufer, N., Liebergall, M., and Donchin, Y. 2010. Preoperative briefing in the operating room: Shared cognition, team work, and patient safety. *Chest*, 137, 443–449.

15 Analysis of the Rate of Interruptions during Physician Rounds

*Yoel Donchin and Meirav Fogel**

CONTENTS

The Relationship between Interference and Errors 217
Physician Rounds in the Internal Medicine Department 218
    Observation as a Research Tool 218
    The Purpose of the Study 219
    Methods and Materials 219
        The Study Population 219
        Sample 219
        Variables 220
    Examples for the Categories of Interference and Their Instigating Sources 220
        Statistical Analysis 220
    Ethical Aspects 221
    Results 221
    The Experimental Population 221
    The Number of Interruptions 222
References 225

THE RELATIONSHIP BETWEEN INTERFERENCE AND ERRORS

In a complex environment, such as that of a hospital, dividing one's attention/multitasking is one of the most important factors that affects performance.[7] A person who frequently shifts his/her attention from one subject to another cannot get a clear picture of the situation around him/her.[7]

    In many areas of work, interruptions in the workplace are known to cause errors. For example, interruptions and distractions are one of the most common causes of pilot errors. In fact, in a study evaluating reports of accidents attributed to crew member error, approximately half the accidents were related to miscommunications due to interruptions, distractions, or involvement in one task at the expense of another.[8] Within the industrial sector, much effort has been invested in researching workplace design; equipment and physical environment that is suitable for the

---

* Meirav Fogel, Yoel Donchin (MD thesis, Hebrew University and Hadassah).

employees in accordance with labor requirements. In contrast, in the medical professions only a number of studies, primarily relating to activities in operating rooms, have been conducted on the rate of interruptions.[8] Physicians and nurses are often interrupted during their work.[11] A study found that emergency room physicians are interrupted at an average rate of 10.3 times an hour.[11] Other studies have found that interruptions and distractions disrupt operating rooms procedures.[12,13] In observations conducted in urological operating rooms, an average of 27 interruptions per hour was found.[14] Furthermore, interruptions are the main causes of errors in the distribution of drugs.[15,16]

Interruptions and distractions at the workplace are a source of work overload, lack of efficiency, increased errors, stress, and employee dissatisfaction.[17] Pressure is also caused by unpredictable or uncontrollable environmental conditions.[18] A study performed in England showed that "constant interruptions at home and at work" was one of the four most influential workday stress factors.

Interruptions do not only affect physicians but also nurses and patients. It was found that work overload and interruptions were a major cause of stress among nurses in the community.[20] As for patients: about 20% of patients surveyed felt that interruptions have a negative impact on medical consultations, 40% felt that it would be better if there were no interruptions, and 18% expressed a negative opinion about interruptions.[21] These studies show that interruptions may be one of the reasons for the stress and dissatisfaction felt by physicians, nursing staff, and patients.

PHYSICIAN ROUNDS IN THE INTERNAL MEDICINE DEPARTMENT

Department rounds represent a focal point for the transfer of information and decision making. It is in the context of the rounds that the staff meets the patient and develops a treatment plan together. The purpose of the physicians' rounds is to receive and process information from the patient and the staff. The rounds are valuable for the purpose of instruction when students participate, and in the past, the discussions regarding diagnosis and treatment were the highlight of the Internal Medicine Department's work. The rounds bring the staff to the bedside of the patient[22] and enable a dialogue between them.

In the past, the physician rounds were conducted with awe and respect, while the department director who led the team granted permission to speak and summarized what was said. The nurses were partners in the rounds. An interruption during the course of the rounds was almost impossible.[23] However, in recent years the rounds have become more *permissive* and do not enjoy the same hallowed aura as in the past.

OBSERVATION AS A RESEARCH TOOL

Observation is an effective method for familiarizing oneself with the goings-on in a department. It is preferable to examining discrete nonroutine events that have already occurred or waiting until a malfunction occurs—alternatives that do not afford preventive measures.[24,25]

In structured observation studies the researcher joins the group being studied and documents its activities systematically. Once the information gathering process is completed, the findings are analyzed through various means. Observations complement existing reporting systems in the documentation of nonroutine events.[27]

THE PURPOSE OF THE STUDY

The purpose of this study was to quantify and characterize, by way of observation, interruptions that the medical staff in the department of internal medicine must cope with during morning rounds, and to utilize this information to develop methods to prevent or reduce the number of errors.

METHODS AND MATERIALS

This is a descriptive observational prospective study (cohort), based on observations on the physician rounds in the Internal Medicine Department. During each observation the interruptions that occurred during the rounds were documented, including the source, duration, and intensity of the disturbance and whether it caused the rounds to be halted. An interruption was defined as *something that prevents the physician from participating in the rounds or to focus on something outside the boundaries of the original task.*

The observations were conducted by three individuals: the person conducting the study, the present coauthor, who was a medical student conducting her thesis work, and the hospital research nurse (who was not a member of the internal department staff). In preparation for the data collection stage, pilot observation studies were conducted, attended also by the supervisor of the pilot study, an expert on safety and human factors. During these pilot observations, each observer collected data separately, after which the data of all three observers were compared and rules were determined for the manner of documentation and its scope.

The Study Population

The participants included the physicians who take part in some of the morning rounds at the Internal Medicine Departments of Hadassah University Hospital, Ein Kerem, Jerusalem. Those present at each of the rounds included the senior physician, interns, nurses, students, and patients (and sometimes their families). The influence of the interruptions was examined by the attending physicians and the specialists only.

Sample

The sample was random and encompassed 24 *grand* rounds in several different Internal Medicine Departments in the hospital. In each round, approximately half of the hospitalized patients in the department (about 15 patients) were examined. The estimate of sample size was based on the expected correlation between the number of interruptions and the number of participants attending the rounds. Assuming a 5% level of significance (one way) and a power of 80%, it would require 24 observations to prove a significantly different correlation coefficient between the number of participants and the number of interruptions (0.8), at correlation coefficient of 0.5.

The sample included rounds conducted by different physicians in different departments to minimize the risk of biasing the sample results. Each sample began when the attending physician joined the round and ended with his/her departure.

Variables

The main variable—interruptions during the rounds.
Each disturbance was characterized according to:

1. Source of the interruption.
2. Duration of the interruption.
3. Intensity of the interruption.

In order to document the interruptions systematically, a coding scheme was developed. The accumulated data was recorded in tables and tested statistically.

EXAMPLES FOR THE CATEGORIES OF INTERFERENCE AND THEIR INSTIGATING SOURCES

Patients	Dementia patient shouts. One patient asks physician a question while the rounds are around another bed. A patient has tried to reach his bed among the round participants.
Family and visitors	Talking: family member asks the physician about his father—the round is not on his father.
Physicians converse	Talking about other patients. Decision is made on a patient in the next room.
Conversation—other than physicians	Nurses, students talk!
Telephone calls	Private calls, calls regarding other patients, phone rang, even if not answered.
Background noise	Respirator alarm, noise from vacuum cleaner, conversation from other rooms.
Other workers	Technicians, physiotherapist.
In and out from the room	Patient, physicians, nurses, food service.

Each interruption was rated on a scale that ranged from 1 to 3 (Table 15.1) according to the reactions it elicited from the staff and the physicians:

1. A stimulus that is liable to be distracting.
2. A stimulus that distracted one of the physicians participating in the rounds and almost caused the rounds to be halted.
3. A stimulus that halted the rounds and became the focus (disturbance) of all the physicians.

Statistical Analysis

Data collection lasted 7 months and the observations were recorded on an Observation Form. The data collection, statistical processing, and diagrams were prepared through

TABLE 15.1
Statistical Summary of the Interruptions Observed over the Entire Sample (All 24 Rounds)

	Number of Interruptions per Round	Number of Interruptions per Hour	Number of Interruptions That Led to the Rounds Being Halted	Duration of the Interruptions That Led to the Rounds Being Halted	Relative Percent of the Time Spent on Interruptions in Relation to the Total Time Period of the Rounds
Mean	79.083	41.760	9.042	07:53	6.63%
Standard deviation	27.42	10.8	5.63	0:06	4.56%
Maximum	132	70.400	23	27:40	18.20%
Minimum	20	22.286	0	00:00	0.00%

the use of Microsoft® Excel 2007 and SPSS® 12.0 for Windows®. The variables for the mean, standard deviation, minimum and maximum values were calculated. Pearson's linear correlation coefficient was employed for the purpose of comparing the variables. Data with a p value ≤5% was considered statistically significant.

ETHICAL ASPECTS

The data collected did not include any medical information or other information about the patients in the medical ward. Prior to conducting the experiment, consent for the study was obtained from the Internal Medicine Department division manager to perform the observations. The researchers were also exempted from the need to obtain the authorization of the Institutional Helsinki Committee.

RESULTS

THE EXPERIMENTAL POPULATION

As mentioned previously, observations were conducted during 24 grand rounds in three different Departments of Internal Medicine in Hadassah Ein Kerem hospital: eight were in Internal Medicine A, eight in Internal Medicine B, and eight in Internal Medicine C. Two to eight physicians participated in each of the rounds, with an average of 4.46 (± 0.68) physicians per round. Two to twelve professionals participated in each round, including physicians, nurses, students, nutritionists, social workers, physical therapists, and so forth, with an average of 6.78 (± 1.22) participants per round. Fourteen of the rounds were conducted by the department head and 10 visits were conducted by a specialist with a "senior position" in the Internal Medicine Department. Each of the rounds lasted from 38–165 minutes, averaging 114.42 (± 32.68) minutes per round.

THE NUMBER OF INTERRUPTIONS

Table 15.2 presents the statistical summary of the disturbances observed. During the morning rounds, on average, an interruption occurred every 90 seconds. During the rounds in which the greatest number of interruptions occurred, one interruption occurred every 0.85 minutes, on average. Approximately 12% of the disturbances caused a halt in the rounds. As a result of the interruptions, an average of 6.63% of the total time period of the rounds was expended on these disturbances.

The most frequent causes of an interruption were movement of people to and from the room (entry and exit) and physicians' discussions that were unrelated to the patient. The causes of interruptions that were less frequent were those that originated from the patient's family members, phone calls made by people present at the round who are not physicians, and background noise.

Most of the interruptions that caused the round to be halted were conversations involving physicians, conversations that were not related to the patient in question, and physicians' phone calls. Interruptions that rarely resulted in the rounds being halted were phone calls of people present at the round who were not doctors, and background noise.

No correlation was found between the duration of the round and the number of interruptions per hour (Pearson correlation = −0.098) or between the duration of the round and the number of interruptions that caused the round to halt (Pearson correlation = −0.091) (Figures 15.1 and 15.2).

TABLE 15.2
Frequency of Interference by Source

Source of the Interruption	The Average Number of Interruptions per Round (A)	The Sum of the Average Intensities of Interruptions per Round (B)	The Relationship between the Sum of the Average Intensities and the Average Number of Interruptions per Round (B/A)
Movement of people to and from the room	38.54	57.38	1.47
Conversations between physicians	10.88	25.71	2.39
Physicians' telephone calls	5.92	13.75	2.29
Patient	4.83	6.63	1.37
Conversations between people who are not physicians	4.83	5.17	1.06
Activity of a paramedical professional in the room	4.04	4.96	1.21
Background noise	3.62	3.83	1.05
Telephone calls by someone other than a physician	3.38	4.54	1.38
The patient's family	3.04	5.5	1.82

Analysis of the Rate of Interruptions during Physician Rounds

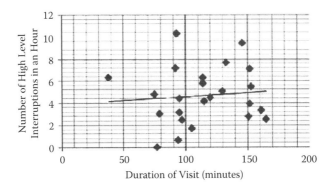

FIGURE 15.1 Number of high-level interferences per hour as a function of visit duration.

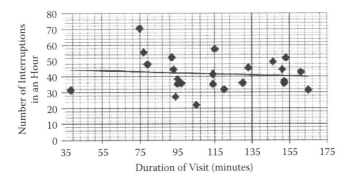

FIGURE 15.2 Total number of interferences per hour as a function of visit duration.

No correlation was found between the number of physicians in the rounds and the number of interruptions per hour (Pearson correlation = 0.37) or the number of interruptions that led to the rounds being halted (Pearson correlation = −0.09). No correlation was found between the number of participants in the rounds (including nurses, students) and the number of interruptions per hour (Pearson correlation = 0.20) or in the number of interruptions that led to the rounds being halted (Pearson correlation = −0.23).

No significant difference was found in the number of interruptions per hour, between the rounds in which the director of the department was the senior physician (42.76 interruptions per hour) and those in which the director of the department was not the senior physician (40.37 interruptions per hour).

One of the objectives of this study was to quantify the number of disturbances during the morning rounds in the Internal Medicine Department. This is the first step in understanding the rate at which physicians are interrupted during the rounds. The sources of the interruptions were examined as well as the reactions of the staff to those disturbances. In this study the relationship between the interruptions and avoidable medical errors was not examined. According to the results of the study, the frequency of interruptions during the physicians' rounds was about 42 per hour on average, or an interruption almost every 1.5 minutes. Some of the interruptions observed caused the round to be halted in order to focus on the cause of the interruption (6.6%).

By comparison, in a study conducted in the emergency room, it was observed that interruptions occurred every 6 minutes, and every 9 minutes the interruption forced the physician to stop the work he/she was doing.[11] In a study of the number of interruptions during urological operations, it was found that interruptions occurred every 2 minutes on average.[14]

These interruptions prevent the completion of thought sequences and/or relaying information during the rounds and do not allow the patient to express him/herself without being interrupted. In this manner, safety is being compromised.[28] The interruption, which adds to the other causes of system failure (lack of manpower, environmental noises), seriously disturb the physician's rounds.

The reactions of the medical staff to the interruptions were rated, as mentioned, on an "intensity of the disturbance" scale. The purpose of the rating was to provide a relative value for each interruption in order to get a better picture of the level of the staff's involvement. Knowing the number of interruptions that occur is important, but it does not indicate their actual influence on the medical staff. In contrast, a rating of interruption intensity indicates how much these interruptions interfere with the continuity of the physician's work during the rounds. This rating can help us see which of the interruptions lead to decreased work efficiency and find the elements that are essential to predicting the risks in patient treatment.

The results reveal that the frequency, intensity, and duration of interruptions vary according to the source of the interruption. Common sources of interruptions were movement of persons to and from the room and conversations between the physicians. In contrast, interruptions caused by the patient's family and telephone calls for nonphysician participants were rare.

We have shown the most frequent sources of interruptions that halted the rounds. For example, movements of people in and out of the room were very frequent, but effectively had a very limited impact on the medical staff and did not interfere with the continuity of the round. In contrast, conversations between physicians or physicians' phone calls were less frequent, but were rated as being more intense and did lead to many breaks in the rounds.

In trying to identify the factors that contribute to interruptions, neither the duration of the rounds nor the number of people participating were found to influence the number of interruptions per hour, whether they were minor interruptions or those that caused the rounds to be halted. The number of interruptions that were observed was also not related to the status of the senior doctor conducting the rounds. This may indicate that the sources of the interruptions relate to the nature of the rounds, the manner in which the department and the hospital function, and the mentality that typifies Israeli hospitals. Apparently, interruptions are such an integral part of the rounds that their rate would be the same whether two or twelve physicians were in charge of it, and whether or not students are present. The same goes for the duration of the rounds—the number of interruptions per hour during the shortest round was similar to the rate of interruptions documented during the longest round.

It should be noted that an interruption is not necessarily something negative, undesirable, or ineffective. Interruptions may include alerting physicians to an emergency or a worsening of a patient's condition, which requires immediate attention. Physicians must indisputably be informed about a deterioration in a patient's condition, even at the cost of interrupting the rounds. Many of the interruptions documented in the study were related to daily work routines. They are generally not related to urgent medical matters, but certainly relevant to work issues. It is likely that had these issues not been tended to during rounds, which are held during the work hours of most of the hospital staff, they most probably would have been postponed to another day. Physicians working in the Internal Medicine Department perform many tasks over the course of the day. However, in the interest of increasing safety, these tasks should not be allowed to intrude upon the rounds.

For example, breaks can be scheduled during the rounds to perform tasks that require immediate attention and the morning rounds can be started early so that there will be enough time to perform these tasks after the rounds are completed. In our opinion, as the awareness of the interruptions and their impact increases, the course of the rounds can be improved and better working conditions can be introduced.

There are many factors that can result in an error in the medical system, and interruptions are one of those factors. With respect to "human error," external factors, such as conflicting task demands or distractions caused by telephones or colleagues, may be implicated.[29]

The results of the study were brought to the attention of the senior doctors in the Internal Medicine Department.

Proposed solutions for reducing the number of interruptions will be delineated in consultation with all participants in the rounds. For example, it was suggested that cell phones be turned off during the entire time the rounds are taking place, except for short periods of time when the participants are transitioning from one room to another. Another suggested solution was to establish a culture of conversation and discussion and to bar, through "friendly" enforcement, participants from entering and exiting the rooms during the rounds. The effectiveness of these proposals will only be seen after they have been implemented.

REFERENCES

1. Etchells E. 2003. Patient safety in surgery: Error detection and prevention. *World J Surg.* August 27(8):936–41.
2. Donchin Y. 2002. Avoiding human errors in the hospital—Mission possible? *Harefuah* May 141(5):453.
3. Gaba D.M. 1989. Human error in anesthetic mishaps. *Int Anesthesiol Clin.* Fall 27(3):137–47.
4. Calland J.F., Guerlain S., Adams R.B., Tribble C.G., Foley E., and Chekan E.G. 2002. A systems approach to surgical safety. *Surg Endosc.* 16:1005–1014.
5. Reason J. 1995. Understanding adverse events: Human factors. *Qual Health Care* 4:80–89.
6. Wickens C.D., Cordon S.E., and Liu Y. 1998. *An Introduction to Human Factors Engineering.* New York, NY: Addison-Wesley (Longman Imprint).

7. Cook R.I. and Woods D.D. 1990. Operating at the sharp end: The complexity of human error. In: Venturino M, (ed.) *Selected Readings in Human Factors*. Santa Monica, CA: The Humans Factors Society 255–310.
8. Dismukes R.K., Young G.E., and Sumwalt R.L. 1998. Cockpit interruptions and distractions: Effective management requires a careful balancing act. *ASRS Directline* December:4–9.
9. Donchin Y., Gopher D., Olin M., Badihi Y., Biesky M., Sprung C.L., Pizov R., and Cotev S. 1995. A look into the nature and causes of human errors in the intensive care unit. *Crit Care Med*. February 23(2):294–300.
10. Brennan T.A., Localio A.R., Leape L.L., Laird N.M., Peterson L., Hiatt H.H., and Barnes B.A. 1990. Identification of adverse events occurring during hospitalization. A cross-sectional study of litigation, quality assurance, and medical records at two teaching hospitals. *Ann Intern Med*. February 1;112(3):221–6.
11. Chisholm C.D., Collison E.K., Nelson D.R., et al. 2000. Emergency department work place interruptions: Are emergency physicians "interrupt-driven" and "multitasking"? *Acad Emerg Med*. 7:1239–1243.
12. Carthey J., de Leval M.R., and Reason J.T. 2001. The human factor in cardiac surgery: Errors and near misses in a high technology medical domain. *Ann Thorac Surg*. 72:300–5.
13. Catchpole K.R., Giddings A.E., de Leval M.R., et al. 2006. Identification of systems failures in successful paediatric cardiac surgery. *Ergonomics* 49(5–6):567–88.
14. Healey A.N., Primus C.P., and Koutantji M. 2007. Quantifying distraction and interruption in urological surgery. *Qual Saf Health Care* April 16(2):135–9.
15. Peterson G.M., Wu M.S., Bergin J.K. 1999. Pharmacist's attitudes towards dispensing errors: Their causes and prevention. *J Clin Pharm Ther*. February 24(1):57–71.
16. Reason J. 2004. Beyond the organisational accident: The need for "error wisdom" on the frontline. *Qual Saf Health Care* 13:ii28–33.
17. Gladstone J. 1995. Drug administration errors: A study into the factors underlying the occurrence and reporting of drug errors in a district general hospital. *J Adv Nurs*. October 22(4):628–37.
18. Kirmeyer S.L. 1988. Coping with competing demands: Interruption and the type A pattern. *J Appl Psychol*. November 73(4):621–9.
19. Cooper C.L., Rout U., and Faragher B. 1989. Mental health, job satisfaction, and job stress among general practitioners. *BMJ* February 11;298(6670):366–70.
20. Leary J., Gallagher T., Carson J., Fagin L., Bartlett H., and Brown D. 1995. Stress and coping strategies in community psychiatric nurses: A Q-methodological study. *J Adv Nurs*. February 21(2):230–7.
21. Dearden A., Smithers M., and Thapar A. 1996. Interruptions during general practice consultations—The patients' view. *Fam Pract*. April 13(2):166–9.
22. Hodgson R., Jamal A., and Gayathri B. 2005. A survey of ward round practice. *Psychiatric Bulletin* 29:171.
23. Kirkpatrick J.N., Nash K., and Duffy T.P. 2005. Well rounded. *Arch Intern Med*. March 28:165(6):613–6.
24. Carthey J. 2003. The role of structured observational research in health care. *Qual Saf Health Care* 12 (Suppl 2):ii13–6.
25. Thomas E.J., Sexton J.B., and Helmreich R.L. 2004. Translating teamwork behaviours from aviation to healthcare: Development of behavioural markers for neonatal resuscitation. *Qual Saf Health Care* 13:i57–64.
26. Rogers S.O. Jr., Gawande A.A., Kwaan M., Puopolo A.L., Yoon C., Brennan T.A., and Studdert D.M. 2006. Analysis of surgical errors in closed malpractice claims at 4 liability insurers. *Surgery* July 140(1):25–33.

27. Singh H., Petersen L.A., and Thomas E.J. 2006. Understanding diagnostic errors in medicine: A lesson from aviation. *Qual Saf Health Care* 15:159–64.
28. Dismukes R.K., Loukopoulos L.D., and Jobe K.K. 2001. The challenges of managing concurrent and deferred tasks. In: R. Jensen (ed.), *Proceedings of the 11th International Symposium on Aviation Psychology*. Columbus, OH: Ohio State University.
29. Rasmussen J. 1994. Afterword. In: Bogner M.S. (ed.), *Human Error in Medicine*. Hillsdale, NJ: L. Erlbaum Associates, pp. 385–393.

16 How Does Risk Management Differ from Accident Prevention?

Yoel Donchin

CONTENTS

What Data Should Be Collected? 233

The first public conference on the subject of errors and accidents that occur in the medical system was held in 1996 in Rancho Mirage, California. The audience was composed of a small group of experts. They were surprised to see the parents of a 7-year-old child who had died after a routine operation to remove a benign tumor in his ear, come up to the stage. The parents were joined by the physicians who treated the child, the operating room nurse, the hospital's lawyers, representatives of the family's insurance company, and the family's lawyers. The discussion was moderated by a well-known TV personality. The story they told left the audience with goose bumps.

The deceased child was originally diagnosed with a cholesteatoma that endangered his hearing. The accepted treatment for such a condition is to remove it. Surgery was conducted using a microscope, which allows better visualization of the delicate structures. A local anesthetic, to which a low concentration of adrenaline was added, was injected into the surgical field to shrink the blood vessels and improve the surgeon's view of the surgical field. During the surgery, the anesthesiologist suddenly noticed a sharp rise in the child's blood pressure. He alerted the surgeon and the surgery was stopped, but resumed when the blood pressure returned to normal. Later, a significant rise in blood pressure occurred again, accompanied by an accelerated pulse. This time the blood pressure did not return to normal and the child began to show signs of hypoxia due to pulmonary edema. The surgery was stopped and the child was transferred to the intensive care unit, where it was diagnosed that he suffers from the results of brain anoxia: he died a few days later.

This case might have ended in the conventional way: the hospital management would tell parents that a higher power was responsible. All the documents had been properly filled out. The surgeon's and anesthetist's signatures were clear and in place. The child was properly prepped for the surgery; the parents had been warned that the child could incur permanent damage for an unforeseen reason. In terms of the hospital risk management system there was no basis for legal action by the family to blame the hospital for their child's death.

However, had this case proceeded in the conventional manner, eventually, another accident would have been inevitable.

However, this time, the situation unfolded in another direction: the room had not been cleaned and the case was not tagged and filed away as just another case of death due to drug hypersensitivity. This time, the hospital's *risk management* nurse took some actions outside the scope of her defined role. She went into the operating room, recorded the names of all who were present during the surgery and gathered all the remaining syringes and solutions from the operating table and nurses' trays. Acting like a police officer at the scene of a crime, she secured the area with a metaphoric yellow tape and collected evidence that could be used in court. Initially, the family had been told that there was no medical explanation for what had occurred; however, it was revealed that the syringe that the surgeon had used to inject the anesthetic did not contain a low concentration of adrenaline, as expected, but pure adrenaline. The injection of this drug into the blood would be expected to result in the same symptoms as were observed in the child. The hospital admitted liability.

It is easy to imagine what must have occurred during the surgery. Both drugs were clear fluids. After the nurse prepared the solutions, she placed them on unlabeled trays and, distracted for one fleeting moment, she drew the adrenaline up into the wrong syringe.

The story spread like wildfire across the globe. After this tragedy, an unlabeled drug would never again be found on an operating tray. The lesson was learned.

The nurse who took it upon herself to investigate the underlying cause of the fatal accident did not work as risk management, but as a safety officer trying to avoid the next accident.

The primary role of risk management is to protect the hospital from financial liability (although the lawyers for medical institutions would deny this vehemently).

Despite the partial and limited overlap between risk and safety management, more often than not, risk management actually encourages activities that defy safety considerations.

In nonhealth related institutions risk management is an integral part of the system. The identification of a potential risk stimulates the system to assess and investigate the situation. A country must examine the dangers it is likely to face (from drought to nuclear attack). A factory owner checks to see if his/her financial status will allow the factory to continue to operate, if sufficient means are available for production, and how to avoid the need to import spare parts from an unfriendly country.

The medical systems imitated this type of risk management and quality control, shaping it into a framework whose main purpose is the prevention of lawsuits by taking steps to deter patients and their lawyers from filing suit against the hospital for malpractice. In the event of a loss, the hospital might have to absorb exorbitant financial losses to cover the costs of compensatory damages and permanent disability. The risk management system in a hospital or clinic represents the legal department of the institution. If an external attorney asks to review a case, the system is automatically activated and an investigation ensues whereby all efforts are made to ensure that the hospital has made every effort to avoid legal liability. This process is comprised of verifying that all necessary documentation has been properly filled out and signed, including the informed consent forms signed by the patients so that they cannot later claim they were not properly apprised of all potential risks.

These departments also perform important preventive actions, such as verifying whether the physicians and nurses complied with clearly delineated administrative procedures and guidelines, whether all forms were signed and stamped, and so forth. Thus, if any deviation from such procedures occurred, the hospital administration is protected if the case goes to trial, and can prove that neither they nor the hospital are to blame but rather the specific person accused had operated contrary to their guidelines. When an accident is revealed, the risk management staff rushes to brief the medical staff on what to say and how to behave in case of an investigation. For example, the lawyers advise the accused client not to say anything, so as to not incriminate him/herself. This may seem a somewhat extreme portrayal of the situation, but it essentially reflects the risk management system in the hospital.

The number of lawyers in the hospital is greater than the number of safety personnel. In fact, there is a great deal of logic in dedicating so many resources for legal protection, as the court judge does not examine the entire system and delve into details such as *why did this or that physician write that?* Rather the judge relies on what he/she is shown and nothing more. The actions of risk management complement the parallel actions of medical quality control, whose purpose is to monitor how well the directives and guidelines are implemented: the rate of infection, strict adherence to hand washing practices, proper use of medications, handling specific situations according to clearly defined procedures, and so on.

The hospital's safety management system is based on the "Swiss Cheese" model: each layer of the system represents a different slice of cheese, while holes in the cheese, randomly distributed among each slice, represent system failures and vulnerabilities, anticipating an opportunity to become manifest. The slices of cheese, or layers of the system, that do not have "holes," represent the final barrier that prevent accidents from occurring.

For example, during the blood transfusion process, the blood type of the donor and the recipient are carefully tested and vital steps are taken to prevent a fatal error. But sometimes, as we have seen during the observations of the blood transfusion process, the unit of blood taken from the blood bank refrigerator for a particular patient is actually intended for use with a different patient. A young physician who is unfamiliar with the blood transfusion procedures may check the packet of blood but neglect to identify the patient, and thus the process deteriorates; the holes in the cheese enable individual mishaps to filter through slice after slice, until they amass into a serious accident.

The risk management system is only activated following the occurrence of a serious accident. Through a process similar to detective work, the system functions to investigate the process step by step, through the holes in the cheese, until it reaches a person who can be held liable or punished. And at this point the task of the risk manager is complete.

In contrast, the safety management personnel attempt to reveal latent flaws in the system and fix them before an accident occurs, in this case, to prevent the possibility of transfusing a patient with the wrong type of blood.

The actions of risk management are similar to those of the police who search for burglars and car thieves by gathering information and following the trail of evidence that lead them to the guilty parties and help bring them to trial. In comparison, the safety personnel attempt to provide sophisticated locks so that break-ins cannot occur at all. They retrace the steps involved in the proper performance of the blood transfusion process, and study and monitor them using methods employed by human engineering. In point of fact, as found in our studies, this is a long and cumbersome procedure, and adding another person to the chain of people already involved, for the supposed purpose of increasing safety, may actually lead to a reduction in the level of responsibility of all involved.

Safety personnel staff believe that mistakes and failures are systemic in nature, and that the blame should not be laid entirely on one individual who represents only one link in the chain. Instead, the entire system needs to be carefully examined in order to find its points of vulnerability: the flaws, the links where malfunction is likely to occur, the holes scattered across the cheese (system). The latent flaws in the system may be technical failures due to poor maintenance of equipment, or the use of high-risk drugs or devices; failures resulting from gaps in knowledge or an illogical situational awareness among the workers; or failures that are based on the interaction between the worker and his/her equipment.

When the examination of risk management personnel reaches the medical staff involved at the "sharp end"—everyone who was present during the mishap—and assign blame, for example, to the anesthesiologist for mixing the wrong medication—they point to the guilty party, and according to an ancient tradition which holds that edification constitutes punishment, fire the wrongdoer and place him/her on trial or make every effort to ensure that he/she will bear the brunt of the responsibility. However, they do not attempt to plug up the *hole in the cheese* or address the circumstances that allowed the problem to occur, and in fact enable failure to occur again. The individual who was punished might learn a lesson from the incident and probably will redouble his/her efforts to be cautious in the future. But unless the flaw in the system is addressed, after the employee is banished, a new victim will arrive on the scene who is just as liable to err as the previous employee was, and will suffer the same consequences when the malfunction eventually recurs.

For example, after investigating a series of mishaps in the hospital pathology department, it was discovered that the source of the problem was not among the pathology personnel, but the manner in which the pathology samples were labeled in the operating room. Labels that were attached to the patient sample were not stuck on properly and fell off, the form accompanying the tissue container was also not well attached, and some of the samples to be tested finally arrived at the lab without an identifying label. How could punishment possibly hope to solve such a problem?

The hospital operating rooms are run efficiently. Efficient room utilization is an important condition for functioning in the system without causing delays between one surgery and the next. Immediately after surgery is performed, a cleaning crew is summoned who is trained to scrub the room from top to bottom. Everything in the room related to the patient who was operated on is cleaned. If a mishap occurred, even one that is serious, the drive to continue on and follow the planned schedule of surgeries outweighs the need to preserve and document what took place in the room. In such cases when error does occur, saving the items would make it possible to discover the nature of the mishap—a bottle containing the wrong concentration of a drug, a magnesium solution that was accidentally placed on the anesthesia cart, and so on. The case of the 7-year-old child described at the beginning of this chapter illustrates the importance of "collecting the evidence."

Even the hospital inpatient departments have a procedure for *treating the deceased*: Immediately after death is pronounced, the nurses disconnect all the plastic tubing, disconnect the ventilator, and turn off the computer monitor. If death was unexpected or an error is suspected to have occurred, this procedure makes it impossible to know what actually happened.

WHAT DATA SHOULD BE COLLECTED?

Personnel—Who was in the room during the incident? Who was the last person proximal to the patient? Did family members visit the patient? All these details should be recorded immediately upon the discovery of the mishap. It is very difficult to reconstruct events after time has passed.

Around the bed—Collect all medical devices. Save and label the syringes. Allow all the automatic syringes to continue to operate. Do not stop the operation of any devices. Attempt to retain the recordings on the screens of the monitors or data from the automatic recording system. Take photographs (with a regular camera)—of the general organization of the room, the syringes, and the body of the deceased—all of which may provide evidence about what had happened. In the operating room all the syringes and needles used by the anesthesiologist are collected as well as the infusion bags, and anything else that can contribute to understanding what happened and be instructive for the purpose of future prevention.

Medical chart—The various laboratory results should be checked, and the times when the patient received blood or drugs should be clarified. Check what the written orders were and which of them were performed.

* * *

When an unusual death occurs in a hospital in Israel, the systems connected to the deceased are not detached, enabling X-rays to be taken to discover the nature of the mishap. In terms of risk management this is an undesirable action—the cause of death was a general failure in the hospital anesthesia department. But in terms of safety it was a lifesaving action, since this made it possible to identify a systemic problem and prevent other deaths from occurring.

17 Reconstruction to Investigate the Sources of an Event in a Medical System

Yoel Donchin

CONTENTS

An Open Heart Surgery 236
    How Is a Reconstruction Performed? 237
Arterial Bleeding 238
The Reconstruction of Events Surrounding a Patient Found Dead in His Bed 241
Summary and Conclusions 244

Dr. Edmond Locard, a physician and lawyer, was the head of the Criminal Investigations Department (CID) in the French city of Lyon, one of the first laboratories for forensic science. In 1910, Locard stated the basic principle "every contact leaves a trace"—any action performed leaves behind traces, which serve as a silent witness to that action. This refers not only to a fingerprint or a hair but also to a scratch on the skin, clothing fibers, paint chips, and more. These are not affected by forgetfulness or excitement—these are concrete evidence. Physical evidence cannot be wrong, it cannot perjure itself or disappear; it is only humans' incompetence in searching and finding evidence which can thwart the ability to use and rely on it.

The role of the Israel Police Forensic Department is to find traces left at the scene of the crime and either establish or rule out a connection between them and the suspect. About 180 police officers work in special laboratories that function for that purpose. The Forensic Department is responsible for a nationwide system that employs approximately 200 technicians to identify and collect crime scene evidence and conduct field tests; and "crime mobile labs," which gather the evidence sent to the laboratories at the National Police Headquarters in Jerusalem to conduct a more thorough examination (according to the Web site of the Israeli Police force).

Airplane accidents are investigated by trained professionals authorized to investigate them. They can reconstruct the stages of the flight until the moment of a crash, based on fragments of the plane and recordings saved in the "black box," constructed to withstand the crash and provide clues as to its causes. Accidents on ships are investigated by naval experts and traffic accidents are investigated by the traffic accident examiners, and sometimes the damaged cars are left untouched at the scene of an accident for long periods of time until the specialists arrive, so as not to obscure skid marks and other data needed to determine the cause of the accident.

In contrast, no carefully conducted investigations are performed on medical accidents, whether they occur from a disconnected ventilator or a fatal error in distributing medication. Instead, an attempt is made to reconstruct the events after the fact on the basis of records and reports. The local investigative committees begin to investigate long after the incident and those involved have often already made up their minds about what occurred.

Yet, medical investigations can also be investigated differently. It would be difficult to preserve or *freeze* the operating room in which a nonroutine event or a death under ambiguous circumstances took place, as if it were a secured crime scene. It would be difficult to shut down an operating room indefinitely and hygienic considerations require that the deceased be immediately transferred to the morgue. But it is possible to examine the source of the mishap and determine if it was preventable by performing an accurate reconstruction, on film, to ascertain all that occurred prior to the error, step by step, action by action. The reconstruction may indicate the real source of the problem and clear suspicion from those who were present but not responsible for it.

Reconstruction of an event may help reveal the true chain of events that led to the problem. However, the hospital's attorneys generally do not permit such a reconstruction, due to their concern that their client, the hospital, might incriminate itself. During the reconstruction, every single person who was in the operating room or in the surgical department at the time of the tragedy is investigated, and the possibility that someone might blurt out something incriminating is too great. Despite the natural tendency to avoid self-incrimination and despite the defendant's right to remain silent, the ethical question which arises is whether such an event should not be seen as a *ticking bomb*? And thus, is it not obligatory to do everything possible to prevent a similar problem from being repeated in the future? Three episodes described below indicate that the specific actions taken undoubtedly protected many patients from harm.

We will demonstrate the method by which adverse events are reconstructed by describing incidents in which a complete reconstruction was performed.

AN OPEN HEART SURGERY

An open heart surgery was about to be successfully completed. The patient, an obese smoker, suffered a myocardial infarction and needed a graft coronary bypass. Two experienced heart surgeons and two anesthesiologists were in attendance at the surgery, which began early in the morning. The surgery lasted 7 hours, during which time the patient was connected to the heart–lung pump. When the heart–lung machine was turned off, an auxiliary device (intra-aortic balloon pump) was used to improve the blood flow in the coronary arteries.

After the patient was off bypass, only one anesthesiologist remained and an attending surgeon performed the routine task of closing the incision. Once the surgery was completed, the patient's blood pressure was observed to have dropped along with a significant slowing of his pulse until cardiac arrest occurred. The anesthesiologist

Reconstruction to Investigate the Sources of an Event in a Medical System   237

checked the system and found that the respirator and the endotracheal tube were disconnected. CPR was immediately performed and the patient's blood pressure and pulse were restored to normal, but the brain had been critically injured. A few days later the patient was pronounced dead.

In addition to the legally required commissions of inquiry, the director of the Ministry of Health also appointed a special investigative committee composed of members of the Center for Work Safety and Human Engineering. A reconstruction attended by all the members of the medical staff was needed to understand the sequence of events that had occurred and despite protests from the hospital's attorneys, permission was granted.

How Is a Reconstruction Performed?

The operating rooms' administrators had retained a written list of all those who participated in the surgery. Anyone present during the final stages of the surgery, as well as the senior surgeons responsible, was asked to attend the reconstruction, held at the exact location of the incident. The hospital was invited to send whichever representatives it wished to observe the reconstruction, along with the attorneys of the medical staff. The reconstruction was videotaped and the actual film was delivered into the hospital's custody. The reconstruction was done within the framework of the hospital quality inspection committee and immunity was provided, so that the film, and anything said on it, could not be used for external legal purposes.

The role of the patient was filled by a resuscitation dummy that was connected to the ventilator and monitoring devices and even covered with sterile drapes. The surgeon and the anesthesiologist were asked to position themselves exactly as they had been during the surgery.

At this point, everyone in the room was asked to say aloud what he/she had seen and what had occurred during the surgery. The other people attending were asked to comment on what was being said and describe what they had seen happen from their perspective. At the time of the accident only the resident surgeon, the anesthesiologist, and the nursing staff were in the room (the senior surgeons who performed the surgery had already left the room) so they alone took part in the actual reconstruction. When the incision was being closed, without any warning, those present in the room noticed a drop in the patient's blood pressure and pulse. The anesthesiologist discovered that the endotracheal tube was disconnected from the anesthesia machine. The circumstances surrounding the adverse event were now revealed. The balloon pump was located at the foot of the patient's bed. There is no technician responsible for the machine and every few minutes the alarm sounds. As a result, the anesthesiologist must occasionally leave his/her position at the head of the patient's bed to stop the alarm.

The senior surgeon requested that the pulse oximeter be removed from the main operating room monitor (the anesthesia/ventilation monitor continued to operate). When the senior surgeon left the room, the oxygen saturation display on the main monitor was not resumed. The remaining monitoring devices did not send out an alarm, as testified by everyone in the room. An examination of the anesthesia chart revealed that during the shift of the anesthesiologists, the end tidal CO_2 was not recorded (in examining the previous charts that were recorded by the same anesthesiologist, it appeared that the required values were not recorded during other surgeries either).

During the reconstruction, it was discovered that the anesthesiologist and the surgeon had disagreed before, indicating that they had a poor relationship and did not communicate well. The reconstruction indicated that the risk and complexity of the operation was augmented by the use of a balloon pump. Nevertheless, the surgical resident was left alone in the room for the skin closure, a fact which caused a considerable delay in the length of the procedure. Both anesthesiologists who participated in the initial stages of the operation were replaced by a single anesthesiologist, who began his shift after a long day of working. An examination of the workstation demonstrated that the anesthesiologist could not see the patient's head, which was covered with sterile drapes. As the surgery continued, these drapes absorbed increasing amounts of blood and fluids, approaching a potential weight of 3 kilograms: this may have factored in the disconnection of the endotracheal tube during this surgery.

The reconstruction revealed a systemic leadership problem in the anesthesiology department, resulting in a situation in which the anesthesiologist was not made aware of the quality of his/her documentation. It was also revealed that the communication between the team members was very poor. Furthermore, the senior surgeons, although they stated several times that "this was a difficult case," did not see fit to stay and help during the closing of the incision. Improper equipment and dangerous workstations only increased the probability that an error would occur.

The conclusions derived from the reconstruction did not find that the specific physician who did not notice the malfunction was at fault. Instead, a systemic correction of the entire system and work methods was recommended. Specifically, that areas of responsibility be determined, that the currently used sterile drapes be replaced with transparent ones that would allow both the anesthesiologist and the surgeon to see the patient and the various tubing, that the supervisory functioning of the anesthesiology department should be examined, and that monitoring devices could not be removed during surgery under any circumstances. These changes were made to improve the procedures for many surgeries and reduce the recurrence of such a failure or one similar to it. Actions such as firing the anesthesiologist or punishing the surgeon would not have been sufficiently effective in preventing a future mishap.

ARTERIAL BLEEDING

A 78-year-old woman was found dead in her hospital bed in the intensive care unit a few hours after she had undergone cardiac catheterization. The nurse who found the woman pulled back the patient's blanket and saw that the catheter placed in the femoral artery had been severed from its attachment to the rinsing system, and as a result, the patient had bled to death.

The easiest and most convenient way of handling the situation would have been to blame the nurse on duty, who was allegedly negligent by not paying attention to the patient. This is the initial scenario that comes to mind and one that is easy to accept, especially when a scapegoat is at hand. Instead, it was decided to reconstruct the event from the moment the patient was brought into the catheterization room until

Reconstruction to Investigate the Sources of an Event in a Medical System 239

PHOTO 17.1 All the involved care team members are gathered at the event room bedside of deceased patient for active reconstruction of the event.

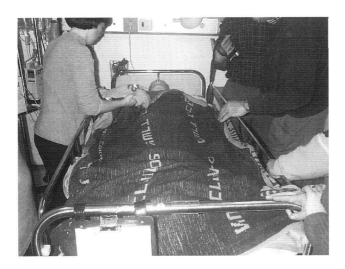

PHOTO 17.2 The bed and the mannequin employed in the reconstruction.

she was transferred to the intensive care unit. A resuscitation doll played the role of the deceased (Photos 17.1 and 17.2).

The same system that was attached to the patient was attached to the doll. Those who attended the reconstruction included the physicians who catheterized her, the nurses from the catheterization room and from the intensive care unit, and any other members of the hospital staff and medical administration who had any connection to the matter. The reconstruction took place in front of the members of the Center for Work Safety and Human Engineering and was videotaped so that participants' faces were not filmed, but only their voices were recorded. During an accurate reconstruction, all stages of the treatment are repeated exactly as they originally occurred. In this case, the patient (resuscitation mannequin) was transferred from the catheterization table to a gurney. The catheterization room nurse passed all the appropriate established forms containing all the information about the patient to the ICU nurse who came to receive her. During the reconstruction, the original form was used.

The patient was transferred to the intensive care unit because the beds in the recovery room, adjacent to the catheterization unit, were all occupied. The gurney was brought into one of the rooms, and the doll was transferred to a bed and connected to the monitoring system, just as had been done for the patient. The nurse who was on duty that night described what happened; how she spoke with the patient, helped her go to the bathroom, supplied her with another blanket, and turned off the light as she left the room. Other patients were in the unit at the same time, two of whom were on ventilators. The staff consisted of two nurses.

The videotape showed the mannequin with the flushing system attached to its *groin*. The 3-way Stopcock is easily disconnected from the system, using as little pressure as that exerted by the weight of the blankets (Photo 17.3) covering the patient (she had complained of feeling extremely cold, which was typical of people coming out of a catheterization). The nurses monitor the devices from a screen located in the center of the unit, but they cannot always see the display, especially when they are busy with another patient in one of the unit's rooms. During the course of the evening, one of the nurses left the unit because the hospital supervisor nurse asked her to help out in another department.

Following the reconstruction, the picture that emerged was different from the one that was apparent immediately after the mishap. The system's points of vulnerability and the preventive actions that could have been taken were revealed. The conclusions of the investigating committee, which had not been previously envisioned, did not lay the blame on any particular nurse. Instead, they revealed the potential dangers faced by the patients in the intensive care unit who were in rooms distant from the staff and not in their direct field of vision.

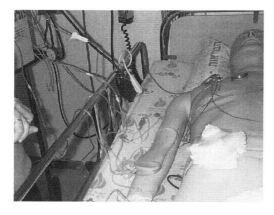

PHOTO 17.3 A blanket's weight was sufficiently heavy to disconnect the three-directional connector.

The conclusions of the reconstruction were as follows:

A. The possibility of the tubing becoming detached is imminent. The catheter is connected to the flushing system by a mere half-turn of a screw and has no locking mechanism, so it can be easily disconnected.
B. There were six patients in beds in the intensive care unit during the event. Usually three nurses work a shift: two in the unit itself and one in the interim unit. Due to the intensity of the workload a nurse was asked to leave the unit and help out in the nearby interim unit. *Only one nurse was present in the intensive care unit* to watch over all the unit's patients, and she was overburdened. At the time of the incident, she was in a room with another patient and was alerted to the bed of the deceased by the monitoring device's alarm, signaling the occurrence of a cardiac arrest.
C. The reconstruction also highlighted problems relating to the structure of the unit itself, the design of the equipment, and the available workforce, all of which threaten the proper functioning of the unit. The structure of the unit was designed to enable privacy and prevent infections and each patient is placed in a small open cubicle. This structure does not allow direct monitoring of all the patients by two nurses, let alone by one nurse.

It is difficult to maintain eye contact when treating one patient while trying to keep the monitoring system in view or directly observe another patient.

The equipment is not sufficient for the needs of the intensive care unit, because if the patients admitted are those who may potentially bleed or whose condition may rapidly change, it is essential to monitor these patients in the currently accepted manner, that is, using a pulse oximeter and a noninvasive blood pressure device, which performs automatically at regular intervals and sounds an alert when detecting values below the predetermined ones. These devices must be an integral part of the equipment included in each intensive care unit room!

There is no doubt that the underlying problem in the intensive care unit relates to a heavy workload that is imposed on an understaffed nursing team, in a department in which the most seriously ill and vulnerable patients in the hospital are kept. Under no conditions should the presence of one nurse alone be permitted in a unit such as this.

THE RECONSTRUCTION OF EVENTS SURROUNDING A PATIENT FOUND DEAD IN HIS BED

A perfectly healthy young man died after shoulder stabilization surgery. A commission of inquiry was established and authorized to reconstruct the chain of events preceding his death.

A mannequin was brought to the operating room at 8 A.M.—the same time that the patient was brought in for his surgery. The surgeon accepted the advice of the anesthesiologist to operate with general anesthesia in addition to a nerve block. The mannequin was covered with a sterile drape and the anesthesiologist performed all the stages involved in providing the local anesthetic, including identifying the anatomical site for the block as well as the direction of its insertion.

After this, peripheral nerve stimulation was performed to confirm that the tip of the needle that would deliver the anesthetic drug to the nerve roots was in the correct position.

The local anesthetic was injected after the patient was under general anesthetic (an unacceptable procedure, to say the least). A catheter was guided through the needle and fixed in place with a sticky transparent sheet, and attached to the body of the mannequin. At the termination of the reconstructed operation, the mannequin's left arm was placed in a neck sling and accompanied by the anesthesiologist and the operating room nurse, was transferred on the gurney to the recovery room and admitted by the nurse. The nurse started the basic monitoring and according to the instructions, provided an attached automatic syringe with anesthetic for pain relief, which could be used to regulate the drug according to the patient's level of pain. This device can be set according to several available programs through an established protocol.

The nurses were asked about the extent of their knowledge and experience with this special type of treatment. They reported that the patient was relaxed upon admission and did not complain of pain. At 1:00 P.M. he was transferred to the orthopedic department, accompanied by an orderly who also brought over the pain relief device. This action was also repeated during the reconstruction, and the researchers followed the gurney on its way into the elevator. Immediately upon being admitted to the department, the patient asked to get out of the bed to urinate, which he was able to do standing. At first he was not allowed to drink but after some time had elapsed he received a meal.

At about 5:00 P.M. the syringe pump for pain relief was disconnected. According to what was recorded on the follow-up sheet the patient did not complain of nausea or vomiting. The surgeon visited the patient before leaving the hospital at 3:00 P.M., and spoke with him. During the reconstruction, the surgeon described his own actions in detail and stated that the patient was fully conscious, alert, and reported some minor pain. The patient was in a room with three other patients who had also undergone surgery. He was not connected to monitoring devices and his blood pressure or pulse was not recorded following the removal of the pain relief device.

In the evening, the anesthesiologist on duty made a routine visit to the patient. For the purposes of the reconstruction, he brought with him the bottle from which he dispensed drugs on the night of the incident and described his actions. He said that the patient rated his pain level at 3 out of 10 so the anesthesiologist decided to add anesthetic through the catheter that had been left in place. The anesthesiologist showed the committee members how he had prepared the drug: he drew up some fluid from the catheter, to ensure that the drug would not be injected into a blood vessel, and then injected 20 ml of the anesthetic solution. The anesthesiologist did not check the patient's blood pressure and pulse rate before the injection or even after and he left the room immediately afterward. He visited the patient again at 11:00 P.M., but that visit was not documented on the chart. At 5:20 A.M. the nurse found the patient dead, after which his death was formally pronounced.

Reconstruction to Investigate the Sources of an Event in a Medical System 243

In terms of the reconstruction and the questions asked at the bedside of the deceased, it is possible to receive a comprehensive picture of the physical hospital system and the mental models of the participants, even long after an event whose results are well known.

The last medical action performed on the deceased was the injection of a drug into the catheter. In the postmortem CT scan, the tip of the catheter appeared to have *migrated* to the spinal cord and the amount of drug injected would have likely caused a total collapse of the general body systems.

But the chain of events preceding the death highlights the failures that led to this final event. The nerve block was first performed in the operating room. The needle was inserted more deeply than accepted and the catheter was also deeply inserted. But considering the report that stated that an electric current passed through the needle caused the expected reaction, and that the injection of 40 ml solution following this did not change the heart rate and blood pressure indices (listed on the chart), it is possible to assume with a high degree of confidence that the catheter was properly placed.

According to the anesthesiologist and staff of the anesthesia department, similar nerve blocks have been performed in the past, but only once or twice in the recent past has this procedure been accompanied by leaving the catheter in place. The reconstruction revealed that the nurses were proficient in operating syringe pumps. They knew that the catheter left for adding anesthetic was not within the scope of their authority. The patient was conscious and did not complain of pain. That evening, in the department, the anesthetic was prepared at the patient's bedside. This was documented, but the anesthesiologist did not relate the injection to the patient's overall physiology—his blood pressure and pulse were not measured either before or after the injection. The sling on the patient's neck had placed pressure on the spot where the catheter was located and may have caused the catheter to migrate.

What have we learned about the system? We have a chain of actions performed by a large number of people: the anesthesiologist who performed the nerve block, another anesthesiologist who then continues the treatment, and yet another who injects the local anesthetic via the catheter. The patient was treated by the operating room nurses, recovery room nurses, and department nurses.

Did any of these professionals know the hidden dangers of using several methods of pain relief? Were each of these professionals familiar with the current stage of the patient's treatment? The answer to both questions is no. When the review committee asked for details on the patient's condition in the department at 10:00 P.M., many forms had to be reviewed, each of which was designated for a specific team. These included a form for pulse and blood pressure records, one for recording drug dispensation, and activities and documentation in the medication cardex.

The recording of the volume of urine passed by the patient could not be found in any of these forms, after having receiving more than three liters of clear liquids during the operation. The physician who came to add local anesthetic did not have a mental model of the patient at the time, nor was he/she cognizant of the predetermined course of action, even though it was a new and *unusual* procedure for the hospital. The communication between the different teams was comprised of what was written on forms that are not sufficiently clear. The extent and manner of information transmission between the different staffs will require a thorough investigation.

Pain relief procedures in hospitals can be very dangerous. The professional literature provides case studies of terminating pain treatment because of the risk involved, and the present case reinforces the need for great caution, because pain relief could cost lives.

The conclusion derived from the reconstruction underscored the need to create an appropriate infrastructure with the introduction of any new treatment regime in the hospital. What is required aside from needles and catheters is to brief the staff and ensure that staff members share their knowledge with each other (nurses, physicians, interns, orderlies). And of course—a thorough explanation must be given to the patient, which might have saved his life if he had said he is not in pain and does not want the injection (all this, of course, is hindsight).

The committee stated the following:

> We cannot assign blame or point an accusatory finger to one person or another. However, from the spirit of what the staff told us and from the analysis of observations conducted from when the surgery began until the patient was pronounced dead, all the stages that the young and healthy patient went through during this treatment must be carefully discussed and studied, from the moment he was brought to the operating room until his release from the hospital.

These findings were delivered to the department and hospital administrations and to the Ministry of Health. The Ministry of Health rejected the committee's conclusions, preferring to assign blame and prosecute those found to be at fault.

SUMMARY AND CONCLUSIONS

Investigating accidents and failures in the health system is not an easy task. These investigations must be launched immediately upon the detection of a mishap, the findings must be gathered, and whatever can be saved should be. In complex cases that involve several teams of staff, an exact reconstruction with the help of a simulation can reveal the reason for the accident, and moreover will point to the *hidden error* and the locations of the "Swiss cheese holes" that must be plugged up to prevent the reoccurrence of similar accidents. An investigation may be done in the department itself, even when a serious accident has not occurred. Reconstructions save lives.

18 Development of a Human Factor Focused Reporting System for Hospital Medical Staff on Daily Difficulties and Problems in Carrying Out Their Work

Ido Morag* and Daniel Gopher

CONTENTS

Results 251
Discussion of Report Results 255
Validation Study of the Centrality of the Major Problems Reported by the
Department Staff 256
General Discussion and Conclusions 257
    The Utility of the System 259
    Raising the Staff's Awareness of the Impact of Human Engineering
    Factors on Patient Safety 259
    Expansion of the Use of the Proposed System 260
    System Management 260
References 261

---

* Doctoral dissertation, Center for Work Safety and Human Engineering, Technion, Haifa, under the supervision of Daniel Gopher. This study was conducted at Rambam Naifa and Hadassah Jerusalem Medical Centers. It was published in *Human Factors* (54, 195–213), by Ido Morag, Daniel Gopher, Avishag Spillinger, Yael Auerbach-Shpak, Neri Laufer, Yuval Lavy, Ariel Milwidsky, Rivka-Rita Feigin, Shimon Pollack, Itay Maza, Zaher S. Azzam, Hanna Admi, and Michael Soudry, "Human Factors Focused Reporting System for Improving Care Quality and Safety in Hospital Wards."

The method of identifying an epidemic is known and recognized in medicine. Whenever there is a sharp rise in the prevalence of a particular disease, epidemiologists go into the field and try to isolate the cause and its source, for example, contaminated food or water. Once they identify the cause, they can prevent further spread of the disease and stop the epidemic. In 1840, there was an outbreak of cholera in London. John Snow, the world's first epidemiologist, discovered that anyone who drank water from the water pump on Broad Street in Soho was infected (one can still see a marker indicating the location of the formerly infected water pump there today). Removing the pump handle, according to the legend terminated the spread of the epidemic.

According to the Public Health Ordinance, a group of laws relating to public health, any physician that diagnoses an infectious disease must report it to the Ministry of Health. The data assembled can then be analyzed by and used to identify possible courses of action and design a prevention program.

This method, the epidemiological method, was initially applied to the epidemic of medical errors as well, an epidemic that has claimed many more lives than either automobile accidents or breast cancer. As publications regarding the scope of these errors in the medical system began to surface, epidemiological methods of investigation were immediately applied; that is, data regarding the *outbreak* of the epidemic were collected. However, unlike reports of infectious diseases, errors and accidents caused by human factors may not always be reported. Physicians and nurses are reluctant to report such events and in most cases information is forthcoming only in very clear-cut cases or in cases involving a lawsuit. Most investigations are conducted in hindsight and do not provide significant value.

Many efforts to improve the quality of care and of patient safety depend almost exclusively on the study of errors and adverse events as the first and primary source of information, and this, as previously mentioned, is of limited scientific value. The process of investigating errors is passive and reactive, in that it cannot take place until after an error has occurred; only then is there an outcry to address the problem. Hence, the researcher does not initiate preventative activities in a systematic fashion, but merely responds to a few isolated outcomes. Moreover, the documentation and reporting of errors are inherently biased and incomplete, since not all errors are reported; therefore, those errors that are reported do not constitute a representative sample of all possible errors but rather a biased sample of the population of medical errors. Error reports are subjective in nature. Partial references and the use of medical records as a source of information are limited, due to the large variation between what is recorded. Even when events are reported, there is no basis on which to compare the population of errors and nonroutine events with the distribution of all the activities being performed in the system being investigated, whether in terms of prevalence, type, or severity. This shortcoming can lead to the misinterpretation and distortion of the impact of these errors and nonroutine events on the operation of the system.

This chapter describes the development of a system that complements the current system of investigating errors and adverse events.[1] This complementary system proposes a systematic analysis of the medical system to identify human factor problems, on the basis of a reporting system in which the department's medical staff voluntarily report problems and difficulties they encounter in their work routine.

Human Factor Focused Reporting System 247

This study examined two main questions: Is it possible to develop an effective tool for a thorough and representative sampling of the problems and difficulties encountered throughout the medical environment, without requiring a significant addition of personnel? And can this tool help identify real problems that significantly impact the department's performance and the quality of patient care?

The study consisted of three main stages:

1. The development of self-report forms through which the medical staff can report the difficulties they face during the course of their routine work. Such reports afford a more focused analysis of the characteristics of the difficulties and obstacles encountered by medical staff, and thus documents the potential for error. This approach is efficient in terms of its use of resources, as it allows for an analysis of the treatment in its entirety, without the need to assign human engineering professionals to observe each stage of the treatment process. In addition, it enables the simultaneous collection of data from different departments by a small team of experts.
2. The identification of the major problems that emerge from the total collection of reports that relate to patient safety and quality of care (*major problems*).
3. The validation of the efficacy of this problem-reporting approach and the veracity of the problems identified as crucial to the quality of patient care.

With this method, there is no need to wait for an accident or adverse event to occur and researchers are not simply responding to a few selected incident outcomes. The reporters provide a detailed description of a problem or difficulty in detail and without fear of accountability, and an investigation thus enables identification of the factors underlying the problem. Since problems and difficulties at work are more common than mishaps and nonroutine events, and since such problems and difficulties tend to recur, the information collected by this system more accurately represents the range of activity in the department and may indicate the severity of the problems and the degree of impact they have on the department's operation. A thorough understanding of the difficulties may encourage proper allocation of resources for the resolution of major problems rather than minor ones. Finally, the method is efficient in terms of resources (in terms of finances and personnel) and involves the entire medical staff in the collection of data.

* * *

Developing a form for use in the hospital involves more than simply drawing lines on a page. The form was designed in collaboration with human factors engineering experts from the Center for Work Safety and Human Engineering at the Technion together with the medical staff from two departments at the Rambam Medical Center in Haifa (the internal medicine department and the orthopedic department). It was based on a preliminary analysis that included a thorough study of the characteristics of the work environment and work routines in these departments.

The preliminary analysis was based on a model of identifying problems and difficulties from the field of human engineering,[2] which identifies gaps between a person's capabilities and the demands of the role that he/she is required to perform, while accounting for the characteristics of the environment the role is performed (such as physical and structural constraints or the availability of information).

The preliminary analysis used to develop the forms included:

1. Interviews with the medical personnel (physicians, nurses) and other department personnel (reception and registration clerks).
2. Observations of routine operations within the department, with special attention dedicated to the transfer of responsibility accompanying shift changes, physicians' rounds, and patient examinations. The observers were present during the staff meetings and the administration of medications.
3. A review of literature on previous studies dealing with human factors engineering in medical systems.

Based on the observations and information obtained from the literature review, five main topics were selected for the reporting system that we feel encompass the array of human engineering problems present in the medical system:

(a) *Work procedures and patterns*—Difficulties originating in policies, work procedures, and collaboration between team members.
(b) *Physical layout*—Difficulties arising from physical infrastructure and spatial layout.
(c) *Administration of medication*—Difficulties in identifying and distributing medications.
(d) *Documentation and record keeping*—Difficulties in transmitting information, documentation, and in the use of forms.
(e) *Equipment and instrumentation*—Difficulties arising from the use of various medical devices.

These five issues were used to develop five types of uniformly structured reporting forms, each of which was divided into three sections.

In the first section, reporters are asked to describe, in their own words, the problem, difficulty, or obstacle that motivated them to fill out a report. This could provide the most important information—awareness that the problem exists.

In the second section, reporters are asked to mark the factors that they believe contribute to the problem or difficulty. Each of the five different forms suggests a variety of possible contributing factors. This section also informs the reporter of the different aspects of human factors engineering in the subject area on which the form is based. The reporter may also propose other factors not mentioned in the form, in his/her own words.

The third section of the form is divided into two parts. In the first part, reporters are asked to suggest solutions that they believe could help reduce the recurrence of the problem and to add any additional details that may help clarify the issue. In the second part, reporters are asked whether they wish to submit an anonymous report or identify themselves in case further clarifications on the reported problems are needed.

The heading on the form is listed as: *Report of a Difficulty, Obstacle, or Nonroutine Event*—the word *error* does not appear anywhere in the form so as not to create the impression that the aim is to receive reports of errors. This title is intended to motivate reporters first and foremost to use the forms to document the difficulties they encounter, and only then to report nonroutine events. Figures 18.1 and 18.2 present examples of a completed form and the second section of each of the five forms.

The study was conducted in the Department of Orthopedic Medicine and the Department of Internal Medicine at the Rambam Medical Center in Haifa, and two gynecology departments at the Ein-Kerem and Mount Scopus hospitals of the Hadassah Medical Center in Jerusalem. Once the form was developed, they were given to senior hospital staff members to obtain feedback regarding two of its components: does the form refer to the full range of issues pertaining to the work environment, and is it sufficiently clear and understandable for staff use.

After receiving feedback on these two issues, revisions were made and the form was put into use.

The study population encompassed the medical staff who submitted the problem reports (physicians, nurses, and auxiliary staff) and the management of the hospitals, all of whom expressed the commitment to encouraging the reporting and providing support at all stages of the study.

The study was conducted at highly disparate inpatient departments that differ with respect to their overall character and the types of patients typically admitted, in two different medical centers with different organizational cultures. This was done in order to check the validity of the proposed system across a wide range of treatment processes and organizational structures.

The problem reports were filled out by a total of 71 nurses and 35 physicians.

The reports were collected by the research team after clearly defining the criteria for selecting the key issues to be addressed:

1. *Major problems*—Problems that arise in numerous reports and relate to treatment aspects that affect the quality of care and the safety of the staff. For example, a shortage of rehabilitation equipment such as wheelchairs and bed frames (52% of all reports in one department); problems in communication and information transfer between physicians and nurses (19% of all reports in one department).
2. *Specific problems*—Problems of a specific rather than generalized nature that affect the quality of care and staff safety; for example, uncomfortable computer workstations in the physicians' room and lack of availability of an on-call physician during the evening hours.
3. *Problems that are unrelated to the field of human engineering and safety*, such as poor maintenance of washrooms and infection-control problems related to the removal of dirty laundry.

All reported problems were sorted into one of these three groups by a team of human engineering professionals. The results were presented to a small team of medical staff comprising the department director and the head nurse. This team was asked to discuss the existence and significance of the issues that were raised.

Work Procedures and Structure

☐	Carrying out a procedure (knowledge, understanding, ability to follow)	☐	Difficulty transferring information among staff members of the same group
☐	Absence of procedures	☐	Work structure in relation to support units
☐	Division of work among different types of staff members	☐	Information exchange with support units (consultants, pharmacy, etc.)
☐	Division of work between members of the same group (e.g., nurses)	☐	Difficulty transferring information between members of different groups
☐	Task load	☐	Other _____

Physical Space and Layout

☐	Department design and space division (e.g., crowding)	☐	Work and preparation areas
☐	Lack of specific rooms	☐	Environmental factors causing danger
☐	Incompatibility of designated rooms	☐	Furniture design causing danger (e.g., patient bed)
☐	Storage space	☐	Other _____

Medication Delivery

☐	Drug information (dosage, effect, sensitivity, drug family)	☐	Positive or negative reactions to other drugs
☐	Drug giving method (including preparation, giving rate)	☐	Drug giving time
☐	Drug recording (place, convenience, copying)	☐	Calculation problem
☐	Drug prescriber's instructions	☐	Drug administration procedure
	Drug nomenclature or commentary	☐	Problematical auxiliary procedure
☐	Similarity in name/packing with other medication	☐	Working surface and drug preparation
☐	Label misleading caption (concentration, units)	☐	Drug storage
☐	Lack of information	☐	Other _____

FIGURE 18.1 The five human factors problem areas and their contributing factors.

Instrumentation

☐	Knowledge of equipment operation	☐	Equipment reliability or calibration failure
☐	Reading displays and indications	☐	Lack of equipment
☐	Ergonomic problems (awkward posture, force, weight)	☐	Equipment area
☐	Unfriendly user operation and use	☐	Equipment organization and arrangement
☐	Operation mistakes	☐	Equipment variety (lack of standardization in performing similar operation)
☐	Maintenance	☐	Other _____

Reporting and Recording

☐	Need to copy (record the same data several times on the same form)	☐	No uniformity in the field names (size, abbreviations, etc.)
☐	Lacking space to record the necessary data	☐	Problems reading the information
☐	No place on the form for definitive patient identity	☐	Problems understanding the information
☐	No identification of time and source of recorded data	☐	Form storage space
☐	Hard to locate the important information out of the total	☐	Other _____

FIGURE 18.1 (Continued) The five human factors problem areas and their contributing factors.

RESULTS

A total of 359 problem reports were received from both medical centers over a 12-week period. The distribution of reports received at each center is presented in Figure 18.3.

There were 241 reports collected from the Hadassah Medical Center during this period (compared to only 51 incident reports received from these two departments during the 5-year period prior to the present study, via their obligatory reporting system). And 118 reports were received from Rambam Medical Center during this period (compared to only 149 reports received in previous years).

Another important difference between the two systems is the source of the report. Previously, reports were received exclusively from the nursing staff, whereas in the proposed system, physicians participated and submitted reports as well; their relative average contribution across all four departments constituted roughly 30% of the total number of reports submitted.

Research Center for Work Safety and Human Engineering, The Technion
Rambam Health Care Campus
Department:_____ Date:_____

Work Procedures and Structure

> Describe in your own words the problem or existing difficulty, time stage, and segment in which it occurs and the nature of resulting difficulties.
> _____
> _____

Mark the factors associated or contributing to the difficulty that was described (more than one factor can be marked):

☐	**Carrying out a procedure** (knowledge, understanding, ability to follow)	☐	**Difficulty transferring information among staff members of the same group**
☐	**Absence of procedures**	☐	**Work structure in relation to support units**
☐	**Division of work among different types of staff members**	☐	**Information exchange with support units** (consultants, pharmacy, etc.)
☐	**Division of work between members of the same group** (e.g., nurses)	☐	**Difficulty transferring information between members of different groups**
☐	**Task load**	☐	**Furniture design causing danger** (e.g., patient bed)
☐	**Storage space**	☐	**Other** _____

Proposed Solutions:
_____
_____
_____

Additional Comments:
_____
_____
_____

> Department:_____
>
> Department Position: _____
>
> Reporter's Name and Telephone number for additional clarification:
> _____
>
> ☐ I am not willing to give this information.

FIGURE 18.2 An example of a work procedure and structure report form. (From Morag I. and Gopher D., 2012, A Novel Reporting System for the Improvement of Human Factors and Safety in Hospital Wards, *Human Factors*, 54, 195–213.)

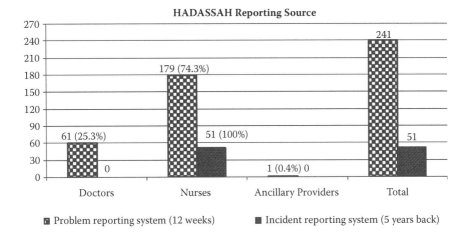

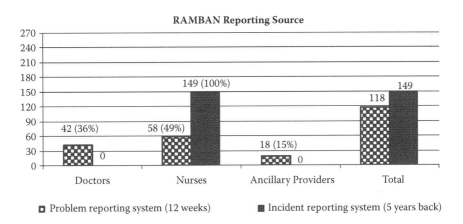

FIGURE 18.3 Frequency and reporting source with the existing and new reporting systems in each medical center.

Table 18.1 presents the type and frequency of reports obtained at both medical centers through each of the two reporting systems. The two reporting systems demonstrated significant differences in the types of problems cited, as well as in the distribution of reporters. Whereas patient falls and errors in medication administration were among the most frequently reported events in the past, there were no reports of patient falls and reports related to the topic of medications were much less frequently reported in the new problem reporting system. This change may indicate that the proposed system encouraged the medical staff to examine their work environment from a new perspective, based on an analysis of human factors engineering and safety.

As this table illustrates, the new reporting system, which included reports from physicians, generated a large database of reports encompassing a wide range of activities in the different departments, and this wide database may more reliably

TABLE 18.1
Frequency and Topics of Collected Reports with the Two Reporting Systems for Each of the Four Wards

Compulsory Incident Reporting System (5 Years: September 2003–September 2007)

	Rambam–Internal	Rambam–Orthopedic	Hadassah Mt. Scopus–Gynecology	Hadassah Ein-Karem–Gynecology
Medication administration	9 (10%)	13 (23%)	19 (58%)	4 (22%)
Falls (from bed and in corridor)	79 (86%)	40 (70%)	7 (21%)	7 (39%)
Data management	4 (4%)	4 (7%)	3 (9%)	3 (17%)
Others	0 (0%)	0 (0%)	4 (12%)	4 (22%)
Total	92	57	33	18

New Problem Reporting System (12 Weeks: During August–September 2007)

Medication administration	5 (8%)	1 (2%)	0 (0%)	3 (2%)
Instrumentation	26 (41%)	23 (43%)	18 (17%)	33 (24%)
Physical space and layout	17 (27%)	14 (26%)	20 (19%)	27 (20%)
Work procedures and structure	11 (17%)	13 (24%)	59 (57%)	72 (52%)
Data recording	5 (8%)	3 (6%)	6 (6%)	3 (2%)
Total	64	54	103	138

Source: From Morag, I. and Gopher, I., 2012, A Novel Reporting System for the Improvement of Human Factors and Safety in Hospital Wards, *Human Factors*, 54, 195–213.

represent the range of problems faced by the full spectrum of medical personnel (physicians and nurses).

After collecting all the reports and classifying them according to the specified criteria, a meeting was held between a team of human engineering professionals and advisory staff from the different departments (the department director and the head nurse) to discuss the key issues identified by the experts. Table 18.2 summarizes the major problems identified in each of the four departments.

As illustrated in the table, different and unique problems were reported in each of the departments. For example, the major topic of the reports from the Rambam Department of Internal Medicine related to the improper manner of providing monitoring that suited the patient's medical status. In another department, problems arose regarding ambiguity in job definitions and the division of responsibility between physicians and nurses. In the absence of a clear consensus among medical personnel on the duties of each staff member and the individual responsible for the patient at each stage of the treatment process, either of the following major problems can occur: overuse—two different staff members provide identical treatment to a given patient and the patient receives unnecessary treatment; and underuse—each staff member mistakenly believes that his or her colleague provided the necessary treatment, and the patient remains untreated. Moreover, physicians claimed that nurses lack sufficient medical knowledge and understanding of medical procedures and

TABLE 18.2
Topics of Reports in Each of the Four Wards

Rambam Medical Center

Internal (64 Total Reports)	Orthopedic (54 Total Reports)
• Quality of ongoing monitoring of patient status (24 reports, 38%) • Documentation and knowledge sharing (13 reports, 20%) • Medication (7 reports, 11%)	• Shortage of support and rehabilitation equipment (wheelchairs, frames) (28 reports, 52%) • Documentation and knowledge access (8 reports, 15%)

Hadassah Medical Center

Gynecology Ein-Karem (138 Total Reports)	Gynecology Mt. Scopus (103 Total Reports)
• Ambiguity in work structure and division responsibilities between doctors and nurses (36 reports, 26%) • Physicians' workload (6 reports, 4%) • Lack of and nonfunctional instrumentation (17 reports, 12%) • Physical setting (23 reports, 17%)	• Communication and information transfer between doctors and nurses (20 reports, 19%) • Lack of and improper instrumentation (19 reports, 18%) • Ambiguity in job definitions and responsibilities between doctors, nursing staff, and aide force (9 reports, 9%)

thus cannot detect adverse events or supervise the physician's work. Conversely, the nurses claimed that physicians were not sufficiently involved in the treatment process and did not provide them with vital information.

DISCUSSION OF REPORT RESULTS

The proposed reporting system was received positively by both physicians and nurses, who willingly completed the appropriate form for any problem they considered important. Thus, it is not surprising that the number of reports received from this system greatly exceeded the number of reports received from the hospital's obligatory risk-management reporting system. The reluctance of medical staff to report errors has been described extensively in the literature. For example, the publications of the Department of Health of the State of New York (1999) indicate that only 16% of all events are actually reported.[3] Differences between the existing system and the proposed system can be attributed to the content of the forms and the motivation of medical personnel to complete and submit a report. In terms of the report's content, errors and adverse events are only the tip of the iceberg of problematic behaviors, and their prevalence is much lower than that of the problems and difficulties faced by medical staff in their routine work. The differences between the contents of the reports filled out in each system are expected, of course, and derive from the different questions and emphases in the two reporting systems. Nevertheless, the question arises: which type of reporting is more informative and/or more effective in preventing accidents and reducing errors—incident reports or reports of difficulties and problems?

A major difference between the existing system and the proposed one is the willingness of the staff to participate in the reporting. In the current system, incident reports are solely the purview of nurses. Although the medical staff recognizes the importance of reporting incidents to prevent recurrence of errors and nonroutine events, reporting is perceived as a threat, especially among the treating physicians, and the personal benefits of doing so are minimal or nonexistent. For example, reporting incidents may potentially call attention to issues of accountability, jeopardize the authority of workers in other roles, and highlight work that was not done, variance according to standard procedures, and so on.

For this reason, it is estimated that only 5% of all errors and nonroutine events are actually reported in the existing systems.[4,5] However, reporting the obstacles that pose difficulties for the medical staff in their daily work prior to the occurrence of a mistake or a nonroutine event is not seen as threatening. Rather, it reflects the staff member's concern, conscientiousness, and awareness of the problematic issues in their work environment. If management supports these reports and the problems are attended to, the staff will benefit from an improved working environment. Under such circumstances it is not surprising that the entire medical staff (physicians and nurses) was willing to participate by producing an extensive number of reports.

Another major issue to consider at this point is the validity of the problems reported by the medical staff. Do they represent actual problems that are meaningful with respect to how the department is run? Do they highlight problems and difficulties that endanger the quality of care? Or perhaps the problems being reported are unsubstantiated and of a personal and/or minor nature.

VALIDATION STUDY OF THE CENTRALITY OF THE MAJOR PROBLEMS REPORTED BY THE DEPARTMENT STAFF

Since each department and the problems therein are different, they were each individually validated using various techniques and assessment methods.

The human engineering personnel together with representatives of the senior medical staff from each department defined the appropriate method of assessment in accordance with the unique characteristics of the department. The method encompassed an assessment procedure and measures that represented the quality of patient care. For example, we will present an analysis of the quality of inpatient monitoring, which was the subject of 38 of the 64 total reports made in the internal medicine department. Patient monitoring was examined by simulating a monitoring regimen initiated upon a patient's admission to the department. The simulation was then compared to the patient's actual monitoring. Two criteria were used to select a profile of nine monitoring indices. These criteria were selected in order to facilitate their retrieval from the patients' medical charts: the frequency at which the indices were monitored during routine care, and the manner in which they were documented in the medical records. The nine indices selected were: (1) heart rate, (2) blood pressure, (3) oxygen saturation, (4) temperature, (5) assessment of heart function by echocardiography, (6) ECG, (7) blood sugar level, (8) urinalysis, and (9) weight.

For the purpose of the simulation, 100 patient charts were sampled from those of patients previously hospitalized in the department of internal medicine with complaints of "general weakness and a sensation of pressure in the chest (angina pectoris)." Six internal medicine specialists, who were not members of the department, were asked to set up an appropriate monitoring regimen for the patients sampled, based on the nine monitoring indices listed above. Their recommendations were compared to the actual monitoring profiles.

These 100 patient monitoring profiles and their admission sheets were then presented to six other physicians, who were asked to assess the quality of the actual monitoring conducted for the patient on a scale from 1 to 5 (1—very poor quality, 5—very good quality). The admission sheets contained all the information needed regarding the patient's condition prior to their hospitalization in the department. If the quality of monitoring rating fell below 5, the physicians were asked to explain the reasoning behind their rating. The doctors were not supplied with information regarding the results of the patient's hospitalization, such as if the patient was readmitted or died during the hospitalization.

The major problem in Rambam's hospital orthopedic department related to the state of their equipment and the shortage of rehabilitation devices. Therefore, a method was developed to examine the impact of this shortage on the quality of medical care and efficacy. But just as the research began, the department was transferred to a new, better-equipped facility so the study was discontinued.

At the Hadassah Medical Center Gynecology Departments, the validation of the main topics of complaint required a different method of assessment. The departments are jointly managed (both were subject to the same management system) and their major problems were expected to be similar. Indeed, problems related to work regulations and informal procedures between the physicians and nurses were identified in both departments; however, the use of the proposed reporting system made it possible to distinguish between different areas of difficulty within the same class of problem. In one department, the difficulty lay with ambiguity in job definitions and the division of responsibility between physicians and nursing staff. In the other department, the major problem related to inadequate communication and information transfer. The topic of ambiguity in defining the role and the division of responsibilities between the physicians and nurses was measured through process diagrams, a tool developed by Sinreich et al. (described in Chapter 5 of this book).

The problem relating to the nature of communication between the physicians and nurses was examined through a unique questionnaire developed for this purpose, designed and completed by the department personnel. A detailed description of the three validation studies and their results can be found in the human factors section (Chapter 9). However, suffice to say that these studies significantly supported the hypothesis that the major complaints identified through these reports are indeed harmful to the optimal functioning of the department and its ability to provide the best possible medical treatment.

GENERAL DISCUSSION AND CONCLUSIONS

The findings demonstrated the central influence of the major problems identified through the reporting system on the three departments. Moreover, the results revealed that the report form developed for the study was indeed effective in identifying the difficulties and problems in the work environment. The validity of the reporting system as a tool to identify problems and difficulties in various clinical environments was supported through the demonstration of its efficiency in two medical centers with two differing organizational cultures (Rambam and Hadassah).

The ability to identify and distinguish between different types of difficulties that stem from the same class of problems testifies to the reporting system's sensitivity and suitability for a medical setting. The system successfully distinguished between two types of problems within the category of *work practices and informal procedures among physicians and nurses* in two jointly managed gynecology departments at Hadassah Hospital with similar practices and procedures; *ambiguity in job definitions and the division of responsibility between physicians and nursing staff* and *inadequate communication and transfer of information.*

The utility of this method was also demonstrated by efficient use of minimal resources in identifying the problems and difficulties. The study results indicated that a self-reporting system for the department's medical staff successfully identified system-wide problems and difficulties. Therefore, this is an effective research method that does not require the continual presence of human engineering professionals, thereby representing only a minimum investment in terms of budget and personnel.

The fact that this system focuses on reporting daily work difficulties and obstacles rather than errors and adverse events encourages the staffs to report a large number of problems in a wide range of topics, without fear of legal or administrative repercussions. Complete, accurate, and detailed reporting was also facilitated by the logical structure of the forms, which enables them to be intuitively completed, and the use of familiar and common terminology.

The well-ordered and straightforward forms include an explanation of the factors that contribute to the emergence of problems in each of five different areas. In addition, these forms help to identify the sources of human engineering problems in health care systems and enable clear and accurate reporting.[8]

On one hand, this system provides the medical staff with insights about the relationship between the daily work difficulties they face and the underlying causes of errors or nonroutine events. It can also underscore the relationship between difficulties and problems that are not directly related to medical practice, yet affect the quality of patient care, such as the transmission of information between different hospital departments or managing patient files. Moreover, the proposed system presented the researchers with a comprehensive understanding of the problems reported, in terms of both general and specific issues related to the work routine and identifying the scope of the factors that contribute to the resulting problems. This type of information can be used as the basis for developing predictive models of the type and range of difficulties in the health care system. Such models may improve the system and significantly reduce the rate of errors and adverse events.

The large number of reports received (359 reports in 2 weeks) can be attributed to the staff's motivation to use the reporting system. In terms of the content, the problems and difficulties cited in the reports are the source and cause of errors and nonroutine events so they encompass many different issues in the staff's daily work.

It is possible that the management's stated commitment to the report process bolstered the staff's participation, since the system was perceived as part of a comprehensive effort to help improve the department. It is possible that the staff perceived the reporting system in the department as a direct channel to senior management. The number of anonymous reports (28%, on average, of the department reports) also testifies to the high level of the staff's motivation, because most respondents did not fear being identified. On the contrary, it is possible they wanted to be identified as making these reports, since they could benefit by being credited with exposing a problem or feeling that they participated in the improvement effort.

The variety of the staff who submitted reports is an important finding, which may also be a testament to the motivation to report. The rate of physicians who actively participated (an average of 32% of the reports received from both medical centers) contrasts sharply with their utter lack of participation in the existing reporting system (as shown in Figure 18.3, which illustrates that only nurses submit reports under the current system). The physicians' reports expand the database to provide a more reliable and representative range of issues by including the full gamut of problems faced by the entire medical staff as a whole (physicians and nurses).

An extensive database can clarify to the decision makers and researchers the types of problems that medical staff must cope with in their daily work, as well as illustrate their prevalence and impact on their performance. In this way, the database may serve as a foundation for identifying the most critical problems in the department's operation. Since the resolution of the problems are, for the most part, limited by resource constraints as well as the lack of awareness of the actual weaknesses in the department, the ability to identify the most salient problems may channel efforts to improve the areas most likely to have the greatest impact on improving the quality of care.

THE UTILITY OF THE SYSTEM

In contrast to other studies that identify system-wide health care problems,[6,7] which require an extensive team of experts to accompany the medical staff throughout the performance of all their tasks, the proposed system uses structured, self-report forms, enabling researchers to examine all elements of the work environment without the need to assign additional professional personnel. In other words, this approach, in which information is gathered from the employees as they work without the need for an external team of experts, is simple, feasible, and economically expedient. Since this system can be applied to many units simultaneously, it can accommodate to a medical center's financial or other relevant constraints.

RAISING THE STAFF'S AWARENESS OF THE IMPACT OF HUMAN ENGINEERING FACTORS ON PATIENT SAFETY

Another important benefit associated with the use of the proposed system is its educational value. Raising the medical staff's awareness of the relationship between human engineering factors (as detailed in the report forms) and the quality of patient care may affect how they perceive their work environment. This awareness may become manifest in the manner in which routine tasks are performed. For example, it may improve the staff's compliance with regulations and work procedures, such as writing medication orders more clearly or increasing their thoroughness when transmitting patient information between staff members. The use of report forms to identify problems in the context of patient care, including aspects not directly related to treatment such as information transfer and communication between different departments, raises the staff's awareness of the influence of these problems on the quality of patient care. In contrast, the current system engages primarily in explicit medical issues, such as the administration of medication.

Understanding the underlying causes of the problems may motivate the staff by instilling the expectation that management will implement the required improvements and changes, such as designing forms that are clear and easy to complete and decipher, improve equipment maintenance, and so on.

* * *

The results of the study were presented to the medical staff and management of the four participating departments at both medical centers, and were very positively received. The staff and management commented on the value of a system that enables continuous improvement. The management of Rambam Hospital adopted the system as policy and plans to introduce it in other departments. The Hadassah Medical Center has already expanded the use of the proposed system in additional departments, such as in pediatric intensive care.

EXPANSION OF THE USE OF THE PROPOSED SYSTEM

The study of the advantages and limitations of the proposed experimental system was limited to four departments. Consistent and comprehensive improvements in the quality of patient care compel that it be implemented in many more departments and units. The study has proved that it is possible to use the proposed system concurrently in different units, and that its operating costs are low. This type of investment is financially beneficial for the hospital since the systematic identification of difficulties and reduced probability of errors and adverse events may reduce the number of hospital claims and the cost of insuring against them.

SYSTEM MANAGEMENT

Having proven that the system is prudent financially and can be effectively used in different units simultaneously, it is important to provide those who collect, analyze, and identify the major problems with the necessary resources to implement appropriate solutions. Operating the proposed system without the benefit of such resources will result in its loss of credibility among the medical staff, and risk the discontinuation of the problem reports, thereby offsetting the advantages of using the proposed system over the existing one.

REFERENCES

1. Gopher, D. (2004). Why it is not sufficient to study errors and incidents—Human factors and safety in medical systems. *Biomedical Instrumentation and Technology*, 28(8), 387–391.
2. Gopher, D. and Weil, M. (1989). Applying human factors engineering approach for thorough investigation of the human factor in workplace accidents—Comparison between four working environments. *HEIS-10-89. Research Center for Work Safety and Human Engineering*, Israel: Technion.
3. Johnson, C. (2007). Human factors of health care reporting systems. In: Carayon (ed.), *Handbook of Human Factors and Ergonomics in Health Care and Patient Safety*. Hillsdale, NJ: Lawrence Erlbaum, pp. 525–560.
4. Cook, R.I., Woods, D.D., and Miller, C., eds. (1998). *A Tale of Two Stories: Contrasting Views on Patient Safety*. National Health Care Safety Council of the National Patient Safety Foundation at the AMA.
5. Cullen, D.J., Bates, D.W., Small, S.D., Cooper, J.B., Nemeskal, A.R., and Leape, L.L. (1995). The incident reporting system does not detect adverse drug events. *Joint Commission Journal on Quality Improvement*, 21, 541–548.
6. Funk, K.H., Doolen, T., Nicolalde, J., Bauer, D.J., Telasha, D., and Reeber, M. (2006). A methodology to identify systemic vulnerabilities to human error in the operating room. *Proceedings of the Human Factors and Ergonomics Society 50th Annual Meeting*, 999–1004.
7. Nemeth, C., Cook, R.I., Dierks, M., Donchin, Y., Patterson, E., Bitan, Y., Crowley, J., McNee, S., and Powell, T., (2006). Learning from investigation: Experience with understanding health care adverse events. *Proceedings of the Human Factors and Ergonomics Society 50th Annual Meeting*, 914–917.
8. Zohar, D. and Luria, G. (2004). Climate as a social-cognitive construction of supervisory safety practices: Scripts as proxy of behavior patterns. *Journal of Applied Psychology*, 89(2), 322–333.
9. Morag, I., Gopher, D., Shpillinger, A., Auerbach-Shpak, Y., Laufer, N., Lavy, Y., Milwidsky, A., Feigin, R.R., Pollack, S., Maza, I., Azzam, Z.S., Admi, H., and Soudry, M. Human factors focused reporting system for improving care quality and safety in hospital wards. *Human Factors*.

19 Patient Safety Climate
Development of a Valid Scale to Predict Safety Levels in Hospital Departments

Yael Livne* and Dov Zohar

CONTENTS

Background 264
Patient Safety Climate 265
Measuring Safety Climate in a Medical Setting 268
The Components of a Nurse's Role 270
Research Methods 271
    Measuring Patient Safety Climate 271
    Measuring the Level of Safety in the Departments 272
Results 273
Conclusions 274
Appendix A: Patient Safety Climate Scale 279
    Hospital Climate 279
    Unit Climate 280
References 281

Patient Dies Minutes after Receiving Improper Medication

By Ron Resnick and Horey Jackie Ha'Aretz
October 24, 2005

A 62-year-old woman died in a hospital minutes after a nurse mistakenly gave her an antibiotic she was allergic to. The hospital reported that the patient received the drug (Penicillin) intravenously, although it had been designated for another patient.

    According to the hospital spokesperson, the drug was administered when the patient was in the department cafeteria. Moreover, he said that the patient's allergy to the drug was duly recorded in her medical chart, and apparently the nurse did not check if the drug was intended for this patient, as required. According to the spokesperson, the "competent and fully-qualified" nurse was sent on forced

---

* Yael Livne. Doctoral dissertation, under the supervision of Professor Dov Zohar.

vacation until the case was investigated. The patient's daughter testified that minutes after the infusion began, her mother started to convulse and lost consciousness. The daughter alerted the medical staff, which brought the mother to the intensive care unit. Efforts at resuscitation failed. The incident was reported by the hospital to the Ministry of Health and an internal committee was established to investigate the affair.

BACKGROUND

Over the last decade, the subject of patient safety has become a growing concern worldwide. According to a report of the American Institute of Medicine, treatment errors are the eighth leading cause of death in the United States.[1] In a recent national survey in the United States, 35% of the physicians and 42% of patients or their families reported that they witnessed a treatment error done to them or a family member.[2]

There is no data on the situation in Israel because it has not been collected, except for that appearing in newspaper headlines. According to a rough estimate based on the data published in the United States, each year 1,000 to 2,000 people die in Israel as a result of medical errors.

A considerable portion of medical system failures originate from blatant staff errors, and can thus be prevented.[3,4] Therefore, substantial research efforts and funding has been directed toward understanding the causes of the errors and implementing various plans to improve the situation. So far these methods have not yielded satisfactory results and it is estimated that since the report on the subject was published in 1999, there has not been any real progress in the promotion of patient safety.[5,6]

Therefore, those dealing with the subject have become gradually aware that the actions currently employed must be accompanied by efforts on the administrative level to change the organizational culture and climate in the health systems. The practice of placing blame and doling out punishments must be replaced by medical accident investigations, which emphasize learning from past mistakes in order to prevent their reoccurrence. In this spirit, we offer an administrative-systematic approach to cope with the phenomenon by adopting the concept *organizational climate* as a theoretical framework for creating a work environment that encourages professionalism and safety in health systems. This approach does not contradict the existing approaches, but complements them and focuses on improving the function of the nurses in their work environment in addition to error prevention.

The case described at the beginning of the chapter, as do many other cases, raises the question: Are there hospitals or departments in which the risk of medical treatment errors is higher than others? Commissions of inquiry set up to clarify the circumstances of unusual events in terms of patient safety concluded that in most cases, proper management would have prevented these errors. They found an unsafe atmosphere, failure to follow procedures, poor communication between the medical care personnel, missing documentation in medical records, insufficient collaboration with the patient and his/her family during treatment, and other factors relating to the professionalism of the treatment provided in the department. In other words, the mistakes are generally related to a system-wide weakness in which safety is not considered the top priority.

We propose to systematically examine the source of the differences between departments and determine what each head nurse, as the department's nurse manager, can do to improve the situation. This chapter is the first attempt of its kind to describe the components involved in professional nursing care and develop a climate questionnaire to measure the perceptions of nurses in hospital departments regarding the department/hospital safety climate.

PATIENT SAFETY CLIMATE

Organizational climate refers to the shared perceptions of people in an organization about the importance of certain aspects of the work environment such as service, safety, and quality. These concepts are based on policy, formal procedures, and informal practices (typical behavior patterns).[7] Policies are the organizational goals defined by senior management, procedures are the methods determined by the management to implement the policy (tactical action instructions), and practices are the actions taken to apply the policies and procedures in each subunit of the organization through the specific directions provided by lower-level managers. The assumption is that decisions on policies and procedures are accepted on an organizational level and applied at the group level through practice.

Climate is commonly referred to as a focused concept, which relates to an organizational strategy or a specific organizational objective (for example, a climate for innovation). Since organizations have many objectives (such as customer service, procedural quality, worker safety), the senior management must develop a set of policies and procedures for each objective.

When the policy is perceived and interpreted by the workers, unique climates are formed for each organizational objective. Therefore, *patient safety climate in the nursing system* describes the importance nurses place on adhering to various aspects of patient safety during their daily work. The source of these perceptions is the nurses' interpretation of the formal policies of the senior management (the hospital nursing management) and the practices applied by the department's head nurse. During the daily interaction with the head nurse, nurses make judgments about the priority he/she attaches to various aspects of the work, especially in situations where there are competing goals (such as treatment speed versus treatment quality). A consistent preference for certain aspects of their role sends a message to the nurses about the importance of these aspects. Namely, it shapes their perceptions about the nature of the climate in the department (the degree of importance of those aspects). Climate can be measured according to two main parameters: level and strength.

Climate level represents the degree of perceived importance of a certain objective relative to other organizational objectives. In our case, this relates to the extent to which the nurses perceive patient safety to be a top priority.

Climate strength relates to the degree of consensus among the nurses regarding the importance of patient safety in their department. Based on research evidence on the correlation between climate level and worker safety, we expect that the level of the climate will affect the level of patient safety in a medical setting. That is, the more importance the nurses relate to patient safety as an objective of their department and their hospital, the more motivated they are to ensure the patients' safety during their work.

It is worthwhile mentioning that the definition of a climate assumes a consensus among the nurses. That is, it is argued that in the absence of a minimal level of agreement, there will be no defined climate in the department. A consensus is formed by consistent messages conveyed by managers to the employees under their supervision. The clearer and more consistent the pattern of the administration's actions is, the stronger the climate that will dictate employee behavior.

For example, in a department in which the head nurse behaves inconsistently in regards to patient safety (for example, in certain circumstances he/she is stricter than in others, or is stricter during the day shift than during the night shift), there is a basis for assuming that the conditions will not support a consensus and that a defined climate will not develop.

On the other hand, it is possible that the nurses will come to agree that the head nurse's behavior actually testifies to a limited commitment to patient safety. In practice, identifying the departments in which the climate is weak (due to lack of consent) is as important as identifying the departments in which the climate is strong but low, since in both cases these are departments in which the probability of errors in treatment is high. The relationship between the level and strength of the climate and organizational results has only been studied in a limited scope and the findings are not clear cut.[8-10] Therefore, one of the goals of our project was to study the relationship between these variables in a medical setting.

Another characteristic of the concept of *climate* is that it may be investigated through several levels of analysis. Organizations with a hierarchical structure may develop different climates according to the organizational, group, or department level, as policies and priorities are determined by the senior management and implemented by intermediate level supervisors and employees. However, since the various organizational units are engaged in different activities and managed by different people, the policies undergo a process of interpretation by junior managers before they are translated into concrete actions. This process is inevitably influenced by personal dispositions, management styles, and supervisors' obligations to their superiors.

Therefore, different organizational units differ in how the policies are implemented, which gives rise to differences in climate perceptions.[11,12] The policies and procedures set by senior management are, therefore, the basis of climate perceptions at the organizational level, while the behavior patterns of those in charge of the different units are the source of climatic perceptions at the group level.

In hospitals the nurses are informed of management policies through procedures, publications, staff meetings, and the like, from which they derive the relative importance that senior management places on patient safety.

Nevertheless, since all head nurses have some degree of autonomy in the implementation of hospital policy, they can emphasize one organizational objective over another, and not necessarily in line with the emphasis intended by the management. Therefore, it is likely to find differences in climate between the hospital and its various departments. For example, nursing management may expect the nurses to expand their professional skills and continuously update their professional knowledge. But the head nurse of a particular department may consistently refuse to authorize nurses to attend professional courses, due to excessive workload. Such gaps between the messages conveyed by management and those conveyed by the head nurse may give rise to differences in climate levels between the hospital and the department.[13,19]

Most studies of safety climate in health systems do not relate to these different levels of analysis.[14–17] Our research confronts this issue through the development of a multilevel measure of patient safety climate, which distinguishes between the messages expressed by the direct nursing supervisor (the head nurse of each department) and by the hospital nursing management. In fact, we have developed two separate climate questionnaires—one for measuring the departmental climate perceptions and another for measuring the hospital-wide perceptions. By relating to both levels it is possible to identify the source of the discrepancies between the emphases placed by managers at different organizational levels and locate the dominant source of influence on the level of patient safety climate in the department.

An important aspect of climate research concerns the distinction between the statements made by management and its actions, that is, between stated organizational policy and its real policy.[18] Stated policy is expressed through written material, procedures, publications, and guidelines distributed to nurses, and so forth. The stated policy is openly declared and documented, while the real policy remains covert and undocumented. Nurses identify it through the typical responses (practices) of the head nurse to changing circumstances during work, and through the everyday decisions he/she makes.

For example, during staff meetings the head nurse may emphasize the need to involve the patient and his/her family in the treatment process and provide them with information that will enable them to be active partners in their treatment (stated policy). But when this causes a delay in the distribution of drugs, she may make a comment to a nurse for spending too much time explaining the side effects of a medication to a patient. In this case, the underlying message conveyed by the head nurse is: There is no need to show the patients courtesy and patience, and providing clear and comprehensive explanations is appropriate only under certain circumstances (real policy). The gap between stated policy and actual policy is revealed, in that the actual commitment of the head nurse to patient safety is not as great as his/her professed commitment. If such a behavior is repeated and becomes habitual, it will result in the emergence of a relatively low climate for patient safety in the department.

The hidden nature of the real policy facilitates the formation of a discrepancy between it and the stated policy, especially since the stated policy is used as a formal backup in case an adverse event takes place. However, the climate perceptions are determined by the real policy only. Thus, it can be said that the climate level reflects the common view (whether there is agreement among the nurses) of the department's head nurse's (and of the hospital's nursing management) actual level of commitment to patient safety, and, therefore, on the importance of strict adherence to their safety under different situations.

The practical importance in measuring organizational climate is the fact that it influences behavior. The relationship between climate and behavior is based on the expectations that employees develop regarding the results of their behavior at work. Workers tend to look for an *organization order* that will enable them to predict which behaviors will be rewarded by the organization.[19] Climate perceptions provide information regarding behaviors that the organization finds acceptable and desirable. They raise workers' expectations that certain behaviors will yield certain results and these expectations guide their behavior.[12] In health care organizations, nurses observe the behavior pattern of the head nurse (What does she notice? What is important to her?)—from which they derive their understanding of the expected results of maintaining patient safety.

This information allows them to make assumptions on whether or not they will benefit from maintaining patient safety, and these assumptions guide their actions in order to achieve desirable results, such as positive feedback, evaluation, recommendations for promotion, participation in courses, consideration related to work scheduling, and so forth. Since the reactions of the direct supervisor within the organization are a major source of reward to employees, the head nurse in the department is primarily responsible for shaping the behaviors of the subordinate nurses.[20,21] When previous experience teaches the nurses that despite the formal declarations, what really matters to the nurse in charge is rapid and efficient treatment, even at the cost of abandoning certain procedures, they will behave accordingly.

Medical errors occur in hospitals when safety considerations are overlooked in favor of speed and efficiency considerations. Priorities such as these by the appointed supervisors contribute to the formation of low patient safety climate and encourage the use of shortcuts that increase the probability of errors. On the other hand, the higher the level of the department's patient safety climate, the greater the probability that the nurses will demonstrate safe behaviors.

The relationship between organizational climate and behavior has been documented in many studies. For example, studies on the service climate in bank branches showed that when bank employees grasped that good service was an important management objective the quality of customer service improved.[22] It was found in the field of safety that, among other things, climate is one of the factors that influence unsafe behavior among workers and worker injuries[23] and climate perceptions predict the occurrence of micro-accidents (behavior-dependent accidents that result in minor injuries) even months after the climate is measured.[11]

MEASURING SAFETY CLIMATE IN A MEDICAL SETTING

Several tools have been developed to measure the safety climate in health systems.[24] A thorough review of the existing measures raises some questions about their validity, because they are too inclusive. First, most of the measures include items related both to worker and patient safety, based on the assumption that these are identical indicators of the department's priorities. However, when all the questionnaire items are summed to reach a single climate score, there is no way of knowing for certain the results it predicts—worker safety, patient safety, or the safety of both of them together.

Second, most of the measures refer to several levels of analysis in the same questionnaire, without distinguishing between the perceptions of climate at the organizational level and at the group level. It is important to mention that due to the complexity and lack of routine characterizing the head nurse's role and the considerable extent of

Patient Safety Climate 269

autonomy afforded to him/her, it is reasonable to assume that each head nurse will implement the management policy differently. Therefore, it is likely that different climates and priorities will develop in different departments as well as differences between the management's priorities and those of a particular head nurse. Therefore, a valid climate measure must include separate scores for the climate at the organizational level (the perceived commitment of senior management in the hospital) and the climate at the group level (the commitment of the head nurses in the departments).

The third concern relates to the fact that, in some of the measures the questionnaire is designed to be filled in by several professional groups simultaneously (nurses, physicians, paramedical staff), ignoring the fact that the unique organizational structure of hospitals is based on several parallel professional hierarchies. It is important to emphasize that an organizational climate stems from the perception of the commitment of the relevant person in charge, as the worker cannot obtain the information he/she needs to learn which behaviors are expected of him/her from the behaviors of a different supervisor.

In fact, the nurses' perception of the climate testifies to the commitment and priorities of the department's head nurse, and of the hospital's nursing management, whereas the perceptions of climate among physicians relate to the relevant superiors in their professional group. Therefore, it is not possible to identify the source of the climate if the same questionnaire relates to professionals from different occupations, because there is no way of knowing which supervisor the respondents' answers are related to. A measure of climate in a hospital should be unique to the profession and separate questionnaires must be dedicated to each professional group.

Another drawback of available safety climate measures in health systems relates to their predictive validity. Most existing climate questionnaires are not validated, hence the degree of correlation between the climate levels revealed by the questionnaire and any outcome related to patient safety was not examined.[24,25] Ideally, predictive validity is achieved by comparing the results of the questionnaire to objective data, such as employee behaviors, patient injuries, or other organizational results.[26] However, even in studies in which outcome variables were collected, most often the data were based on self-reported frequency of medical errors or unsafe conduct collected through the same questionnaire, which measured the climate.[24]

Study designs such as these may distort the results due to single source bias, which stems from the fact that both the climate data and the safety data come from the same subjects, so that the respondent's perceptions of climate may influence his/her perception about the organization's safety status. Thus, it is likely that the correlations between the predictor variable (climate) and the outcome variables will be exaggerated. In this context, it is worth mentioning a study that examined the predictive validity of safety climate through hospital reports on safety events.[27] This study is limited in that the events reported usually do not reflect the true scope of errors in treatment, since many errors are not officially reported to the hospital authorities and are not documented. Therefore, in our project we have taken an objective view on the safety behavior of nurses as a measure that approximates the level of safety in the department.

THE COMPONENTS OF A NURSE'S ROLE

The success of nurses in maintaining patient safety is not only due to compliance with safety instructions and meticulously following procedures. From conversations with nurses in hospitals, we discovered that patient safety (as opposed to worker safety in other industries) is an integral part of quality of care and in fact stems from the level of professionalism of the nurses. In order to promote the safety of patients, nurses must continuously stay up to date with the professional knowledge, communicate effectively with the other caregivers, work collaboratively with the nursing staff in the department, and foster caring and concern for the patients—due to the fact that ensuring patient safety is part of the broad spectrum of nursing practice and is strongly associated with various aspects of professionalism in nursing.

For example, the component of caring includes, in addition to humane and personal treatment, also teaching patients about the treatment they are receiving, providing explanations about their drugs and drug side effects, updating the family, and more.

Moreover, nurses who develop close and personal relationships with patients can provide them with better care and avoid making mistakes. Similarly, teamwork involves not only a routine update on the patients' condition and professional consultation, but also alerting others to situations that are dangerous without fear of inappropriate responses, and reporting "near mistakes" without fear of punishment. Based on a comprehensive literature review and interviews with nurses in different roles, we identified three basic components in the role of the nurse: professional development, patient orientation, and teamwork.

Professional development—Advances in clinical knowledge and the development of innovative medical technologies generate periodic changes in the current knowledge base and methods of patient care. These changes require caregivers to commit to continuously study and update their professional skill set beyond the level achieved through formal on-the-job training.

Familiarity with up-to-date procedures and medications would allow nurses to provide patients with the best treatment, and to avoid errors resulting from lack of knowledge. Therefore, health literacy is an essential component of professionalism that affects the quality of care and patient safety. It has been found that inadequate knowledge is one of the risk factors associated with errors. Nurses who continually update their knowledge of drugs make fewer medication errors than those who do not.[28-31] Therefore, investing in the professional development of nurses is essential for successful performance.[32,33]

Patient orientation—This component of nursing encompasses caring behaviors (such as emotional support, maintaining the dignity and privacy of the patient, providing a comprehensive explanation of the treatment) and behaviors directly related to patient safety (such as compliance with safety procedures and avoiding dangerous shortcuts). The unique nature of health care, in contrast to other service industries, is characterized by people taking care of other people in times of need and stress. Therefore, stable, trusting relationships between a patient and his caregivers can be critical to managing an illness.[34]

Offering emotional support is recognized by nursing staff and patients as an essential part of nursing practice.[35] Patient care (caring) also means providing clear instructions on the medications administered to him/her, thereby reducing the risk involved in taking them. Providing comprehensive and understandable information to the patient allows him/her to be more involved, develops his/her ability to make informed decisions about treatment, and impedes the occurrence of potential mistakes.[36] A nurse who fosters a therapeutic-supportive relationship with his/her patients usually demonstrates a high level of concern for them, listening to them, uncovering details about their potential drug sensitivities, and demonstrating interest in their response to treatment. He/she responds effectively to patients' needs and, therefore, the chance of making an error is low due to the personal relationship between them.[37]

Teamwork—Nurses spend a large amount of time coordinating patient care across nursing shifts, different professional groups (physicians, nurses, paramedical staff), and between hospital departments (admitting patients from the emergency room, sending patients for tests performed out of the admitting department, etc.). Continuity of treatment depends on cooperation among the nursing staff and effective communication with other caregivers.

Optimal patient care and ensuring his/her safety requires the accurate transfer of information in real time, providing constructive feedback and sharing essential knowledge and skills.[33] Good communication between caregivers enables not only the continuous updating of the patient's status and professional consultation, but also alerting relevant persons regarding potentially dangerous situations. Studies show that communication, collaboration, and interaction between nurses and physicians improve the quality of care. Conversely, poor communication was identified as one of the risk factors associated with medication errors.[38,39]

As a result from the direct and ongoing relationship between nurses and patients, the nursing staff is in a key position in terms of their ability to reduce errors within the health care system.[38] Nurses can immediately detect inadequacies in the treatment provided and take corrective action. Therefore, the nurse is able to impede medical errors and prevent patient harm. It is for this reason that we chose to focus on the nursing staff and measure patient safety climate as a source of influence on nurses' performance and impetus for behavioral changes.

Therefore, the purpose of this study was to develop a measure of patient safety climate unique to nursing and validate it on the basis of objective outcome data. We also sought to examine the effects of patient safety climate at the organizational and group levels and examine the association between the level and strength of climate as predictive factors of patient safety.

RESEARCH METHODS

MEASURING PATIENT SAFETY CLIMATE

Based on the considerations described above, we have developed two climate questionnaires that relate to the two relevant levels of analysis: the organizational level (hospital) and the group level (the department). The questionnaires are designed to be completed by nurses in the inpatient departments and they relate to the nurses' perception of the importance of patient safety in the hospital and in their department. Each questionnaire includes items related to the three components of the nurse's role: professional development, patient orientation, and teamwork. Factor analysis and internal reliability tests have confirmed the existence of three subscales corresponding to these three components.

The following are some examples of items in the climate questionnaires. In relation to professional development in the hospital subscale: "The hospital's nursing management promptly offers professional courses dealing with subjects in which there is deficiency (including expensive courses)," and in the department subscale: "In our department I am encouraged/required/expected to keep up to date on new professional knowledge." An example of an item in the climate questionnaire that deals with patient orientation in the hospital subscale is: "The hospital nursing management puts a lot of effort in improving patients' safety in the hospital (preventing errors)," and in the department questionnaire subscale: "In our department I am encouraged/required/expected to provide [for] the patient's needs, even if not directly asked to." Finally, the following is an example of an item in the climate questionnaire relating to teamwork in the hospital subscale: "The hospital nursing management encourages nurses to report problems in working relations with physicians in the department." An example of this in the department subscale is: "In our department I am encouraged/required/expected to share the workload with other team members."

The development of the questions was based on a preliminarily stage which consisted of identifying senior management activities and head nurses' practices, who represent the standard for assessing the real hospital and departmental priorities by nurses. The complete questionnaire appears in the appendix to this chapter.

Climate level was calculated by averaging the responses to the questionnaire items on a scale of 1–5, so that a high score represents a high climate level. Climate strength was calculated using the homogeneity index R_{wg},[40] so that the greater the homogeneity within the group, the stronger the climate.

The questionnaires were completed by 955 nurses from 69 departments in three urban hospitals in Israel. The response rate was 72%. The answers were anonymous and the participants were promised absolute confidentiality. The hospital management authorized participation in the study in exchange for obtaining the study findings and conclusions after the data was processed.

MEASURING THE LEVEL OF SAFETY IN THE DEPARTMENTS

About 6 months after completing the questionnaires, we conducted observations of the behavior of the nurses in the departments, according to two measures of safety: medication safety and emergency safety. Five observations were conducted in each department at random times, during both morning and evening shifts. We focused on measuring the nurses' work processes, not their results. Assuming that certain behaviors increase the probability of medical errors, the nurses' behavior served as an approximate measure of patient safety, since it allows for the detection of inadequate work processes that may result in adverse outcomes.

In order to comply with methodological standards, we had to limit the content of the observations to objective indicators only and avoid relying on subjective observer judgments. Therefore, the observations did not relate to *soft* behaviors, such as the nurses' decision-making processes, teamwork, effectiveness of departmental communication, and other components of professional practice that could affect patient safety.

Observations were made according to a checklist developed in consultation with experts in the field of nursing and related to two categories of behavior: medication safety and emergency safety. We chose these two categories to encompass the range of nursing behaviors relevant to both routine and emergency situations. Actually, we measured the rate of safety behaviors demonstrated by nurses during their work from the sum total of all observed behaviors. All the deficiencies that we referred to during the observations were classified by experts as inadequacies derived from unprofessional or unsafe behavior, that is, behavior-dependent deficiencies.

Medication safety was measured using a list of items relating to the proper storage of drugs, their expiration dates, documentation in nursing records, identification of the patient prior to administering the drug, and adherence to safety rules during the distribution of drugs. These criteria represent routine activities that nurses perform frequently enough to be statistically analyzed.

Emergency safety is related to nurses' preparedness to nonroutine situations. We chose this measure to be used in our observation because some of the nurses' errors derive from lack of alertness toward unexpected situations.[37] As a result, the extent to which the department is prepared for emergency situations could affect the safety of its patients. Thus, emergency safety was measured using items that deal with the continuous maintenance of the resuscitation cart, such as updating the list of its contents, and matching the contents of the list to the supplies on the cart, and so forth.

RESULTS

First, the predictive validity of the climate questionnaires was examined. Predictive validity refers to the questionnaires ability to predict the behavior of the nurses concerning the maintenance of patient safety. Table 19.1 presents the means and standard deviations of the study variables and the correlations between them.

The findings indicate that the department level climate predicts the nurses' behavior both in the category of medication safety as well as in emergency safety.

Similar results were obtained for the patient safety climate at the hospital level. In other words, the nurses' safety behaviors are influenced by their perceptions of the importance placed by the department head nurse and the hospital nursing management on patient safety.

It was also found that the effect of the climate at the departmental level on the nurses' safety behaviors depends on the climate level at the organizational level (see Figure 19.1). In other words, when nurses perceive that the hospital's nursing management assigns high priority to the safety of patients, they are less influenced by contradictory messages they receive in the department. However, when the hospital nursing management's attitude regarding patient safety is perceived as lax, the impact of the department climate increases.

TABLE 19.1
Descriptive Statistics and Intercorrelations among Study Variables

Variables	Mean	SD	1	2	3	4	5	6
1. UC level	4.09	.27	—	.67[a]	.62[a]	.16	.10	.01
2. HC level	3.16	.35		—	.39[a]	.26[b]	.01	.13
3. UC strength	0.79	.12			—	.21[c]	−.03	.20
4. Medication safety	0.95	.06				—		.04
5. Emergency safety	0.68	.24					—	

Source: From Zohar, Livne, et al., 2007, Healthcare Climate: A Framework for Measuring and Improving Patient Safety, *Critical Care Medicine*, 35, 1312–1317.

Note: UC, unit climate; HC, hospital climate. UC level and HC level are measured on a 1–5 scale and represent the mean score of all units; UC strength is the mean R_{wg} of all units; medication safety and emergency safety are the % of safe conditions out of all sampled conditions.

[a] $p < .001$
[b] $p < .05$
[c] $p < .8$

Another factor that influences the relationship between the department level climate and safety behavior is the climate strength, that is, the degree of consensus among the nurses regarding the importance of patient safety in their department. As shown in Figure 19.2, the effect of climate at the departmental level on the nurses' behavior is greater when there is widespread agreement on the importance of the patient's safety in the department (strong climate). Thus, when the nurses are of similar mind regarding the departmental priorities, a high climate leads to greater safe behavior than a low climate. However, when the climate is weak (for example, due to inconsistent messages from the head nurse), each nurse works according to his/her perception and understanding of the situation and the rate of safety behavior in the department is low.

The results presented above indicate that the climate of patient safety can consistently predict the level of safety in the departments, as reflected in the safety behaviors of the nurses, both with respect to routine and emergency situations. Overall, an impressive statistically significant result was obtained regarding the questionnaires predictive validity ($p < 0.01$). This indicates that the chance of getting similar results in repeated tests exceeds 99%.

CONCLUSIONS

This study applies the concept of organizational climate, which is commonly investigated in manufacturing and aviation industries, to the context of health care. Unlike business organizations, in which the cost of mistakes is expressed primarily in financial terms, the cost of errors in health organizations is much higher.

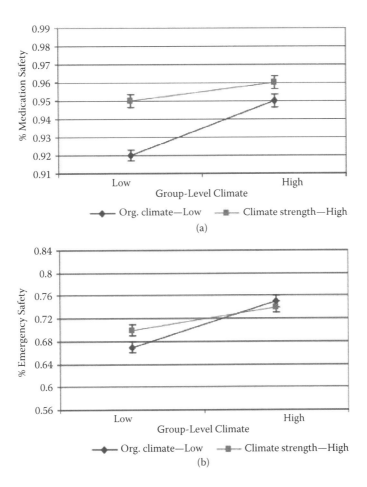

FIGURE 19.1 The effect of department safety climate on safe medication administration (a) and safe behavior in emergency conditions (b), as a function of safety climate at the organizational level. (From Zohar, Livne, et al., 2007, Healthcare Climate: A Framework for Measuring and Improving Patient Safety, *Critical Care Medicine*, 35, 1312–1317.)

The differences in settings require two significant adjustments in the conceptualization of climate. First, instead of examining climate effects on worker safety, the dependent variable in this study is the patient. First, instead of examining the effect of climate on employee safety, we studied its effect on the safety of patients. In other words, we assumed that the employees' climate perceptions affect their safety behavior not only with regard to themselves but also toward their "clients." This assumption also underlies other studies that examined safety climate in health care organizations.[27,41]

Secondly, we realized that the emphasis on patient safety is an inherent component of quality of care and integral to the other aspects involved in the nurse's role. The level of safety in the department is not affected by the degree of the nurses' compliance to safety instructions to the exclusion of all else, but is also affected by

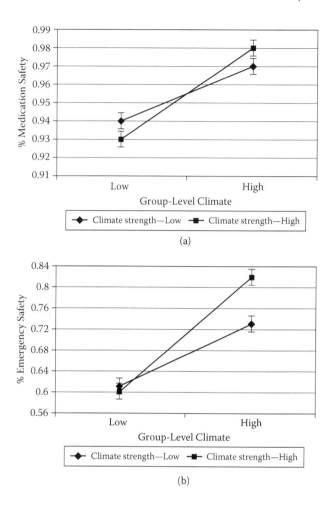

FIGURE 19.2 The effect of department safety climate levels on safe medication administration (a) and safe behavior in emergency conditions (b). (From Zohar, Livne, et al., 2007, Healthcare Climate: A Framework for Measuring and Improving Patient Safety, *Critical Care Medicine*, 35, 1312–1317.)

factors such as their professional knowledge, the quality of their communication with other professionals, and their level of concern for the patient. Therefore, in addition to the safety components, the measure that we developed referred to the perceived importance of these other aspects of the nurse's work that also contribute to patient safety.

The major purpose of the study was to develop a measure of patient safety climate and validate it using two objective outcome variables. Indeed, the results showed that the perceptions of nurses regarding the policy of the senior management and the practices of the head department nurse predict safe behaviors in the areas of medication and emergencies. In other words, we were able to validate the two climate questionnaires using criterion data collected independently through objective observation. Therefore, the measure we developed enables us to predict the probability of errors occurring in each department and identify departments and hospitals in which either a higher or lower than average probability of error exists.

On the basis of this study's findings it is possible to initiate a process of enhancing patient safety as a preventative step, before rather than after the occurrence of a medical error. Since the head nurses are the main source for the development of climate perceptions, any intervention must be based on changing their practices. An intervention program could be developed that begins by measuring the baseline of patient safety climate as well as the actual level of patient safety. Later, workshops on the development of patient safety climate should be conducted for head nurses, providing them with feedback on their behaviors. The program would conclude after remeasuring the patient safety climate level and the levels of safety. A similar intervention program for implementing a safety climate in industrial plants in Israel led to a significant improvement in the climate perceptions by the workers, after which the rate of safety behaviors increased and the rate of injuries decreased.[42]

Our results indicate the importance of hospitals' nursing management and of the department's head nurse in shaping nurses' behaviors regarding patient safety, and therefore in improving the safety levels in the departments and in the hospital overall. The behavior of the head nurse affects the basic probability that treatment errors will occur in his/her department, although a particular human error made by a specific nurse may be affected by other factors as well. In this respect, the practices of the head nurse constitute an underlying factor in increasing or decreasing the probability of medical errors and accidents.

The data presented here elucidates the issue of patient safety from the managerial perspective and offers a way of using tools from the field of management to cope with it. As previously noted, this approach does not preclude other approaches for improving patient safety and it is certainly possible to combine it with existing efforts to complement the traditional methods of dealing with this issue. The data demonstrates three management parameters that affect the safety performance of nurses in hospital departments:

- Those in charge must consistently ensure that the patient is at the center of nursing practice through compliance with safety procedures and the development of a caring attitude toward the patient, even during uncomfortable situations.
- The practices of supervisors should reflect a continuing investment in the professional development of nurses.
- Practices that reflect a consistent cultivation of teamwork and effective communication between the caregivers.

These findings have great practical importance, because they make it clear that strict adherence to safety procedures alone is not enough to ensure the patient's safety. Rather, it also requires an investment in the professional development of nurses and the cooperation of the entire department working as a coordinated team. By considering separately the various components of the nurse's role in the climate questionnaire, it is possible to identify the professional dimensions that are perceived by nurses as receiving high and low priority by the management.

One of our findings relates to the influence of climate strength on the relationship between the climate level in the department and the behavior of the nurses. We have seen that climate level has a greater effect on the safe behavior of nurses when it is also strong, rather than weak. Because climate strength depends on the clarity and consistency of the messages delivered by the head nurse, these results emphasize how vital the role of the head nurse is, not only in prioritizing professional treatment but also in routinely propagating *coherent and consistent messages, stressing the high priority of quality nursing*. Therefore, an intervention program to improve patient safety should also relate to climate strength during the measurement stage, as well as while training and providing feedback to the head nurses.

The findings support the hypothesis that organizational processes at different levels of analysis influence one another. Specifically, we have seen that the head nurse has a major role in influencing the nurses' behavior and the level of safety in the department, especially when the hospital's nursing management does not emphasize these issues. The combined effect of the climate at the department and the organizational level in predicting the nurses' safety behaviors highlights the need to consider the climate at different levels within the organization, especially in light of researchers' tendency to limit their study to one level of analysis. It is even recommended that the sample size be increased in the future to allow comparison not only between hospital departments, but also between hospitals, using more complex statistical analyses.

In conclusion, our study focused on the nursing system and measured patient safety climate among nurses in hospitals. However, despite the nurses' important role in ensuring patient safety, they are merely part of the caregiving team in any given department. They work alongside physicians, technicians, social workers, and others. The findings support the existence of a patient safety climate and its ability to predict the behavior of nurses, paving the way to measure patient safety climate in hospitals among other professional groups, especially among physicians. Other such climate measures will enable a more detailed and deeper understanding of the factors that contribute to patient safety in a complex medical environment.

APPENDIX A: PATIENT SAFETY CLIMATE SCALE

Hospital Climate

To what extent do you agree with the following descriptions of your hospital? (There are no right or wrong answers—circle the most suitable number in each sentence, in your opinion.) (1 = Least agree; 5 = Most agree.)

The Hospital's Nursing Management:

1	Publishes a lot of information concerning professional courses for nurses.	1	2	3	4	5
2	Holds plenty of professional meetings (lectures, conferences) for nurses.	1	2	3	4	5
3	Often stresses the importance of communication between nurses and doctors.	1	2	3	4	5
4	Poses the improvement in nurse's professionalism as a major goal for the hospital.	1	2	3	4	5
5	Promptly offers professional courses dealing with subjects in which there is deficiency (including expensive courses).	1	2	3	4	5
6	Demands a high level of nursing professionalism in the hospital (above the minimum).	1	2	3	4	5
7	Requires each nurse manager to claim responsibility on the safety of patients within her/his ward.	1	2	3	4	5
8	Takes into account internal evaluations when deciding on nurses' promotion.	1	2	3	4	5
9	Grants the nurse manager a lot of power and authority to supervise nurses' professionalism.	1	2	3	4	5
10	Provides nurses with the required means to learn while at work (library, Internet).	1	2	3	4	5
11	Seriously considers nurses' suggestions to improve caring.	1	2	3	4	5
12	Allocates large amounts of money to professional training programs for nurses.	1	2	3	4	5
13	Puts a lot of effort into improving patients' safety in the hospital (preventing errors).	1	2	3	4	5
14	Conducts audits on nurses' caring for patients in the hospital.	1	2	3	4	5
15	Tries to improve communication between nurses and doctors in the hospital every year.	1	2	3	4	5
16	Refers to the issue of patients' satisfaction in every meeting between hospital management and nurses.	1	2	3	4	5
17	Invests in the improvement of nurses' teamwork.	1	2	3	4	5
18	Asks nurses to report on problems in working relations with doctors in the ward.	1	2	3	4	5
19	Distributes patients' satisfaction reports to the wards.	1	2	3	4	5
20	Stresses the importance of a multiprofessional teamwork (nurses–doctors) in the ward.	1	2	3	4	5

UNIT CLIMATE

The following questions refer to the condition in your ward. Circle the most suitable number in each sentence, in your opinion. (1 = Least agree; 5 = Most agree.)

In Our Ward I Am Encouraged/Required/Expected to:

1. Know how to deal with families' tough reactions. 1 2 3 4 5
2. Think ahead prior to every action (continuously fight the routine). 1 2 3 4 5
3. Get constantly updated on new professional knowledge. 1 2 3 4 5
4. Share information with team members (not to withhold information). 1 2 3 4 5
5. Check the patient's sheet for relevant sensitivities before providing any type of medication. 1 2 3 4 5
6. Demonstrate personal resourcefulness and be able to learn on my own. 1 2 3 4 5
7. Communicate with doctors only through the nurse manager. 1 2 3 4 5
8. Comprehensively inform patients about their treatment. 1 2 3 4 5
9. Make sure every patient has comfortable surroundings (light, temperature, noise, etc.). 1 2 3 4 5
10. Report every near-miss event (even if no harm done, and nobody witnessed). 1 2 3 4 5
11. Ask for explanation on every medication, so I understand its purpose and effects. 1 2 3 4 5
12. Inform a colleague when a patient complains about him/her. 1 2 3 4 5
13. Provide for the patient's needs, even if not directly asked to. 1 2 3 4 5
14. Examine every near-miss event in order to learn from it. 1 2 3 4 5
15. Welcome criticism from doctors on professional issues. 1 2 3 4 5
16. Propose ideas to make a more comfortable environment for the patients. 1 2 3 4 5
17. Always give medications on time (even during busy hours). 1 2 3 4 5
18. Notice any patient's irregularities (even if he/she is not under my responsibility). 1 2 3 4 5
19. Consider my colleagues' needs when coordinating vacations and absences. 1 2 3 4 5
20. Point out any unprofessional conduct by nurses (even if not exceptional). 1 2 3 4 5
21. Place nursing professionalism as a genuine top priority (not just as a slogan). 1 2 3 4 5
22. Invest time in clear guidance to patients (make sure they understand). 1 2 3 4 5
23. Insist that doctors rewrite unclear instructions before I carry them out. 1 2 3 4 5
24. Direct team members' attention to medications which are not commonly being used in the ward. 1 2 3 4 5
25. Approach doctors with professional questions (not necessarily related to a specific patient). 1 2 3 4 5
26. Attend as many professional courses as possible. 1 2 3 4 5
27. Frequently check my noting in the kardex. 1 2 3 4 5
28. Share the workload with other team members. 1 2 3 4 5
29. Seek information whenever a question arises. 1 2 3 4 5
30. Significantly contribute to the development and maintenance of good working relations within the team. 1 2 3 4 5

REFERENCES

1. Kohn, L.T., Corrigan, J.M, and Donaldson, M.S., eds. (1999). *To Err Is Human: Building a Safer Health System*. Washington, DC: Institute of Medicine, National Academy Press.
2. Blendon, R., DesRoches, C., Brodie, M., Benson, J., Rosen, A., Schneider, E., Altman, D., Zapert, K., Herrmann, M., and Steffenson, A. (2002). Views of practicing physicians and the public on medical errors. *The New England Journal of Medicine*, 347(24), 1933–1940.
3. Leape, L.L., Brennan, T., Laird, N., Lawthers, A., Localio, R., Barnes, B., et al. (1991). The nature of adverse events in hospitalised patients: Results of the Harvard Medical Practice Study II. *New England Journal of Medicine*, 324, 377–384.
4. Andrews, L., Stocking, C., Krizek, T., Gottlieb, L., Krizek, C., Vargish, T., et al. (1997). An alternative strategy for studying adverse events in medical care. *The Lancet*, 349, 309–313.
5. Wears, R.L. and Berg, M. (2005). Computer technology and clinical work: Still waiting for Godot. *Journal of the American Medical Association*, 293(10), 1261–1263.
6. Brennan, T.A., Gawande, A., Thomas, E., and Studdert, D. (2005). Accidental deaths, saved lives, and improved quality. *New England Journal of Medicine*, 353, 1405–1409.
7. Reichers, A.E. and Schneider, B. (1990). Climate and culture: An evolution of constructs. In: B. Schneider (ed.), *Organizational Climate and Culture*. San Francisco: Jossey-Bass, pp. 5–39.
8. Gonzalez-Roma, V., Peiro, J.M., and Tordera, N. (2002). An examination of the antecedents and moderator influences of climate strength. *Journal of Applied Psychology*, 87(3), 465–473.
9. Lindell, M.K. and Brandt, C.J. (2000). Climate quality and climate consensus as mediators of the relationship between organizational antecedents and outcomes. *Journal of Applied Psychology*, 85, 331–348.
10. Schneider, B., Salvaggio, A.N., and Subirats, M. (2002). Climate strength: A new direction for climate research. *Journal of Applied Psychology*, 87, 220–229.
11. Zohar, D. (2000). A group-level model of safety climate: Testing the effect of group climate on micro-accidents in manufacturing jobs. *Journal of Applied Psychology*, 85, 587–596.
12. Zohar, D. (2003). Safety climate: Conceptual and measurement issues. In: J.C. Quick and L.E. Tetrick (eds), *Handbook of Occupational Health Psychology*. Washington, DC: American Psychological Association, pp. 123–142.
13. Hackman, J.R. (2003). Learning more by crossing levels: Evidence from airplanes, hospitals and orchestras. *Journal of Organizational Behavior*, 24, 905–922.
14. DeJoy, D.M., Murphy, L.R., and Gershon, R.R.M. (1995). Safety climate in health care settings. In: A.C. Bittner and P.C. Champney (eds.), *Advances in Industrial Ergonomics and Safety VII*. New York: Taylor & Francis.
15. Felknor, S.A., Aday, L.A., Burau, K.D., Delclos, G.L., and Kapadia, A.S. (2000). Safety climate and its association with injuries and safety practices in public hospitals in Costa Rica. *International Journal of Occupational and Environmental Health*, 6, 18–25.
16. Singer, S.J., Gaba, D.M., Geppert, J.J., et al. (2003). The culture of safety: Results of an organization-wide survey in 15 California hospitals. *Quality Safety Health Care*, 12, 112–118.
17. Sorra, J. and Nieva, V. (2003). *Psychometric Analysis of the Hospital Survey on Patient Safety* (Final Report to Agency for Healthcare Research and Quality). Washington, DC: AHRQ.
18. Argyris, C. and Schon, D. (1996). *Organizational Learning II: Theory, Method, and Practice*. Reading, MA: Addison-Wesley.

19. Schneider, B. and Reichers, A.E. (1983). On the etiology of climates. *Personnel Psychology*, 36, 19–39.
20. Luthans, F. and Kreitner, R. (1985). *Organizational Behavior Modification and Beyond*. Glenview, IL: Scott, Foresman.
21. Stajkovic, A.D. and Luthans, F. (1997). A meta-analysis of the effects of organizational behavior modification on task performance, 1975–95. *Academy of Management Journal*, 40, 1122–1149.
22. Schneider, B., White, S., and Paul, M.C. (1998). Linking service climate and customer perceptions of service quality: Test of a causal model. *Journal of Applied Psychology*, 83, 150–163.
23. Hofman, D.A. and Morgeson, F.P. (1999). Safety related behavior as a social exchange: The role of perceived organizational support and leader member exchange. *Journal of Applied Psychology*, 84, 286–296.
24. Gershon, R., Stone, P., Bakken, S., and Larson, E. (2004). Measurement of organizational culture and climate in health care. *Journal of Nursing Administration*, 34(1), 33–40.
25. Pronovost, P.J., Weast, B., Holtzmueller, C.G., Rosenstein, B.J., Kidwell, R.P., Haller, K.B., Feroli, E.R., Sexton, J.B., and Rubin, H.R. (2003). Evaluation of the culture of safety: Survey of clinicians and managers in an academic medical center. *Quality Safety Health Care*, 12, 405–410.
26. Podsakoff, P.M., MacKenzie, S.B., Lee, J.Y., and Podsakoff, N.P. (2003). Common method biases in behavioral research: A critical review of the literature and recommended remedies. *Journal of Applied Psychology*, 88, 879–903.
27. Katz-Navon, T., Naveh, E., and Stern, Z. (2005). Safety climate in healthcare organizations: A multidimensional approach. *Academy of Management Journals*, 48, 1073–1087.
28. Fink, J.L. (1983). Preventing lawsuits: Medication errors to avoid. *Nursing Life*, 3(2), 26–29.
29. Fuqua, R.A. and Stevens, K.R. (1988). What we know about medication errors: A literature review. *Journal of Nursing Quality Assurance*, 3(1), 1–17.
30. Leape, L.L., Bates, D.W., Cullen, D.J., et al. (1995). Systems analysis of adverse drug events. *Journal of the American Medical Association*, 274(1), 35–43.
31. Rainbow, J. (1984). Six legal safeguards versus drug errors, *Nursing Life*, 4(1), 56–58.
32. Gibson, F. and Soanes, L. (2000). The development of clinical competencies for use on a pediatric oncology nursing course using nominal group technique. *Journal of Clinical Nursing*, 9, 459–469.
33. Page, A. (2003). *Keeping Patients Safe: Transforming the Work Environment of Nurses*. Washington, DC: Institute of Medicine, National Academy Press.
34. IOM (Institute of Medicine). (2001). *Crossing the Quality Chasm: A New Health System for the 21st Century*. Washington, DC: National Academy Press.
35. Bulechek, G., McCloskey, J., Titler, M., and Denehey, J. (1994). Report on the NIC project: Nursing interventions used in practice. *American Journal of Nursing*, 94(10), 59–66.
36. ANA (American Nurses Association). (1998). *Standards of Clinical Nursing Practice*. Washington, DC: ANA.
37. Benner, P., Sheets, V., Uris, P., Malloch, K., Schwed, K., and Jamison, D. (2002). Individual, practice, and system causes of errors in nursing. *Journal of Nursing Administration*, 32(10), 509–523.
38. Mitchell, P. and Shortell, S. (1997). Adverse outcomes and variations in organization of care delivery. *Medical Care*, 35(11), NS19–32.
39. Shortell, S., Zimmerman, J., Rousseau, D., Gillies, R., Wagner, D., Draper, E., Knaus, W., and Duffy, J. (1994). The performance of intensive care units: Does good management make a difference? *Medical Care*, 32(5), 508–525.

40. James, L.R., Demaree, R.G., and Wolf, G. (1993). RWG: An assessment of within-group inter-rater agreement. *Journal of Applied Psychology*, 78, 306–309.
41. Flin, R. and Yule, S. (2004). Leadership for safety: Industrial experience. *Quality Safety Health Care*, 13, ii45–ii51.
42. Zohar, D. (2002). Modifying supervisory practices to improve sub-unit safety: A leadership-based intervention model. *Journal of Applied Psychology*, 87, 156–163.
43. Zohar, D., Livne, Y., Tenne-Gazit, O., Admi, H., and Donchin, Y. (2007). Healthcare climate: A framework for measuring and improving patient safety. *Critical Care Medicine*, May 35(5):1312–1317.

20 Beyond Fatigue
Managerial Factors Related to Resident Physicians' Medical Errors

Zvi Stern, Eitan Naveh, and Tal Katz-Navon

CONTENTS

Active Learning...286
Method..287
Autonomy...288
Organizational Voice..289
Standardization...290
References..291

Physicians, after completing their studies in medical school, begin a 5-year training period, known as a residency, to become expert physicians in their field. During the residency period, residents work long, consecutive hours including night shifts in which they have only a few hours of sleep. Therefore, residency is considered a difficult and stressful period that is challenging both physically and mentally. However, the long working hours are part of the traditional method of training young physicians and are seen as essential to their learning process.

Resident physicians are the main workforce in teaching hospitals. They bear great responsibility despite being inexperienced and are only just beginning their professional path. The tremendous responsibility on the one hand and lack of experience and knowledge on the other creates fertile ground for errors as is evident from studies published on this topic.[1,3] In recent years, studies have demonstrated a significant correlation between residents' long working hours and sleep deprivation, and the rate at which they make errors. For example, research has shown that the rate at which serious medical errors are made by residents who worked long shifts was higher by 35.9% than among the residents in the intervention group who worked shorter shifts, and the rate of the residents' diagnostic errors was 5.6 times higher.[4] Moreover, it was found that long shifts and sleep deprivation also threaten the residents themselves, since at the end of a shift of 24 hours or more their driving skills are similar to drivers with a blood alcohol level of 0.04%–0.05%,[5] and the probability of their getting into a car accident when driving home is over double the average level.[6]

As a result of these studies, the residents' work hours have been limited in the past few years. In some countries a general restriction has been set, limiting the scope of residents' work hours to between 60 and 80 hours a week. In Israel, a physician is on duty for 45 hours a week and the resident works an extra four to eight shifts a month. The average workweek of a resident in Israel amounts to 69 hours. However, despite this limitation there has been no actual change in the rate of the errors made by residents.[1,7,8] In this chapter, based on a series of studies we conducted, we point to several organizational and managerial factors that may reduce the number of errors among resident physicians.

ACTIVE LEARNING

Active learning refers to the accumulation of skills by active work experience. It refers to learning new skills and improving existing ones, active experimentation, and testing out new ways to perform the work. This is in contrast to passive learning, in which professional studies occur outside the workplace, such as in classrooms. Active learning is essential to the residents' learning process since their work is characterized by high uncertainty levels and their tasks are complex, varied, and nonroutine. Passive learning of work procedures cannot encompass all possible situations; therefore, a more flexible type of learning is needed, through active hands-on experience. For example, although residents can simulate an operation using a specialized apparatus in order to gain practice, they must also experience participating in a real operation, much like pilots who train on a flight simulator, but must actually fly real aircraft in order to learn.

An active learning climate is the degree to which the organization or department emphasize active learning. A high climate of active learning in the department is essential to the resident's learning and to the development of his/her knowledge and skills. However, when the active learning climate is high, the learner explores his/her environment actively and tries out new options, sometimes by trial and error, which inevitably leads to the commission of errors. The cost of these potential medical mishaps or errors is very high because it can be a matter of life and death. Thus, it is necessary to find the middle ground between the need for a high learning climate that advances the residents' skills and abilities, and the fact that active learning entails errors and mishaps that can cause potentially serious injury.[17]

In order to integrate a resident's active learning and avoid errors as much as possible we suggest two systematic-managerial factors that can reduce the number of medical errors caused by a high active learning climate.

The first factor is the department's safety climate. Safety climate is defined as the degree to which the members of an organization communally perceive the importance of a safe work environment versus the importance of achieving other organizational goals, such as productivity or efficiency.[9,12,13] Studies have found a correlation between safety climate and ensuring employee safety behaviors and a lower rate of accidents among workers.[15]

The second factor is the degree to which the employee's direct manager is concerned about safety. The senior management and the direct manager influence the employees' behaviors by setting a personal example and by formally and informally rewarding desirable behaviors. The department managers' attitudes are particularly significant in promoting safety, since they come into daily contact with the residents, supervise safe behaviors, admonish residents if safety rules are not followed, and identify and resolve potential hazards.[16]

Managers who place importance on safety and act safely themselves influence their subordinates throughout the organizational hierarchy. Positive attitudes of managers and supervisors toward safety are reflected by high degrees of safety awareness at different levels of the organization. However, when managers do not convey safety importance, we can expect that this will in turn affect the attitudes and performance of more junior employees, and safety will not be viewed as important.[15,16] Indeed, studies find that if the management, especially the direct manager, take a strong stance in supporting safe behaviors this correlates strongly and significantly to a reduced number of work accidents and to the employees' awareness that they are responsible for safety.

METHOD

The study was conducted in two teaching hospitals, each of which treats over 100,000 patients annually. The study was based on questionnaires distributed to all the residents in both hospitals, totaling 123 residents from 25 departments (surgery, anesthesia, cardiology, gastroenterology, orthopedics, OB/GYN, emergency medicine, and pediatrics). The response rate was 80%. Two-thirds of respondents worked in 15 departments in one hospital and the remaining third in 10 departments in the other hospital. The number of respondents in each department ranged from 3 to 13. Seventy percent of respondents were male. The average age of the residents who participated in the study was 32.5 (SD 3.93); 68.6% were in the first to third years of their residency and 31.4% were in their fourth to fifth years.

The active learning climate was assessed with five items adapted from Ames and Archer (1988) and from Migdley et al. (1998),[27] such as "To what extent in your department are you encouraged to try new things at work?" and "We search for new ways to improve professionally." The safety climate was assessed with six items, such as "In our department, in order to do the work, some aspects of safety must be overlooked," and "When there is pressure at work, it is preferable to work as quickly as possible even if safety is compromised." Finally, the commitment of the direct supervisor to safe behaviors was assessed with five items, such as "Our department manager supervises us more closely after someone in the team violates safety instructions."[16]

To examine within-department agreement we used the r_{wg} agreement index and intraclass correlations (ICCs) were used to justify aggregating individual responses to the department average. The active learning climate, safety climate, and the department supervisor's commitment to safety produced sufficiently high measures to justify aggregating these variables to the department level. Thus, an average score was calculated for each department for each of the three scales, by averaging the ratings of the residents and assigning each resident his or her department mean score.

Treatment errors were defined as "any error in the performance of an operation, procedure, or test; in the administration of treatment; in the dosage or method of using a drug; and in general, any inappropriate treatment that resulted in an accident, that is, harm to the patient."[1] Based on this definition we asked four expert physicians to compose a list of potential treatment errors that were preventable, severe, and identifiable. The final list included 12 potential errors.

A senior nurse was asked to count the number of treatment errors from the list made by each resident who completed the independent variable questionnaire over the last 3 months.[1] Each resident was evaluated by a different nurse who worked alongside the resident and was personally familiar with his/her work. In order to check inter-rater reliability, we asked 20 additional nurses to evaluate the same residents. Four control variables were also measured. The first one was the year of residency, which ranged from 1 to 5 (average 2.81, SD = 1.35). This variable controlled for the participant's level of experience and knowledge. The literature on treatment errors among residents emphasizes the critical influence of the resident's fatigue on his/her tendency to err. The level of fatigue is associated with the learning strategies used for complex tasks, especially among people who are not familiar with the tasks. Therefore, the second control variable was the number of night shifts that the resident reported working during the last 3 months (mean 18.37, SD = 5), as a proxy to the resident's level of fatigue. In order to effectively partial out confounding variables related to the hospital and the department and that may influence the findings, we dummy coded for the hospital and the residents' departments.

Results of hierarchical linear models (HLM) demonstrated, as expected, that the higher the department learning climate was, the greater was the number of the residents' errors. But the number of medical errors due to the active learning climate decreased by approximately 50% when there were both high learning climate and a high safety climate in the department or when the residents' supervisor was highly committed to safety. The safety climate and the commitment of the supervisor to safe behaviors reduced the number of errors resulting from a high learning climate.

AUTONOMY

Employee autonomy refers to the degree of freedom and discretion that employees have to plan, schedule, and carry out their jobs as they see fit.[18] Traditionally, physicians have always emphasized their professional autonomy.

Studies show that employees' autonomy is positively associated with work outcomes because it allows the employees who are most knowledgeable about the task to make decisions relating to that task. Moreover, role autonomy has motivational value, since it stimulates a sense of responsibility for the work and its consequences, whether they are positive or negative results.[18,19] However, more recent studies found that the relationship between the worker's autonomy and work performance is not necessarily positive and therefore, a question arises as to what is the correct level of autonomy to grant an employee?

The work of residents is complex and diverse, requiring autonomy and discretion. Without autonomy residents cannot complete their tasks independently, but need to wait for instructions from their superiors. For example, a problem may emerge if a situation requires an immediate decision and the supervisor is not available. On the other hand, residents are novices in the first stages of learning their profession and often lack the knowledge and experience required to deal with various situations. Therefore, granting too much autonomy may result in errors, because the novice resident needs regular guidance and supervision. Thus, it is crucial to find the appropriate degree of autonomy for residents, allowing them to use their judgment to the extent necessary to make decisions and yet provide the essential guidance and supervision.

Following the research method described above, we asked residents to assess the degree of autonomy given to them within their department. The level of perceived autonomy was measured with four items based on Hackman and Oldham,[18] such as "in your daily work, to what extent do you decide on the order of doing your tasks?" "Do you plan your work?"

Results of hierarchical linear model analysis, as expected, demonstrated a curvilinear relationship between the residents' degree of perceived autonomy and the number of errors committed when the active learning climate in the department was low. In contrast, when the departmental learning climate was high, the higher the perceived degree of autonomy was, the lower was the number of errors (from 8 errors on average to almost zero errors; data per resident over a 3-month period).[20] In other words, autonomy improves the residents' performance when they are also learning, experimenting, and accumulating knowledge. In the absence of learning, too much autonomy may harm their performance (from approximately zero errors when learning to about four errors when there is no learning and the level of autonomy is high; data per resident over a 3-month period).

ORGANIZATIONAL VOICE

The term *organizational voice* refers to the degree that employees feel it is useful to express their opinions to senior managers. When employees in an organization feel they are entitled to express their opinions freely, discuss problems with the supervisors, offer solutions and seek help, the organizational voice level is high.[21] Whereas *organizational silence* is characterized by employees' perception that expressing their opinion within the organization and sharing information, ideas, and concerns is not worth the effort. In this case, employees believe that their words would be ignored and would not be likely to effect a change. Moreover, expressing their opinions might place them at risk so they are afraid to talk. Thus, they prefer to remain silent and keep the information to themselves.

An employee decides whether to speak up or remain silent based on organizational cues. For example, if the management of the organization is perceived as willing to listen, has an open-door policy, and is supportive and genuinely interested in improvement, workers will tend to express their opinions, bring up new ideas, and ask questions. Whereas if the organizational environment is perceived by the employee as threatening, and he/she believes that proposing new ideas will not be beneficial, and might even prompt anger, punishment, or ridicule, the employee will undoubtedly prefer to keep silent about his/her ideas and opinions.

A high organizational voice enables improvements in work processes and safety, since employees' ideas, suggestions, and knowledge can help identify risks and hazards, and facilitate improvements by implementing various courses of action. For example, if residents believe that asking for the help or consultation of a senior physician on-call during the night shift will be met with a dismissive response, or be interpreted as a lack of knowledge, they will refrain from contacting senior doctors, thus risking errors in medical treatment. On the other hand, if residents feel comfortable in calling, consulting, directing attention, raising ideas that can lead to improvement and discussing and learning from mistakes, the information derived from these types of communication may contribute to improvement and error prevention.[20,22]

In the study described above, we also asked the resident physicians to assess the degree of the organizational voice in the department. Organizational voice was assessed with six items based on LePine and Van Dyke (1998), such as "In your department, to what extent can residents feel free to come to senior doctors with any question?" and "Residents feel free to address the department managers when they find a problem in the department's work processes."

Results of hierarchical linear model analysis demonstrated that when the active learning climate in the department was high, a high organizational voice led to a reduction in the number of errors (approximately seven errors when the organizational voice was low to one error when the organizational voice was high; data from residents over a period of 3 months). When the active learning climate was low, the organizational voice had no impact on the number of errors.[20]

STANDARDIZATION

Standardization refers to structured organizational processes defined by practices, procedures, protocols and manuals that state the manner in which work is to be performed, and are based, in part, on past learning. Standardization determines the actions required and the order of executing them so that the job is safely completed.[23] Formally, organizations react to safety breaches by refreshing safety procedures and adding new procedures and implementing the standards for work processes. Standardized work processes have a positive effect on patient safety in hospitals because they promote the transfer of knowledge between physicians on the proper way to perform various activities.

Residents may perceive the hospital safety procedures as being different than what the organization formally intended, for example, residents may perceive in different levels the degree of detailing involved in safety regulations, their degree of relevance to all components of their work, and the degree to which they contribute to or hamper their work.[16] These procedures may be perceived as impeding work, impossible to implement, too detailed, and a nuisance or burden that takes up too much time and human resources. Residents may alternatively believe that there are actually too few safety procedures that do not cover all aspects of the work or do not provide sufficient guidelines for safe behavior. As such, we concluded that what is needed is an optimal level of safety procedures, which allow residents to employ procedures in their daily work but also allows them to maintain the required autonomy, thus preventing over-rigidity of work processes.

We also propose that standardization is an effective tool for coping with the negative effects of fatigue on the resident's work. Working according to imposed work practices prevents shortcuts and "cutting corners." For example, an article in the *New England Journal of Medicine (NEJM)* (2009) found that checklists prevented a large number of errors in the operating rooms.[24] Moreover, when the level of fatigue is high, the cognitive performance decreases, resulting in confusion, forgetfulness, and erroneous judgments. Thus, working "by the book" or according to checklists will prevent errors. In a profession in which judgment is a key component, physicians occasionally oppose the institution of standards that limit their discretion in certain situations. But it seems that when residents are fatigued, the standardization of work processes may actually prevent errors.[25]

In summary, we presented several studies conducted on the relationship between managerial/systemic factors and the number of errors committed by resident physicians. These are factors that the organization has control over and can influence change in the hospital. Although there is no doubt that long work hours and resident fatigue are the basis for errors[1,7,8] our research shows that errors are also caused by the overall set of system-wide factors in the residents' environment. Organizational and managerial factors may mitigate or amplify the negative impact of fatigue on the resident physicians' work. Given that budgetary constraints do not *allow a further reduction of the residents' work hours, hospitals may be helped by the information presented in this chapter in their campaign to reduce residents' medical errors.*

REFERENCES

1. Landrigan, C.P., Rothschild, J.M., Cronin, J.W., Rainu, M.D., et al. 2004. Effect of reducing interns' work hours on serious medical errors in intensive care units. *New England Journal of Medicine*, 351, 1838–1848.
2. Leape, L.L. and Berwick, D.M. 2005. Five years after *To Err Is Human: What Have We Learned? JAMA*, 293, 2384–2390.
3. Stern, Z. 2003. Opening Pandora's box: Residents' work hours. *International Journal for Quality in Healthcare*, 15, 103–105.
4. Landrigan, C.P., Czeisler, C.A., Barger, L.K., Ayas, N.T., Rothschild, J.M., and Lockley, S.W. 2007. Effective implementation of work-hour limits and systemic improvements. *The Joint Commission Journal on Quality and Patient Safety*, 33, 19–29.
5. Arnedt, J.T., et al. 2005. Neurobehavioral performance of residents after heavy night call vs. after alcohol ingestion. *Journal of American Medical Association*, 294(9), 1025–1033.
6. Berger, L.K., et al. 2005. Extended work shifts and the risk of motor vehicle crashes among interns. *New England Journal of Medicine*, 352(2), 125–134.
7. Gaba, D.M. and Howard, S.K. 2002. Fatigue among clinicians and the safety of patients. *New England Journal of Medicine*, 347, 1249–1255.
8. Lockley, S.W., Cronin, J.W., Evans, E.E., and Case, B.E. 2004. Effect of reducing interns' weekly work hours on sleep and attentional failures. *New England Journal of Medicine*, 351, 1829–1838.
9. Zohar, D. 2000. A group-level model of safety climate: Testing the effect of group climate on microaccidents in manufacturing jobs. *Journal of Applied Psychology*, 85, 587–596.
10. Vaughan, D. 1990. Autonomy, Interdependence, and Social Control: NASA and the Space Shuttle Challenger. *Administrative Science Quarterly*, 35, 225–257.

11. Schein, E.H. 2004. *Organizational Culture and Leadership*, 3rd ed. San Francisco: Jossey-Bass.
12. Schneider, B., ed. 1990. *Organizational Climate and Culture.* San Francisco: Jossey-Bass.
13. Schneider, B., White, S.S., and Paul, M.C. 1998. Linking service climate and customer perceptions of service quality: Test of a causal model. *Journal of Applied Psychology,* 83, 150–163.
14. Zohar, D. 2002. Modifying supervisory practices to improve subunit safety: A leadership-based intervention model. *Journal of Applied Psychology*, 87, 159–163.
15. Naveh, E., Katz–Navon, T., and Stern Z. 2005. Treatment errors in healthcare: A safety climate approach. *Management Science*, 51(6), 948–960.
16. Katz–Navon, T., Naveh, E., and Stern, Z. 2005. Safety climate in healthcare organizations: A multidimensional approach, *Academy of Management Journal*, 48, 1075–1089.
17. Katz-Navon, T., Naveh, E., and Stern, Z. 2009. Active learning—When more is better? The case of resident physicians' medical errors. *Journal of Applied Psychology*, 94, 1200–1209.
18. Hackman, J.R. and Oldham, G.R. 1976. Motivation through the design of work: Test of a theory. *Organizational Behavior and Human Performance*, 16, 250–279.
19. Langfred, C.W. and Moye, N.A. 2004. Effects of task autonomy on performance: An extended model considering motivational, informational, and structural mechanisms. *Journal of Applied Psychology*, 89, 934–945.
20. Stern, Z., Katz-Navon, T., and Naveh, E. 2008. The influence of situational learning orientation, autonomy and voice on error making: The case of resident physicians, *Management Science*, 54, 1553–1564.
21. Morrison, E.W. and Milliken, F.J. 2000. Organizational silence: A barrier to change and development in a pluralistic world. *Academy of Management Review*, 25, 706–725.
22. Naveh, E., Katz-Navon, T., and Stern, Z. 2006. Readiness to report medical treatment errors: The effects of safety procedures, safety information, and priority of safety. *Medical Care*, 44, 117–123.
23. Naveh, E. 2007. Formality and discretion in successful R&D projects. *Journal of Operations Management*, 25, 110–125.
24. Haynes, A.B., Weiser, T.G., Berry, W.R., Lipsitz, S.R., et al. 2009. A surgical safety checklist to reduce morbidity and mortality in a global population. *New England Journal of Medicine*, 360(5), 491–499.
25. Stern, Z., Katz-Navon, T., Levtzion-Korach, O., and Naveh, E. Resident physicians' level of fatigue and medical errors: The role of standardization. *International Journal of Behavioral and Healthcare Research*, 1(3), 223–233.
26. Katz-Navon, T., Naveh, E., and Stern, Z. 2007. The moderate success of quality of care improvement efforts: Three observations on the situation. *International Journal for Quality in Health Care*, 19, 4–7.
27. Migdley, C., Kaplan, A., Middleton, M., Maehr, M.L., Urdan, T., Anderman, L.H., Anderman, E., and Roeser, R. 1998. The development and validation of scales assessing students' achievement goal orientations. *Contemporary Educational Psychology*, 23, 113–131.

21 Gentle Rule Enforcement

Ido Erev and Dotan Rodensky

CONTENTS

The Value of Gentle Rule Enforcement 294
Hand Washing, Gloves, and Gentle Rule Enforcement in Hospitals 295
How Does It Work? 297
Summary and Conclusions 297
References 298

The relationship between the legal and medical systems resembles a conversation between two people who do not speak each other's language. The lawyer sees the problem and looks for a guilty party, whereas the physician and human factors engineer look for the source and cause of the problem, in terms of human behavior, to find a way to prevent similar failures in the future.

The law tries to eliminate problems and errors in medical settings through handing out an appropriate punishment. Medical malpractice that results in injury requires that the victim be compensated for damages; toward that end, the culprit must be found and punished. The implicit assumption is that physicians must be deterred from negligent behavior.

The focus on punishing malpractice that results in injury implies that only a small fraction of the malpractice incidences can be punished. This implication is a product of the fact that most incidences do not lead to injuries, and the fact that the cost of legal punishment is very high. Thus, only a small part of the malpractice incidences can be detected, and only a small part of the detected incidences can be punished. The legal system addresses this "low probability of punishment" problem by using severe punishments.

The "severe punishment" solution can be justified based on classical research in economics. One justification is Becker's[1] analysis that assumes risk aversion: Assuming that people are risk averse the magnitude of the punishment is more important than its probability. A second justification uses Kahneman and Tversky's[5] experimental analysis. Their results and descriptive model suggest oversensitivity to rare events (such as low probability punishment) and oversensitivity to losses.

However, empirical research suggests that there are good reasons to doubt the effectiveness of the severe punishments solution in the context of typical medical systems. The best-known problem of the focus on severe punishments involves the fact that this focus leads hospitals to employ an expensive risk management apparatus that functions to prevent the legal system from proving negligence. Yet, the investment in "safety management" is minor.

A related problem involves the fact that the hospitals' reactions to the risk of legal consequences includes firing employees, experienced doctors and nurses who were previously successful in their roles who were discharged before they could apply what they had learned about the error they committed (labeled as *negligence* by the plaintiff, otherwise there would be no cause for action).

A third reason to doubt the effectiveness of the severe punishment solution comes from recent experimental studies of the effect of experience on choice behavior. This research reveals large differences between decisions that are made based on a description of the incentive structure and decisions that are made based on experience. People tend to exhibit oversensitivity to low probability events when they rely on a description of the incentive structure (as in the situation examined by Kahneman and Tversky),[5] but to exhibit the opposite pattern in decision from experience (see review in Hertwig and Erev).[4] Experience leads people to behave as if they believe that "it won't happen to me."

One study that demonstrates this pattern focuses on a choice between the status quo (0 with certainty) and a gamble that leads to "a gain of 1 Shekel (about $0.25) in 90% of the cases, and a loss of 10 otherwise (probability of 10%)." When subjects were presented with a description of the gamble and were asked if they want to play it, only 25% said "yes." Most subjects preferred the status quo and avoided this bad gamble (it is bad in the sense that the average payoff from playing it is negative). However, when subjects were presented with this choice for 100 trials with immediate feedback after each trial, they tend to prefer the bad gamble: The gamble is selected in about 60% of the trials. That is, experience leads people to behave as if they underweight the rare loss.

Additional research[6] (Yechiam et al.) shows that planning decisions are similar to decisions from description. When people are asked to plan their 100 decisions in advance, they tend to exhibit high sensitivity to rare losses. However, experience often leads them to deviate from their plan and prefer the bad gamble.

In summary, the current analysis suggests that the use of severe punishments to reduce malpractice in medical settings can be costly and ineffective. The cost involves the fact that it leads hospitals and insurers to follow counterproductive defensive policies, and the limited ineffectiveness is a result of the tendency of the personnel to underweigh rare punishments.

THE VALUE OF GENTLE RULE ENFORCEMENT

The shortcomings of severe punishments imply that, in certain settings, gentle rule enforcement can be more effective than a deterrence policy that relies on severe penalties. Gentle rule enforcement is likely to be particularly effective when two conditions are met: (1) the probability of detecting rule violation is high, and (2) the subjects of the rule plan to obey it[2] (Erev et al.). Severe punishments are unnecessary in these settings. It is often enough to remind people when their behavior deviates from their own plan. Assuming that the reminder is unpleasant to the receiver (and the workers prefer to avoid it), this gentle method can eliminate rule violation.

We first tested this optimistic prediction in studies conducted in 11 factories in Israel. The results suggest that it was easy for managers to accept the ideas underlying gentle rule enforcement. Senior managers were aware of the shortcomings of severe punishment (specifically, they note that the middle-class managers do not want to use severe punishment), and therefore they supported the recommended program. The study was conducted over a period of 3 years. The predictions were substantiated—and the studies resulted in an increased percentage of safe behaviors relative to the period prior to the experiment.

HAND WASHING, GLOVES, AND GENTLE RULE ENFORCEMENT IN HOSPITALS

In 1856, Ignaz Philipp Semmelweis argued that maternal mortality is caused by physicians who carry some unknown factor (it was only a few years later that Pasteur discovered the real cause). To get rid of this factor Semmelweis ordered the doctors to wash their hands properly.

However, the determination that the physicians' hands were contaminated did not take hold. Ignaz knew that his suggestion could save lives but his ideas were rejected and he went insane. He was admitted to a mental asylum where he was murdered by his therapists. Today no surgeon dares operate without carefully and extensively scrubbing his/her hands and donning sterile gloves and a sterile gown. The risk of infecting the surgical wound is a clear and present danger. In contrast, in nonsurgical hospital departments, the staff is not careful to wash their hands, a problem exacerbated by the fact that they may have to wash their hands dozens of times during the course of one working day yet a sink and a faucet are not always readily available.

Many different methods have been used to try and regulate medical staff's compliance to procedures such as washing their hands and wearing gloves, without significant results. One of the methods tested was eliminating the requirement to wash hands with soap and water; instead, hospital workers were given small vials of antiseptic solution. This method has proven itself in several places, but only when it was integrated into specific educational programs in which the entire medical staff was rallied to fight infection. Based on the analysis described above, and in light of the positive findings in the studies in industrial plants, an experiment was conducted in several departments of a hospital in northern Israel (Erev et al.).[3]

The purpose of the experiment was to examine the effect of gentle and mutual enforcement on a number of tasks performed by medical staff:

- Wearing a glove while drawing blood samples from a patient.
- Changing gloves after each procedure.
- Washing hands whenever required before and after a procedure on a patient.

The study was initiated in August 2005 by observing the current status with respect to these three tasks. After obtaining consent from the department managers and primary nurses, the experiment began.

The first step in implementing behavioral change is to present the idea of gentle rule enforcement to the staff, and assure participants that it is not a compulsory directive, rather a proposal for collaboration. The staff of five hospital wards was presented with the objectives of gentle rule enforcement, the target behaviors, and the principles and methods of implementing it. The presentation was given during the routine morning department meetings, in which current management issues are generally discussed. For the most part, the entire nursing and medical staff were present, including the night staff, who completed their shift, personnel arriving for the morning shift, and the head nurses of the hospital. After presenting the issue, we allowed the team members to express their opinions. Physicians said that wearing a glove while drawing blood interferes with the delicate tactile sensation needed to insert the needle properly.

The department head explained the importance of wearing the glove in protecting both the patient and the staff member taking the sample. The nursing staff's questions were answered by the head nurses. It was clear to all that the issue of the convenience or inconvenience of the glove was not relevant when withdrawing blood samples from a patient with an infectious disease, such as hepatitis, and since the immunological status of most patients is unknown, it is appropriate to protect and be protected—this was the spirit of the responses. In the four other departments (skin, cardiology, rheumatology, and outpatient) which served as the controls, the subject was presented in a lecture on the importance of these procedures, but enforcement was not included.

The enforcement process. Enforcement among the physicians was performed by the department head and his deputy, assisted by senior physicians or a senior physician who volunteered for this, and by having colleagues provide each other with reminders when any of them detected a failure to follow instructions or a deviation from the correct procedure. For example, doctors were asked to remind their coworkers when they forgot to change gloves between blood collection rounds. With respect to the nursing staff, the head nurse or his/her deputy enforced the proper, safe procedures and sometimes the nurses and their colleagues engaged in mutual enforcement. For example, nurses were encouraged to remind their coworkers when they forgot to replace gloves between patient treatments.

With the implementation of the program, observations were also conducted according to a structured form developed for this purpose. The staff observations were performed by members of the Center for Work Safety and Human Engineering from the Technion, who visited each department several times a week during all operational hours. The Technion team measured the percentage of safe glove use by physicians when taking blood samples (including switching gloves between one patient and another) and the safe use of gloves by the nursing staff when changing an infusion. (During the year, the Technion team was summoned to the hospital administration to explain the abnormal increase in the use of protective gloves.)

Every month participants received an update on the status of the program, including the data of the department itself as well as the data from the other departments. In addition, the teams participated in discussions during which they were updated on the progress of the program.

The results of the project, summarized in Figure 21.1, indicate a significant improvement in safe work behaviors by the medical and nursing staff. The graph indicates improvement in safe behavior in each of the departments. In one department, the percentage of safe behavior performance among the medical staff was 55% before the start of the project, which increased to over 80% 2 months after the project was initiated. Subsequently, there was an even further increase in compliance

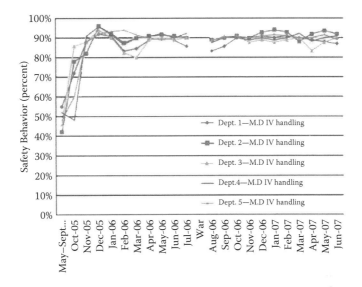

FIGURE 21.1 Percent of safe IV handling by physicians in the four experimental groups of gentle compliance. (From Erev, Rodensky, Levi, et al., 2010, The Value of Gentle Enforcement on Safe Medical Procedures, *Quality and Safety in Health Care*, 19(5), 1–3.[3])

to 83–90%, and the rate finally stabilized at approximately 90%. Similar results were also obtained in the other internal medicine department (adhering to safe behaviors improved from 43% to 90%). The observations revealed a significant improvement in adherence to the protocol and donning gloves when preparing an infusion. (The observations were suspended for 2 months due to the Second Lebanon War.)

HOW DOES IT WORK?

High visibility—In order that gentle rule enforcement is effective, the behavior being enforced must be highly visible, meaning that the vast majority of breaches of the enforced behavior can be viewed and recorded.

High probability—Gentle rule enforcement (admonitions when the desired behavior is breached) should be carried out with high probability; that is, it is important to provide reminders for the majority of violations of the enforced behaviors. This is particularly important at the initial stages of the project when the implementation of the principles of gentle rule enforcement is commencing.

Perceived as worthwhile—The enforcement should be conducted in a manner that is gentle rule, mutual, and effective. The enforced behavior should be perceived as advantageous by the workers. The data described above indicate that a large portion of the safety problems meets these criteria.

SUMMARY AND CONCLUSIONS

After a suitable period of preparation, it was decided to begin implementing gentle rule enforcement in the hospital, as was done in the industrial plants. The nursing and medical staff implemented the gentle rule methods of enforcing the safe use of protective measures. The results show a significant 30–40% improvement rate among medical staff in adopting safety measures when drawing blood, and in the use of safe methods by the nurses to prepare and set up intravenous treatment for their patients. Projected results indicate that implementing this method can also help reduce the rate of cross-infection in hospital departments.

REFERENCES

1. Becker, G. (1968). Crime and punishment: An economic analysis. *Journal of Political Economy*, 76, 169–217.
2. Erev, I., Ingram P., Raz, O., and Shany, D. (2010). Continuous punishment and the potential of gentle rule enforcement. *Behavioural Processes*, 84(1), 366–371.
3. Erev, I., Rodensky, D., Levi, M., Hershler, M., Adami, H., and Donchin, Y. (2010). The value of gentle enforcement on safe medical procedures. *Quality and Safety in Health Care*, 19(5), 1–3.
4. Hertwig, R. and Erev, I. (2009). The description–experience gap in risky choice. *Trends in Cognitive Sciences,* 13, 517–523.
5. Kahneman, D. and Tversky, A. (1979). Prospect theory: An analysis of decision under risk. *Econometrica*, 47(2), 263–291.
6. Yechiam, E., Barron, G., and Erev, I. (2005). Description, experience and the effect of rare terrorist attacks. *Journal of Conflict Resolutions*, 49(3), 430–439.

22 Human Engineering and Safety in Health Care Systems—What Have We Learned?

Daniel Gopher

CONTENTS

The Importance of Viewing the System and the Process as a Whole 300
Technology Comprises More Than Just Monitors, Computers, and Imaging Systems 301
Multiprocess and Multistage Work Is the Norm 301
There Are Many Different Elements That Contribute to Workload 302
The Work of the Team Has Many Dimensions 302
Sharing of Information and Informal Communication 304
Data Collection and the Development of a Valid Database for Scientific Work 305
Concluding Remarks and a Look to the Future 309
What Should Be Done and What Can Be Done? 310

In the introduction to this book we presented the argument that the increasing number of medical errors, as well as the increased load and decreased efficiency in the functioning of medical staff members, are caused by faulty human factors engineering and a lack of organizational planning of the personnel's roles and medical work environments. We emphasized the fact that the rapid development of health care systems, treatment techniques, and their increasingly greater sophistication were accompanied by a significant increase in the technological complexity of the systems, cumulative amount of medical knowledge, and need for teamwork. However, they were not accompanied by a concomitant investment in the planning and adaptation of these new systems and work processes for the medical staff members who work within them. Therefore, medical personnel are forced to work in a hostile work environment, which not only does not support them, but rather increases the difficulty of their job performance.

The various chapters of the book illustrate these claims, broaden them, and cover a wide array of topics and problems that describe the different dimensions and aspects of such discrepancies, their significance, and how they interfere with the medical staff's ability to function. All in all, these difficulties primarily manifest in an increased probability of errors occurring. In each case, the steps that were taken or that could be taken to improve the situation are described, as well as their influence on job performance. In many cases, these modifications are relatively simple and low in cost; however, they have an important and significant impact.

It is worthwhile to again emphasize that the aspects and issues that this book deals with do not include the examination of medically related judgments, medical justifications for the diagnoses provided, or treatment recommendations, nor with the skill used in executing the medical intervention. Rather it describes the other factors and aspects that may help or interfere with the medical actions taken, such as the manner in which information is presented, its integrity, quality, the organization of the work environment and workstations, work procedures, how information is shared, the coordination and integration of teamwork, and the like.

This book enables the understanding of the vital nature of the contribution made by human factor ergonomics and the recognition, support, and encouragement it provides to the medical team member specialists, enabling them to demonstrate their abilities and knowledge optimally. It also provides the opportunity to understand why the planning of the system and its work structure requires the collaboration of professionals from different disciplines—not just doctors—whose concern is the optimal design of complex work environments that integrate technical systems and the people who operate them. The personnel involved in the different projects described in this book include professionals in ergonomics, industrial engineers, cognitive psychologists, industrial designers, social psychologists, and organizational behavior professionals.

The book and the projects it describes provide conclusive evidence that human factors engineering and the planning of a user-compatible system reflects a manner of thinking, a worldview, and the use of a variety of methods and manners of collecting data that must accompany and be integrated within every technological development process that unites technological systems with a human user. Without such an integration, the risk of decreased functional effectiveness, increased workload, and the potential for errors and malfunctions increases.

In examining the book and its chapters, a number of general topics and conclusions are raised.

THE IMPORTANCE OF VIEWING THE SYSTEM AND THE PROCESS AS A WHOLE

One of the most important insights that emerges from the different projects that were presented in this book is the need to view and analyze the medical system and the medical process or job comprehensively. The surgery or the surgical process (Chapter 4), dispensing medication (Chapter 6), treatment of a neonate (Chapter 7), or medical supervision in the intensive care unit (Chapter 3) are not limited to the direct medical action (e.g., surgery, administrating medication) and are not confined to it; rather it encompasses a wide spectrum of activities, interventions, and accompanying processes, of which they are an integral part, which can assist and contribute to its successful administration or interfere with its effectiveness and cause harm to the patient. The registration processes, transmission of information, preparation, and coordination are an integral part of the medical intervention process just as much as the surgeon's scalpel and the imaging device. Only a comprehensive view of the entire system and the manner in which the different components are integrated can enable an overall understanding of the system and how to suitably plan each of its components. It is not possible to design a single element without considering its place in the overall process and its integration into it. This argument is equally true for the elements involved in the process of preparing a single drug as well as for the design of the medication rooms and the complete physical complex of the outpatient department.

The lack of comprehensive and integrated planning of health care systems and hospitalization environments is one of the central problems in human factors engineering in medicine. It would be inconceivable not to design a pilot's cockpit or the process control system in an engineering plant as a uniform work environment, whereas health care systems are built mostly as a patchwork, according to local restrictions and budgetary considerations, which do not take into account the impact of these considerations on the overall nature of the job and its performance. The absence of a comprehensive systematic view of the system and its functioning is blatant. This problem becomes apparent in almost all the chapters of the book and in all the work environments it describes.

TECHNOLOGY COMPRISES MORE THAN JUST MONITORS, COMPUTERS, AND IMAGING SYSTEMS

The medical work environment has abundant resources and relatively simple technological devices whose proper planning and effectiveness are just as important for the success of the medical treatment as are the complex technological systems in the technological forefront, such as surgical systems, imaging devices, and sophisticated monitoring devices. Health record forms (intensive care, radiotherapy, neonatal units), labels for medications (magnesium, medication room, labels), storage containers (magnesium, medication rooms), automatic syringe pump systems, hoses, connectors, and work surfaces (magnesium, intensive care, neonatal units) are also technological systems and represent an integral part of the medical care process that also influences the safety of medical treatment and its effectiveness. The importance of planning them properly is no less important than investing in the planning of sophisticated medical care systems. Their impact on the success of the medical treatment is equivalent to its contribution.

MULTIPROCESS AND MULTISTAGE WORK IS THE NORM

As a general rule, the tasks of the medical team member—such as a physician or a nurse—are multistage and multiple process tasks, as we have seen in the different chapters. Even apparently simple tasks such as dispensing medication (magnesium, medication room) or documenting the patient's chart (radiotherapy) encompass multiple steps and stages and are not at all simple. Multistep or multistage tasks that are important for the orderly progression and integrity of the task are those that require special training, supervision, and practice. Special work patterns and methods of planning and monitoring the organization of the process and ensuring its integrity are required. Designing the radiation therapy form and the information it contains (Chapter 10) and preparing pre-operative briefings (Chapter 14) are two examples of the planning and organization needed to assist in the execution of multistep and multistage tasks.

THERE ARE MANY DIFFERENT ELEMENTS THAT CONTRIBUTE TO WORKLOAD

Another theme that pervades most of the medical roles and work environments is the claim that the workloads are too taxing. The most frequently heard request is for added nursing and medical personnel. The book does not focus on workload issues except in the context of the neonatal unit project presented in Chapter 7; however, the various aspects and factors that contribute to the increasing workload appear in many of the other chapters and projects outlined in the book. The main component influencing the workload is, of course, the number and scope of the activities required of the medical staff member to perform his/her job. As a result, there is a direct correlation between the ratio of medical staff members—physicians, nurses, and assisting personnel—to the number of patients in a unit (referred or hospitalized). Indeed, the number of patients that each medical staff member in Israel is responsible for is greater than that which is accepted in other advanced countries. Nevertheless, it should be recognized that a corresponding and no less important element that influences the workload of the worker is the difficulty and complexity of the tasks required. Many other factors serve to increase the workload on medical personnel, such as workstations that are not planned properly, a lack of consistency, disparity between workstations designated for a single purpose, a lack of adequate resources for documenting information and the difficulty in locating them, spatial congestion, a lack of organized work procedures, and the difficulty of coordinating and transferring responsibilities between staff members. All of this leads to the conclusion that it is possible to cause a substantial reduction in the work and mental load imposed on staff members even within the limits of the current available staff. The neonatal unit project (Chapter 7), medication rooms (Chapter 9), the intensive care unit (Chapter 3), emergency rooms (Chapter 5), surgical rooms (Chapter 4), the process of dispensing medications (Chapter 6), and the main difficulties in the hospital departments that were revealed through the problem report forms (Chapter 18)—all emphasize an array of topics and problems that, if dealt with, could improve and greatly increase the efficiency of the medical staff member's job and reduce his/her workload.

THE WORK OF THE TEAM HAS MANY DIMENSIONS

The main feature of the activities performed in a modern health care system is the crucial importance of thorough and coordinated teamwork. As we have seen in the various chapters of the book, the work performed by these teams should be synchronous, as in the surgical room or during the doctors' rounds, or serial, as in the hospital unit's medical team (ICU, neonatal unit) during the surgical process or in the emergency care unit (Chapter 5). Efficient, synchronized staff work requires the staff to act as one, in a coordinated fashion and with each member being aware of the other member's job and actions. To describe the situation in which staff members do not work in synchrony, we sometimes use the image of a relay race, in which the stick that is passed from hand to hand represents the allocation of responsibility and transfer of information. Just one failed transfer of the stick may cause the entire process to fail.

As reflected in the various chapters of this book, effective and efficient activity of the medical staff is characterized by two main dimensions:

A. The cognitive-functional dimension.
B. The social-organizational dimension.

The cognitive-functional dimension: The cognitive-functional dimension of effective teamwork incorporates a number of variables and job demands. The central requirement is to coordinate the activity according to defined roles and to apportion responsibility between the team members. Each member of the team must be aware of his/her role and responsibility within the context of the activity being performed. Furthermore, each of the team members must also be cognizant of the roles and areas of responsibility of the rest of the team members. Only with this type of shared knowledge can each of the team members perform his/her role in an optimal manner, help the other team members, and be helped by them. This information is vital for both parallel and serial work teams. When this is absent, various problems in the coordination of the joint work arise, as we have seen in the chapters of this book (the surgical process, the surgical room instructions, neonatal units, how physicians and nurses think, medical teams). One of the primary ways to ensure this type of coordination is through joint situational awareness regarding the patient and the medical process. The importance and nature of joint situational awareness were discussed in detail in the pre-operative briefing project (Chapter 14) and in the discussion about the thinking patterns of physicians and nurses (Chapter 18). Creating and continuously updating joint situational awareness has an important role in the task performance of each of the staff members on one hand, and in ensuring the coordinated activity of the entire team on the other. This leads to the awareness of the importance of communication, of sharing information between team members, and documenting it.

An interesting challenge of coordinating the work of the team arises from the fact that most teams include members whose education and professional backgrounds differ, as for example the physicians and the nurses. These differences lead to disparities in professional language, concepts, and patterns of communication that must be bridged in order to ensure efficient communication and teamwork. Problems in teamwork that arise from differences in education and language are salient, recognized, and harmful to the joint teamwork of doctors and nurses (see chapters: intensive care [Chapter 3], surgical room briefings [Chapter 4], neonatal units [Chapter 7], how doctors and nurses think [Chapter 13]).

Another important element in ensuring the effectiveness of teamwork is the management and organization of the task, namely, to appoint an agreed upon process manager to supervise, manage, and coordinate the activity as a whole.

The social organizational dimension: It is important to remember that all hospital departments and medical units are social organizations. That is, as in other organizations, there is an organizational culture governing its conduct and values, which affects the behavior of the individuals in the groups and guides them. Three chapters of this book deal with this issue. The findings of the studies emphasize that the organizational culture regarding safety (Chapter 19), the encouragement of learning and advancement (Chapter 20), and the detailed regimen imposed when procedures are not followed (Chapter 21) have a significant impact on safe behavior and the prevention of errors. It is important to remember this additional aspect of teamwork in a social environment, and its influence on the members' behaviors.

SHARING OF INFORMATION AND INFORMAL COMMUNICATION

As already stated, one of the main benefits of modern medicine is the ability to measure, examine, and collect vast amounts of up-to-date information about the patient's condition, his/her medical history, and the concurrent treatment processes he/she is undergoing. This richness is one of the system's strengths but it may pose a serious obstacle, as it requires adequate documentation and information sharing among the caregivers and team members. One of the most challenging stumbling blocks of modern health care is the accumulation of large amounts of informal and undocumented information that is not shared between medical team members involved in the same treatment process. The problem of transmitting informal information is clearly reported in the chapters that deal with the intensive care unit (Chapter 3), the development of charts for radiotherapy treatments (Chapter 10), the surgical process (Chapter 4), pre-operative briefings in the surgical rooms (Chapter 14), neonatal care units (Chapter 7), the development of a new system for reporting problems (Chapter 18), the thinking processes of the doctors and nurses (Chapter 13) as well as in other chapters. The general principle is that all relevant information about the patient and the medical process must be documented and accessible for all medical personnel involved in the medical process. Despite this, every treatment process has a relatively large pool of informal information that accumulates and is not documented or accessible and only part of it gets transmitted properly between the team members. This is the Achilles heel of modern treatment processes, and the three main reasons for it are:

1. The lack of an appropriate and simple means of documenting and retrieving information (Chapters 3, 7, 10, 15).
2. Team members forget to transmit information that they are privy to (Chapters, 3, 7, 10).
3. A team member has information but is not aware of its importance to the rest of his/her team in order to perform their roles (Chapters 7, 10, 13, 14, 18).

These three reasons are not mutually exclusive and together they contribute to the fact that information vital to the success of the treatment and the judgment of the caregivers is not passed along to them and made available to them when making decisions.

One of the important challenges that human factors engineers in the health care system must meet is to accrue the experience needed to solve this problem by using information technologies, developing work procedures, increasing situational awareness, and developing methods of documentation and retrieval.

DATA COLLECTION AND THE DEVELOPMENT OF A VALID DATABASE FOR SCIENTIFIC WORK

The systematic collection of information and creation of an appropriate database are the cornerstones of every suitable scientific study and successful intervention. In the introduction and in other places in the book, the argument is presented that the investigation of medical errors and accidents cannot be used as a basis for the systematic and appropriate information needed to guide a scientific researcher, whose aim is to improve the health care system, reduce the number of medical errors, and increase patient safety. In the studies presented in the various chapters of the book, a variety of methods have been used to collect information and describe the problems in the system from the perspective of human factors engineers. This section presents a summary of these methods and of the ideas underlying them. It is important to emphasize that the main goal in collecting data and in this type of analysis is the understanding of a unit's work processes or the processes of the treatments provided, and the assessment of the interactions and gaps between the role requirements. In addition, these techniques aim to familiarize human factors engineers with the technical and physical environmental features, work procedures, and the abilities of the staff members who fill these roles. The methods used for this purpose and for the data collection are:

1. Interviews
2. Observations
3. Task analysis, analytical analysis
4. Simulations
5. Reconstructions
6. Development of problem report forms
7. Questionnaires

1. *Interviews*—In most cases, the research teams were composed of graduate students and human factors experts in areas outside of medicine, who were not previously familiar with the work environments or processes that were being examined. The main goal of the interview was to provide the research team with an initial familiarity with the unit, its work processes, issues and problems—as they were perceived by the unit's staff and administrators. The interviews provided an immediate impression of the unit, the opportunity to learn the relevant concepts, and to provide the experts with a familiarity of the framework of each of the units examined. These

interviews were not for the purpose of gathering data and did not entail systematic formal or quantitative data gathering.

2. *Observations*—Systematic observation was one of the main research methods adopted in many of the studies cited in the book (for example, Chapters 3, 4, 7, 14, 21). In all cases, the observation was stratified, and an effort was made to ensure an adequate representation and the appropriate frequency distribution of the main variables to be investigated, such as work shifts, types of surgeries, days per week, patient categories, workstations, and so forth. Based on the results of the observations, measures of quantities, frequencies, and distributions, judgments regarding the severity and degree of risk were formed.

One of the important discoveries resulting from the observations, beyond the description of the actual types of work processes, was to note any nonroutine or adverse events during the course of the work as an alternative and complementary method of recording errors and accidents. Unlike accidents, which are rare, adverse events are far more common and allow for the assessment of the quality of the process as a whole even if an actual mistake did not occur.

The argument is made that "near miss" events such as these are markers for potential mistakes and raise the probability of them occurring. For example, in the chapter that discusses presurgical briefing (Chapter 14), nonroutine and adverse events were grouped into nine categories, as follows:

   a. Teamwork
   b. Information availability and management
   c. Lack of knowledge and skill
   d. Procedures and regulations
   e. Deviations from routine
   f. Work schedules
   g. Human engineering and safety
   h. Equipment
   i. General

Adverse and nonroutine events in each of these categories may reflect the nature of the system's functioning and increase the probability of error related to the patient's health.

It should be noted that an observation, as outlined above, is a long and arduous process that demands the dedication of many, irregular hours for the collection of data and its analysis. But this is the most sensitive method of detecting problems and evaluating their sources and severity.

3. *Task and analytical analysis*—Another way of systematically analyzing the work environment and the work processes originates from the fields of cognitive psychology and human factors engineering; using methods of task analysis to perform an analytical analysis of the work environment and the work process involved in the performance of a particular task. Analyses such as these are illustrated in the description of the surgical process (Chapter 4), dispensing magnesium (Chapter 6), the neonatal unit (Chapter 7), and the studies conducted to investigate the department of emergency medicine (Chapter 8) and the format work procedure in the Gynecology

Human Engineering and Safety in Health Care 307

Department in Hadassah Ein Kerem (included in Chapter 18). In addition, process analysis tools, a technique adopted from the field of industrial engineering, accompanied by a quantitative model to describe the differences between descriptions of work processes obtained from different staff members, were also used. Task analysis and the accompanying analytical tools enable a direct relationship to topics and aspects of cognitive psychology and human factors. This type of mapping makes it possible to define the severity of the problem and the outlining of possible solutions.

4. *Simulations*—In some of the studies described in the various chapters, controlled simulations were conducted to enable the evaluation and systematic study of the medical work process. In Chapter 9, which describes the planning and construction of a room for dispensing medications, a simulation study was conducted of preparing a prescription for three drugs, to compare the new medication rooms to the old ones. In a study that compares the thinking patterns of doctors and nurses in the intensive care unit, described in Chapter 3, a 90-minute simulation study was conducted of the process involved in the medical and nursing admissions and treatment of a patient to the ICU.

In Chapter 18, which describes the development of problem report forms regarding the human factors related difficulties and risks involved in performing physicians' and nurses' clinical work, a simulation was conducted of how patients admitted to the Internal Medicine Department were monitored. The examination compared the recommended monitoring regimens of senior physicians from a sample of 100 admission charts of patients who were previously hospitalized to the actual monitoring of these patients after their admission to the department.

In each of the above three cases, simulations enabled an investigation and a controlled, orderly assessment of how a complex medical process is managed in its natural environment and context. Moreover, the research team was given the opportunity to observe the process and to compare performance measures. The simulation procedure is an effective, controlled, and reliable technique for gathering data about the personnel's actions during a complex and multivariable process. Its main shortcoming and weakness lies in the fact that it requires a great deal of effort as well as a thorough understanding of the processes, which is not always possible.

Of course, it is also true that the participants in the simulation are aware of the fact that they are being observed and inspected. However, based on our experience with many such simulations, we believe that this has a negligible influence. This results from the fact that the participants are not aware of which variables are being examined and there are no clear rules for "correct" behavior. Thus, the participants soon forget that they are being observed and behave naturally.

5. *Reconstruction*—This is a research method based on simulation, which is conducted when an accident, adverse event, or error actually occurs. This technique and its advantages are described in detail in Chapter 17. This technique has three prominent advantages:

1. It enables all of those who participated in the event to reenact their tasks, and as such, it assists them in retrieving facts from their memory of the event and in reconstructing episodes and processes that are not usually included in the verbal report.

2. It enables the members of the medical staff to interact with one another and thus to clarify, compare, and verify versions that are generally difficult to compare.
3. It allows the researchers to get an impression of the overall situation and see aspects of physical and interactive details and processes that are not available through verbal questioning and are generally overlooked.

However, reconstruction as a method of investigating an event, is very difficult to perform and involves many careful preparations. Moreover, it is usually very difficult to obtain the consent of the event's participants for a full reconstruction, due to conflicts of interest and legal considerations.

6. *Development of Systems for Reporting Difficulties and Problems*—All of the techniques for the research and data collection described above have something in common. They must be carried out and executed by a team of professionals and require the active involvement of the staff members in observations, running the simulation, studying the process, and collecting the data.

This fact places constraints on the process of data collection and the development of an appropriate database in large systems such as hospitals, which contain a great number of units and work environments. A solution to this problem is offered in Chapter 18, which describes a method of data gathering based on the reports of the staff about human factors related difficulties and problems they face when performing their daily tasks. This reporting system offers the medical staff members a method of identifying, describing, and reporting on focal problems prior to the occurrence of error or adverse event. These reports are collected from each unit over time, and are examined, categorized, and assessed by a team of human factors engineers. They can be used as a sensitive guide in detecting the main focus of the reporting unit's major problems without requiring a detailed and prolonged analysis of the unit and its activities. A professional team such as this can simultaneously evaluate a large number of units and departments in the hospital and concentrate their professional energies on a limited number of topics that are featured in a large number of complaints, as well as on the importance of effective and safe work.

Ostensibly, the need to build this type of database would require the same effort as in investigating accidents and errors and in developing a necessary reporting system for adverse events. Moreover, an incident and adverse events database is limited and is not suitable to function as a reliable, systematic, and representative source of information, as discussed in several of the book chapters.

The proposed reporting system may provide the appropriate database for the guidance of valid and general corrective efforts.

7. *Questionnaires*—As a general rule, the use of questionnaires for the purpose of collecting data and analyzing work processes is a technique that has limited usefulness in the field of human factors engineering. This is due to the fact that the responses to the questions reflect the respondents' opinions and awareness of the subject area that the questionnaire focuses on, and that their responses are subjective by nature. On the other hand, if the topics, variables, and perspectives that human factors engineers deal with are, for the most part, not within the scope of the medical staff member's expertise and knowledge; he/she will not be aware of what these topics and variables are and thus cannot assess their influence on the quality of his/her performance and

on his/her workload. Therefore, the responses and the opinions provided in the questionnaires on these matters are expected to have only marginal value.

The situation is completely different when dealing with a culture of safety and in assessing the system's support of active learning and presenting criticism.

These aspects, covered in Chapters 19 and 20, deal directly with the subjective perceptions of the unit's personnel and medical staff, and derive directly from their opinions and interpretations of the system's policy and the attitude of the administration. In this situation, the personnel is familiar with the matter being examined and the subjective variables revealed in the responses on the questionnaires are of greater value; they can be studied, gathered, and conceptualized through the information provided by the questionnaires. The chapters describing the studies that deal with the subjects of safety culture and active learning do indeed make extensive use of the questionnaires and show their effectiveness in predicting relevant behavior patterns.

In addition to what is stated in Section 6 above, the use of questionnaires is also a method of collecting data from a large population and of processing them with relatively little effort.

CONCLUDING REMARKS AND A LOOK TO THE FUTURE

The general objectives common to all the studies described in the book are to improve the quality of medical care, prevent errors, and maintain patient safety. Hence, the two main concluding questions in the book are: What is its overall message concerning these objectives? And what should be the policy and main thrust of the effort in attempting to cure the system's problems?

The book covers many aspects and elements of the modern medical system and provides a detailed description of their impact on the ability of the system to function and on the quality and potency of the medical care. The general argument advanced in the Introduction, regarding the hostile environment and unsuitability of the medical system from the perspective of human factors, has been illustrated in its many dimensions, colors, and shapes through the description of the projects and problems discussed. It is important to now emphasize that these issues and problems are not unique to medical systems, but are common to many complex military and industrial systems, in control plants and in air, sea, and land transportation systems.

The medical system is unique in that even when the error that transpires is not large, it comes at a very high cost. Moreover, the attention given to the overall design of the system and to the problems discussed is relatively limited in comparison to that of other systems. Thus, from the perspective of human factors engineering the health care systems are behind the times and neglected in comparison to industrial areas and military systems. Part of the problem stems from the excessive attention given to medical negligence and malpractice as the main source of errors and impaired quality of the medical care.

However, the projects presented in this book have shown that even if some degree of negligence exists, the major source of reduced quality of care and safety is the lack of proper design of the medical systems and the work environments. The major message conveyed in the book is the urgent need to change priorities and the focus of our attention. The key for the issues should be searched for in the places in which they are and not in the comfortable, limited area of incident spotlights.

We must recognize the fact that the overfocusing on medical negligence and on medical staff accountability diverts attention from the main issue and blocks the efforts to reach the real source of the trouble and solve the main problems that impact on quality of medical care and its effectiveness (see also Chapter 16, "What Is the Difference between Risk Management and Safety Management?" and Chapter 17 "Use of Simulation to Clarify the Source of an Event").

The preoccupation with medical malpractice impedes the attempts to address the real sources of the problem, from three different vantage points:

A. Attention is being focused on a secondary issue and not on the main problem.
   This is reflected in the topics that are investigated and the work topics, as well as the allocation of resources and development of complex and expensive legal and insurance systems, that thrive upon and reinforce the "blame culture" while diverting the public's and the professionals' attention from the main issues contributing to the problem.
B. The targeting of negligence, personal responsibility, and the risk of lawsuits directly interferes with the ability of human factors engineers and those researching the system to obtain unbiased, comprehensive, and objective information.
   The importance of this type of information for the benefit of scientific and professional functioning was discussed at length in several places in the book.
C. Substantial financial resources are invested in defensive medicine, legal protection, and insurance against tort claims. Were just a fraction of these sums to be diverted toward solving the primary problems, the quality of medical care would improve dramatically.

WHAT SHOULD BE DONE AND WHAT CAN BE DONE?

The main points discussed in the previous section relate to the direction in which, we believe, action should be taken, and the changes needed to bring about a real improvement in the medical system and quality of medical care:

1. Change the priorities, foci of attention and efforts, and concentrate on the problems of design and planning of the system and how it should operate rather than on medical malpractice.
2. Invest energy into planning health care systems from a broad, comprehensive perspective, with the collaboration of multidisciplinary experts which includes professionals from a variety of relevant fields of knowledge: human engineering, cognitive psychology, industrial engineers, system engineers, industrial designers, organizational behavior, and of course, medical personnel. This type of planning can be based on professional experience, learning from it, and then applying the many solutions that have already

been developed and implemented in a wide range of systems and work environments.
3. The sharing and responsibility of senior management's role in developing an overall policy of how the medical system should be operated must be emphasized, refined, and clarified and should include active involvement in managing the processes/supervising them and in problem solving.

Taking concrete steps according to these guidelines will, in our opinion, create a real improvement in the quality of health care and its efficiency. They will lower workload, reduce errors, and raise positive attitudes and emotions. They can also be expected to reduce the overall economic cost of treatment and thus, it will be possible to improve the quality of medical care given to a wider segment of the population without increasing the overall financial investment. We hope that this book and the studies in it will act as an impetus for change in the desired and correct direction.

23 Summary
The Physician's Point of View

Yoel Donchin

Until the discovery of ether's anesthetizing properties in 1846, every surgery involved excruciating pain and, for good reason, many people refused to abandon their bodies to the mercy of the surgical knife. Nearly 160 years later, patients waiting for surgery still fear adverse consequences, due to the possibility of injury due to a medical error or negligence. These concerns increase with every news story about cases such as that of a surgical patient whose healthy leg was actually amputated due to an error in marking the leg, or a patient who received a drug that she was allergic to, resulting in her death. Although these events are rare, they affect the level of anxiety of a patient arriving at the hospital. Often medical mishaps have another injured party aside from the one whose body was injured. This person is the physician or nurse who, by his/her own efforts, led to the mishap. They are hurt as a result of the widespread publicity about the incident, which is often accompanied by legal charges in which they stand accused. And they, no doubt, are aware of this danger.

The first physicians who dealt with anesthesia and stood at the patient's bedside during surgery were among the first to realize that without strict safety rules to ensure the well-being of the patient, the patient could not receive the benefit of safe and reliable medical services. Anesthesiology organizations turned to basic research to learn which factors cause mishaps. The research began by analyzing information obtained from data collected on thousands of mishaps that had occurred during the anesthesia process. This study was conducted by Jeffrey B. Cooper, of Massachusetts General Hospital, among the world's most premier medical centers.

Based on the results of Cooper's research, simple technical changes were introduced in 1980 that produced an alert to warn the physician whenever the oxygen flow decreased below a certain critical value. Pressure sensors were developed to instantly reveal whenever there was a break in the tubing that conducts the anesthetic gases to the patient. Following the remarkable decline in the number of casualties resulting from errors in anesthesia, an institution was established, under the auspices of the U.S. anesthesiologists' organization, whose main function was patient safety. However, a different type of mishap began to emerge, mishaps originating from "human error."

The old adage "to err is human" is obviously not a substitute for the need to investigate this *human factor*. Studies of this kind require skills in the different aspects of human behavior, including human weaknesses and limitations. It is almost as if the physician's hospital routine is determined with complete disregard for human capacity. Supposedly, a physician does not need a minimum number of hours of sleep, has a tremendous memory for thousands of details relating to a large number of patients, and possesses a host of other superhuman attributes. Yet within the huge gap between the demands made on the physician or the nurse and their abilities to meet these demands, there is wide room for unwanted errors and malfunctions.

Over the many years that I worked in the Intensive Care unit I was witness to many errors; some were discovered after the fact and some were discovered in time. I was helpless without tools to help me investigate or prevent such errors or influence the dedicated and highly motivated staff. Looking from within the system, as a physician in charge of a unit which houses the most critically ill hospitalized patients, was not possible. The need to solve (literally) burning existential problems overshadowed our ability to look in the mirror and admit: We are working in a system that is difficult for the patient and caregiver alike. Assistance was needed from outside the system, for a different viewpoint on occurrences inside the hospital, from the patient's entry into the hospital, his/her admission to the system, through the various stages of diagnosis and treatment that the patient undergoes, and until his/her departure from the hospital. Such a perspective was obtained by opening the medical system to collaboration with the investigators from the Center for Work Safety and Human Engineering at the Technion.

It happened by chance that I encountered the term *human factors engineering*, calling to mind the structure of the cockpit, or the work done by people along a conveyor belt. The possibility of applying the vast body of knowledge of human engineering to medical systems never occurred to me.

The first study, described in Chapter 3, was the first turning point that afforded me a different way of viewing the hospital and the unit in which I worked. The human engineering investigators, who in turn had never experienced the daily life of a hospital in general, and the Intensive Care unit in particular, were now spending time in the unit 24 hours. They recorded, described, analyzed, and showed us an image that we had refused to see up until that point, and we faced a powerful sight, through which we first realized how close we come to the brink, how easy it would be for us to take the wrong drug from the stockroom and bring it to the wrong patient.

The researchers helped us understand the essence of poor communication but also how to repair and improve it when the entire staff works together, recognizing the need to prevent the mishap. We realized that safety comes at a certain price: to conduct briefs, act according to a checklist to guide us, and help us overcome the limitations of our human memory.

At the same time that the studies were begun in Israel, news reports and articles on medical errors began to appear in the medical journals and in the mass media, including the potential dangers they pose to an inpatient. Most of the articles dealt with errors: classifying, recording, and reporting them. The studies in this book, some of which have been published over the years in the medical literature, deal with human behavior and the need for the person who operates the medical system to adjust and adapt the system's work environment. This adaptation must start with appropriate education in medical school, through teaching programs designated to highlight the complexity of human behavior, so that those who will need to make quick decisions and work under pressure within a few years' time will be aware of the risks involved. Much as a novice pilot learns the safety rules from the very first day of the piloting course, so must the next generation of physicians be educated.

The chapters in this book illustrate to physicians and nurses the possibility of collaborating with a field that is seemingly distant from that of classical medicine, and yet may bring about great benefits. From the dozens of studies cited in the book it can be concluded that a medical environment must examine itself—the chapter headings of the book can serve as a guideline and a checklist for action: Are the procedures proper and appropriate? Is the physical environment and organization of the workstation suitable for its intended function? To what extent can these factors contribute to the prevention of the next error? What are the thinking patterns of the medical personnel? Is there an understanding between the various teams in the unit in which I am working? How well do we communicate and how can the communication be improved? What is the organizational environment in which we operate? Is it possible to improve the ability to prevent failures by making changes in the organization? How can the safety culture in the microclimate that I am working in be measured and improved? How and what can we learn from a single, rare failure and how can it be studied?

Medical ergonomics is a field encompassing new and unique knowledge that is currently being developed. This field is vital, since the medical systems deal with organisms and not mechanisms. In many situations, medical science is not capable of achieving a speedy and accurate diagnosis. The medical system is neither a cockpit nor an assembly line at a dairy plant or in a weapons factory. Our system is full of surprises and anatomical diversity that cannot be predicted. All these facts demand the creation of Intensive Care units that are not hostile to the user; to prepare forms (even now in the computer age) that can provide all the necessary information in a clear and selective manner; to create a mutual understanding of the thinking patterns among all the teams; and to utilize mental aids (checklist, briefs) to boost our memory capacity, which we cannot afford to rely on exclusively.

This book is a joint product of multidisciplinary research: medicine with human factors engineering, a much-needed approach.

Index

A

Abnormal events, communication, 38
Acquired knowledge, 205
Ambiguity in job definitions, 58
Arterial bleeding (reconstruction), 238–241
    blame, 238
    catheterization room, 239
    device monitoring, 240
    doll, 239
    initial scenario, 238
    intensive care unit, 240
    reconstruction conclusions, 241
    scapegoat, 238
    underlying problem, 241
    videotape, 240

B

Black box problem, 121
Blame culture, 310
Briefing, operating room, 207–215
    briefing, 210–212
        information included, 211
        parts, 210–211
        poster, 212
        shared situational awareness, 211
    development of briefing, 210
        events, 210
        observation, 210
    frequency of errors, 207
    Hand Out, 207
    impact of briefing on number of nonroutine events during surgery, 212–214
        classification of nonroutine events, 212–213
        quality of briefings, 213
        staff members, opinions of, 214
        study finding, 212
        superficial manner of conducting briefing, 213
        systemic problems, 213
        teamwork, 212, 213
    reason for briefing, 208–210
        confirmation bias, 209
        fallacy, 209
        memory associations, 209
        overconfidence effect, 209
        staff performance, 209
        staff task, 208
        shared situational awareness, 208
    summary, 215
    understanding of the situation, 208

C

Central venous pressure (CVP), 165
Climate; *See* Patient safety climate
Clothing; *See* Garment design, for operating room staff
Cognitive processes, tools available to investigate, 196; *See also* Thinking patterns and communication, physicians and nurses in intensive care unit
Communication; *See also* Thinking patterns and communication, physicians and nurses in intensive care unit
    abnormal events, 38
    caregiver, 271
    departmental, 273
    informal, 304
    information derived, 290
    information transfer and, 110–114
        information nodes, 111
        physicians' rounds, 112–113
        shift transfer of nursing staff (handover), 112
        summary, 113–114
    irregularities, 31
    problem of, 12, 37
    problems during surgery, 50
    team, 40, 52
Confirmation bias, 209
Crime mobile labs, 235
CVP; *See* Central venous pressure

D

Daily changes log, 152
Database
    development, 305–309
    mental model, 124
    unique, 142
Defensive medicine, 310
Doctor's rounds; *See* Physician rounds, rate of interruptions during
Documentation, abnormal events, 38
Drug labeling; *See* Labeling of drugs and medical tubing, designed stickers for

317

Index

E

Emergency medicine department, mental models, 55–78; *See also* Human–computer interaction, improvement of emergency medicine unit efficiency through
  describing and quantifying mental models, 74–77
    comparing process charts, 75–76
    example of comparing two process charts, 76–77
    relationship intensity, 75
  discussion, 73–74
    flowcharts, 73
    overlapping of process flowcharts, 74
    similarity index, 73
    workers' perspective, 73
  index for gap assessment between different mental models, 61–62
  industrial engineering system approach, 56–58
    human factor, 57
    linear programming, 58
    literature claim, 58
    Management's Operation Policy, 58
    perception, 57
    process flowcharts, 58
    queuing theory, 58
    system definition, 56
  mental model, 58–59
    predictive and explanatory attributes, 59
    synchronization, 59
    ways of thinking, 59
  mental models of the team, 59–61
    clinical procedure, 60
    gaps, 61
    human error, events referred to as, 61
    multifactor system failure, 61
    operating rooms, 59
    patient information, sources, 60
    teamwork, 60
  process flowchart operations similarity index, 63
    general similarity, 63
    operations similarity, 63
    transition similarity, 63
  use of similarity index for performance assessment, 64–72
    calculation of similarity indices among team members under proposed approach, 72
    calculation of similarity indices between team and management according to proposed approach, 69–72
    description of emergency medicine department selected for field experiment, 64
    internally ill patients, 67
    management policy model, 64–65
    mental models of the medical team in emergency medicine department, 66–68
    minor trauma, 72
    mobile internally ill patients, 72
    orthopedic patients, 72
    process flowcharts, 70
Employee
  banished, 232
  behaviors, 266, 269, 286
  capability, 82
  cards, 134, 135
  dissatisfaction, 218
  functioning at workstation, 6
  inner subjective global concept, 57
  opinions, 289
  reduced burden, 6, 177
  safety, 275, 286
  slipping on polished floor, 14
  tag, 181, 188, 189, 191
  work garment, 176
Errors; *See* Medical errors, history of; Medical errors, types and causes of in intensive care
Event sources, reconstruction to investigate, 235–244
  airplane accidents, 235
  arterial bleeding, 238–241
    blame, 238
    catheterization room, 239
    device monitoring, 240
    doll, 239
    initial scenario, 238
    intensive care unit, 240
    reconstruction conclusions, 241
    scapegoat, 238
    underlying problem, 241
    videotape, 240
  black box, 235
  crime mobile labs, 235
  hidden error, 244
  open heart surgery, 236–238
    changes made, 238
    heart–lung pump, 236
    monitoring devices, 237
    oxygen saturation display, 237
    patient, 236
    reconstruction, 237–238
    resuscitation dummy, 237
    systemic leadership problem, 238

Index

videotape, 237
workstation examination, 238
patient found dead in his bed, 241–244
   anesthesia, 241
   anesthesiologist, 243
   catheter, 243
   communication, 243–244
   mannequin, 241
   nerve block, 243
   pain relief procedures in hospitals, 244
   peripheral nerve stimulation, 242
summary, 244
ticking bomb, 236

F

False alarms, 106
Fatigue; *See* Resident physicians' medical errors, managerial factors related to
FDA; *See* U.S. Food and Drug Administration

G

Garment design, for operating room staff, 175–194
   environmental conditions, 186–188
      cellular phone, 187
      clothing, 186–187
      humidity, 186
      identification tag, 187
      lighting, 186
      pager (beeper), 187
      pens and markers, 187
      temperature, 186
      ventilation, 186
   objectives of project, 176–177
      design preferences, 177
      ergonomic perspective, 177
      laundering of clothes, 177
      microclimate, 177
      protection provided, 176
      work outfit, 176
   project, 178–186
      data collection, 178
      design of operating room clothes, 182–185
      observations, 178, 179
      occupational analysis, 186
      personal details, 185
      privacy, 178
      questionnaire, 179–185
      use of operating room clothes, 180–182
   summary, 193–194
   summary of questionnaire's findings, 188–193
      clothing style preferences, 190
      comfort of operating room clothes, 188, 191
      design preferences for operating room clothes, 190–191
      differences between different professional groups in patterns of use and design preferences, 191–193
      differences between men and women in patterns of use and design preferences, 191
      habits regarding use of the operating room clothing, 188–190
      objects carried in operating room clothes, 189
      selected color preferences, 192
      types of physical activities, 192
GDP; *See* Gross domestic product
Gentle rule enforcement, 293–298
   best-known problem, 293
   gamble, 294
   hand washing, gloves, and gentle rule enforcement in hospitals, 295–297
      behavioral change, 295
      enforcement process, 296
      experiment, 295
      increase in compliance, 296–297
      methods tested, 295
      project results, 296, 297
   how it works, 297
      high probability, 297
      high visibility, 297
      perceived as worthwhile, 297
   "low probability of punishment" problem, 293
   medical malpractice, 293
   negligence, 294
   "severe punishment" solution, 293
   summary, 297–298
   value of gentle rule enforcement, 294–295
      optimistic prediction, 294
      senior managers, 295
Gross domestic product (GDP), 2
Gross national product (GNP), 1
Guilt, accusation of, 19

H

Hand Out, 207
HCI; *See* Human–Computer Interaction
Hierarchical linear models (HLM), 288, 289
Hospital
   improvement of emergency departments, 120
   medication room design, 134
   microcosm of, 24
   most common accident in, 14
   nineteenth-century, 13
   orderly, 34

pain relief procedures in, 244
risk management system, 20
safety; *See* Patient safety climate
Veteran Health Association, 20
Hospital medical staff, human factor focused reporting system for, 245–261
   administration of medication, 248
   discussion of report results, 255–256
   documentation and record keeping, 248
   equipment and instrumentation, 248
   general discussion, 257–260
      ambiguity in job definitions, 58
      daily work difficulties, 258
      expansion of the use of the proposed system, 260
      motivation to report, 259
      raising staff awareness of the impact of human engineering factors on patient safety, 259–260
      system management, 260
      system utility, 259
      validity of reporting system, 258
   human factors problem areas, 250
   major problems, 249
   method efficiency, 247
   outbreak of epidemic, 246
   physical layout, 248
   preliminary analysis used to develop forms, 248
   problem reports, 249
   results, 251–255
      ambiguity in job definitions, 254
      database, 253
      difference between systems, 251
      meeting, 254
      reports collected, 251
      type and frequency of reports, 253, 254
   self-report forms, 247
   shortcoming, 246
   specific problems, 249
   study population, 249
   study questions, 247
   study stages, 247
   topics, 248
   validation study of centrality of major problems reported by department staff, 256–257
      communication, 257
      criteria, 256
      indices, 256
      patient charts, 256–257
   work procedures and patterns, 248
   work procedure and structure report form, 252
Human–Computer Interaction (HCI), 121
Human–computer interaction, improvement of emergency medicine unit efficiency through, 119–130
   black box problem, 121
   component tasks, 121
   conceptual model, 124–129
      accepted rules of interface design, 126
      central activity sequence, 124
      change of parameters, 127
      data alignment, 129
      design details, 128–129
      disregarding of recommendations, 126
      flexibility between subsystems, 125
      navigation format, 127–128
      parameters panel, 128
      principles, 124–125
      reason for presenting so much information on one screen, 126–127
      screen format and interactive screen areas, 125–126
   first line of defense, 119
   hospital challenge, 120
   Human–Computer Interaction, 121
   mental model of the system, 123–124
      active model, 123
      collection of files, 123
      database, 124
      gap, 123
      modifications, 124
   simulation, 120
   summary, 129–130
   task analysis, 122–123
      context of systems' use, 122
      effects of change, 123
      organization and methods departments, 122
      primary users in system, 123
Human engineering
   components, 8
   key function of, 4
   link between medicine and, 20
   neonatal care units; *See* Neonatal care units, human engineering and safety aspects in
   source of knowledge for, 9
Human engineering and safety in health care systems (lessons learned), 299–311
   concluding remarks and a look to the future, 309–310
      attention diverted from main issue, 310
      blame culture, 310
      concluding questions, 309
      defensive medicine, 310
      medical malpractice, 310
      negligence, 309
   data collection and development of valid database for scientific work, 305–309
      development of systems for reporting difficulties and problems, 308
      interviews, 305–306

Index

methods, 305
"near miss" events, 306
nonroutine events, 306
observations, 306
questionnaires, 308–309
reconstruction, 307–308
simulations, 307
task and analytical analysis, 306–307
difficulties, 299
elements contributing to workload, 302
importance of viewing the system and the process as a whole, 300–301
lack of integrated planning, 301
medical intervention process, 301
spectrum of activities, 300
modifications, 300
multiprocess and multistage work, 301–302
sharing of information and informal communication, 304–305
teamwork, 302–304
cognitive-functional dimension, 303–304
dimensions of medical staff activity, 303
shared knowledge, 303
social organizational dimension, 304
technology, 301
what should be done and what can be done, 310–311
change of priorities, 310
planning of health care systems, 310–311
senior management, 311
worldview, 300
Human factors engineering, 314
Human factors and safety in health care, 1–12
cognitive components, role perception, mental models, and safety climate, 11–12
communication, 12
mental model, 12
variables, 12
data collection for evaluation of functional problems and performance of medical systems, 9–11
absence of reference base, 10
limited representation, 9–10
passive and reactive approach, 11
quality of reported information and its completeness, 10
wisdom in hindsight, 10–11
health care in the age of computers, the information revolution, and artificial intelligence, 1–3
burgeoning workload, 2
cutting corners, 2
medical carelessness, 3
medical paradox, birth of, 2
quality of life, 2
human factor characteristics in contemporary medical systems, 5–11
data collection for the evaluation of functional problems and performance of medical systems, 9–11
design of devices and unit systems, 5–6
design and layout of individual workstations, 6
design and planning of large workspaces, 6–7
intervention paths, 5
multifunctional activities, 7
physical and engineering components of the workstation and its surroundings, 5–7
recording of information, access to information, and transfer of information, 7
teamwork, 8
work procedures and work patterns, 7–9
human factor engineering components in working environments and medical systems, 3–5
historical background, 3
human engineering, key function of, 4
root problem, 3
unsuitability, implications, 4
physical and engineering components of the workstation and its surroundings, 5–7
design of devices and unit systems, 5–6
design and layout of individual workstations, 6
design and planning of large workspaces, 6–7
recording of information, access to information, and transfer of information, 7
work procedures and work patterns, 7–9
activities are multistage with many variables and components, operating in a complex technological environment, with a wealth of information, 7–8
effective teamwork, 8–9

I

ICU; *See* Intensive Care Unit
Index and intraclass correlations (ICCs), 287
Industrial engineering system, 56–58
human factor, 57
linear programming, 58
literature claim, 58
Management's Operation Policy, 58
perception, 57
process flowcharts, 58
queuing theory, 58
system definition, 56
Information map, 146, 153

Intensive care, medical errors in; *See* Medical errors, types and causes of in intensive care
Intensive Care Unit (ICU), 23, 196, 303; *See also* Thinking patterns and communication, physicians and nurses in intensive care unit
Internal medicine department, physician rounds in; *See* Physician rounds, rate of interruptions during
Ionizing radiation, 140, 177

J

Job
  analysis, 126
  definitions, ambiguity in, 254, 258
  demands, 5, 82, 141, 303
  description, 36
  performance, 300
  requirements, 23
  training, 270

K

Knowledge
  acquired, thinking patterns and, 205
  base, inner, 203
  management, 41
  requirements, tasks, 84–85
  shared, 303
  source of for human engineering, 9

L

Labeling of drugs and medical tubing, designed stickers for, 163–174
  analysis of results, 165–169
    average performance duration of tasks, 168
    bed order, 166, 167–168
    interaction, 166
    labeling method, 166–167
    questionnaire, 169
    results, 166–169
    simulation, 169
  bag label, 165
  central venous pressure, 165
  efficiency of labeling, 164
  error identification, 165
  examples of new stickers, 170
  experimental process, 171–174
    background data, 171
    laboratory simulation, 171–174
    opinions, 173
    questionnaire, 173
  experimental tasks, 165
  glossy label, 164
  ideal print, 164
  peripheral vein identification, 165
  pump identification, 165
  spaghetti, 164
  syringe identification, 165
  total parenteral nutrition, 165
Linear programming, 58
Log
  anesthetist's, 51
  daily changes, 143, 152
"Low probability of punishment" problem, 293

M

Magnesium sulphate dosage (medication administration problems), 79–94
  agenda, 81–82
  analysis of task requirements and possible errors at each stage, 83–86
    attentiveness, alertness, and concentration requirements, 85–86
    knowledge requirements, 84–85
    memory requirements, 85
    requirements for information processing and decision making, 85
    sensory requirements, 83
  conclusions, effects on drug administration processes, and recommended solutions, 90–93
    bottle labels, 92
    calculation program, 91
    calculation tool, 91
    case analysis, 90
    control of administration, 93
    drug information bank, 91
    engineering solutions, 91–93
    exclusive preparation of drugs in the pharmacy, 91
    manual spreadsheet, 91
    notification of a rare drug instruction, 93
    procedural–organizational solutions, 93
    training improvements, 93
    uniform registration of instructions for administering drugs, 93
  discussion, 86–90
    ability to obtain information on the drug from outside sources, 87–88
    accuracy of drug dosage calculations, 86
    cognitive load, 86
    crowded environment, 86
    error probability, 87
    labels on drug packages, 88–90
    magnesium ampoule volume, 90
    nurse's general knowledge of drug and dosage calculations, 87
    unfamiliar drugs, 88
    written statement, 89

Index

findings, 83
nurse task description (administration of magnesium sulphate), 83
task stages, 83
frequency of adverse events, 80
method, 81
study objective, 81
tools, 82–83
analysis and prediction of errors and analysis of factors affecting performance, 82–93
cognitive function analysis, 82
human factor engineering method to detect and analyze work safety problems, 82
solutions proposed, 83
Management event, 43
Management's Operation Policy (MOP), 58
Medical ergonomics, 315
Medical errors, history of, 13–21
accusation of guilt, 19
anonymous reporting of mishaps, 20
child's death, 18
diagnosis, 14
legal ramifications, 18
manufacturing production line, comparison to, 21
medical event, 14–16
medical negligence, 17
movement for enhancement of patient safety, 19
nineteenth-century hospitals, 13
paradox, 14
parallel paths, 20
paramedic, 16
production line, 13
risk management system, 20
unprecedented events, 18
upheaval-prone surgery, 14
Medical errors, types and causes of in intensive care, 23–28
activities recorded, 25
data collection, 26
environmental conditions, 23
error definition, 25
microcosm of hospital, 24
mistaken diagnosis, 25
morning shift, 24
questions, 28
shift changeovers, 27
Medical malpractice
preoccupation with, 310
victim compensation, 293
Medical paradox, birth of, 2
Medical tubing, labeling of; *See* Labeling of drugs and medical tubing, designed stickers for

Medication administration problems; *See* Magnesium sulphate dosage (medication administration problems)
Medication room design, human factors contributions to, 131–137
comparative evaluation of new medication room, 135–136
differences in performance, 136
discussion, 137
error prevention, 131
height of work platform, 133
height of work surfaces, 132
labels on medication packages, 132–133
lighting and ventilation conditions in work area, 133
location of medication workstation, 132
nurse preparing medications, 133
parking bays for drug carts, 134
poor planning of work area, 134
project stages, 131–132
recommendations, 134
refrigerator doors and drug storage cupboard, 133
self-report questionnaire, 135
storage capacity of medicine cabinet, 134
Memory associations, 209
Memory transfer, 37
Mental models; *See* Emergency medicine department, mental models
Mental stress, 102
Microclimate, 177
Mobile internally ill patients, 72
MOP; *See* Management's Operation Policy

N

"Near miss" events, 306
Neonatal care units, human engineering and safety aspects in, 95–117
cannon fodder, 96
characterization of NICU, 111
communication and information transfer, 110–114
information nodes, 111
physicians' rounds, 112–113
shift transfer of nursing staff (handover), 112
summary, 113–114
vicious circle, 111
findings and discussion, 101–110
additional cognitive load of drug preparation (midsize NICU), 108
average number of operations per hour, 104
cognitive load, 102, 105
common load characteristics, 103–106
environmental conditions, 102–103, 106
false alarms, 106

incubator cable cluster, 106
influence of environmental conditions (small NICU), 109–110
mental stress, 102
multidimensional concept, 101
number of operations per unit time, 101–102, 103–105
number of operations per unit time (large NICU), 107
nurses' qualification and accreditation, 105
summary, 110
unique load characteristics for each of the three units, 107
workload, 101
heavy workloads, 97, 103
historical background, 96
incubators, 97
intensive care, 98
intermediate care, 98
large NICU, 115
    advanced information technologies, 115
    revised work schedules, 115
manpower shortage, 97
mature baby, 98
method, 99–101
    interviews, 100
    mapping the unit, 100
    observations, 100
    questionnaires, 100–101
    research sample, 99
midsize NICU, 108, 115–116
    attention demands, 108
    decision making, 108
    processing and interpreting written material, 108
    recommendations, 115
    search for and location of information, 108
multiplicity of devices, 98
neonatology, 96
proposals for improvement, 114
routine work, 98
small NICU, 109, 116
Neonatal intensive care unit (NICU), 107
Nurses
    knowledge of drug and dosage calculations, 87
    medication preparation, 133
    qualification and accreditation, 105
    role in patient safety climate, 270–271
        component of caring, 270
        emotional support, 271
        feedback, 271
        patient-orientation, 270
        professional development, 270
        teamwork, 271
    task description (administration of magnesium sulphate), 83
    thinking patterns; *See* Thinking patterns and communication, physicians and nurses in intensive care unit

O

Open heart surgery (reconstruction), 236–238
    changes made, 238
    heart–lung pump, 236
    monitoring devices, 237
    oxygen saturation display, 237
    patient, 236
    reconstruction, 237–238
    resuscitation dummy, 237
    systemic leadership problem, 238
    videotape, 237
    workstation examination, 238
Operating room briefing, 207–215
    briefing, 210–212
        information included, 211
        parts, 210–211
        poster, 212
        shared situational awareness, 211
    development of briefing, 210
        events, 210
        observation, 210
    frequency of errors, 207
    Hand Out, 207
    impact of briefing on number of nonroutine events during surgery, 212–214
        classification of nonroutine events, 212–213
        quality of briefings, 213
        staff members, opinions of, 214
        study finding, 212
        superficial manner of conducting briefing, 213
        systemic problems, 213
        teamwork, 212, 213
    reason for briefing, 208–210
        confirmation bias, 209
        fallacy, 209
        memory associations, 209
        overconfidence effect, 209
        staff performance, 209
        staff task, 208
    shared situational awareness, 208
    summary, 215
    understanding of the situation, 208
Operating room and operating process, 29–53
    assumption, 31
    classification of abnormal events, 40–43
        deviations from the norm, 42
        equipment, 42
        general notes, 42
        human engineering and safety, 42

Index

knowledge and its management, 41
lack of knowledge and expertise, 41
procedures, 41
teamwork, 40–41
timetables, 42
cognitive ability of surgeon, 30
description of the process, 33–38
   communication, problematic, 37
   elective surgery, 33
   hospital orderly, 34
   information-rich process, 38
   memory transfer, 37
   multistage process, 37
   numerous locations, 37
   numerous staff, 36–37
   nursing control room, 34
   procedure flowchart, 34–35
   process team members and main roles, 36
discrepancies, 31
observation dates, 32
operating room, 30, 33
pelvic fracture fixation, 43–51
   blood sample test results, 50
   communication problems, 50
   lead apron, 43
   senior surgeon, 49
   X-ray image, 49
procedures, 41
   compliance with actionable sequence, 41
   compliance with procedures, 41
   lack of procedures, 41
results of observations, 38–43
   abnormal events, documentation of, 38
   classification of abnormal events, 40–43
   general data, 38–39
   management event, 43
   management problems versus safety problems, 43
surgery, incidence rate of complications, 30
teamwork, 40–41
   absenteeism, 41
   communication, 40
   coordination, 40
   inexpert requests, 40
   team changeover, 40–41
   team discipline, 41
timetables, 42
   changes in the operations program, 42
   delays, 42
   work plans, 42
troubleshooting, 51–53
   human engineering problems, 51–53
   inventory, 51
   patient charts, 52
   team members, 52
Operating room staff, garments for; *See* Garment design, for operating room staff

Organizational climate, 264, 265
Organizational voice, 289–290
   assessment, 290
   hierarchical linear models, 290
   open-door policy, 289
   organizational cues, 289
   organizational silence, 289
Orthopedic patients, 72
OSHA; *See* U.S. Occupational Safety and Health Administration
Overconfidence effect, 209

P

Patient safety climate, 263–283
   background, 264–265
      blatant staff errors, 264
      organizational climate, 264
      proposal, 265
      system-wide weakness, 264
   components of nurse's role, 270–271
      component of caring, 270
      emotional support, 271
      feedback, 271
      patient-orientation, 270
      professional development, 270
      teamwork, 271
   measuring safety climate in a medical setting, 268–269
      assumption, 268
      drawback of safety climate measures, 269
      nurses' perception of climate, 269
      parallel hierarchies, 269
      questionnaire items, 268
      single climate score, 268
   outcome variables, 276
   patient safety climate, 265–268
      assumptions, 268
      climate level, 265
      climate perceptions, 267
      climate strength, 265
      focused concept, 265
      nursing system, 265
      organizational climate, 265
      organizational structure, 266
      stated policy, 267
   patient safety climate scale, 279–280
      hospital climate, 279
      unit climate, 280
   preventative step, 277
   professional knowledge, 276
   quality nursing, priority of, 278
   research methods, 271–273
      example, 272
      measuring the level of safety in the departments, 272–273
      measuring patient safety climate, 271–272

medication safety, 273
probability of medical errors, 272
results, 273–274
climate strength, 274
departmental level, 273
hospital level, 273
questionnaires predictive validity, 274
Physician, resident; *See* Resident physicians' medical errors, managerial factors related to
Physician point of view (summary), 313–315
adage, 313
anesthesiology organizations, 313
errors witnessed, 314
existential problems, 314
human factor, 313
human factors engineering, 314
medical ergonomics, 315
medical literature, 314
Physician rounds, communication and information transfer, 112–113
Physician rounds, rate of interruptions during, 217–227
physician rounds in internal medicine department, 218–225
duration of rounds, 224
ethical aspects, 221
examples for categories of interference and instigating sources, 220–221
experimental population, 221
factors contributing to interruptions, 224
frequency of interference by source, 222
methods and materials, 219–220
number of interruptions, 222–225
observation as research tool, 218–219
permissiveness of rounds, 218
purpose of study, 219
results, 221–225
sample, 219–220
sources of interruptions, 223, 224
statistical analysis, 220–221
study population, 219
variables, 220
relationship between interference and errors, 217–218
Physician thinking patterns; *See* Thinking patterns and communication, physicians and nurses in intensive care unit

Q

Quality assurance, 151
Quality of life, technologies to improve, 2
Questionnaires
data collection using, 308–309
garment design for operating room staff, 179–185
labeling of drugs and medical tubing, 169
medication room design, 135
neonatal care units, 100–101
Queuing theory, 58

R

Radiotherapy chart, user-centered design of, 139–162
cognitive problems integrated in new design, 147–153
calculations table, 152
chart does not provide format and structure for calculating and monitoring processes, 151
chart location, 148
complete and explicit presentation of information, 147–149
daily changes log, 152
lack of consistency in displaying data, 150
lack of designated fields for documenting information, 148
lack of human factors principles to improve information legibility, 150–151
leaving cells blank creates a dangerous ambiguity, 149
legibility and organization of information, 149–151
modularity and flexibility, 152–153
"not required" situation, 149
problems in identifying changes made during course of the treatment, 152
problems in old chart and solution in new chart, 154–159
supporting quality assurance and means of verification, 151–152
course of study, 144–147
characterization of content and form requirements of new chart, 146
design of new chart, 146–147
detailed cognitive analysis of existing radiotherapy chart, 145–146
information maps, 146
introduction of new chart for use in radiotherapy unit, 147
task analysis of data, 145
exemplary account of event, 140–141
approach, 141
joint team, 140
report, 140
questions and challenges in developing and implementing new chart, 153–160
information map, 153
investment, 153
presentation of information, 153
testing of new chart, 160

Index

radiotherapy (radiation treatment), 141
radiotherapy unit and radiotherapy chart, 141–144
   chart contents, 143
   cognitive artifacts, 144
   collaborative work, 144
   information created, 142
   near mistakes, 144
   process stages, 142
   staff members, 142
   unique database, 142
   summary, 160–161
     cognitive resources needed, 161
     investigation of radiotherapy centers, 161
     requirements for new design, 161
Reconstruction, investigation of event sources, 235–244
   airplane accidents, 235
   arterial bleeding, 238–241
     blame, 238
     catheterization room, 239
     device monitoring, 240
     doll, 239
     initial scenario, 238
     intensive care unit, 240
     reconstruction conclusions, 241
     scapegoat, 238
     underlying problem, 241
     videotape, 240
   black box, 235
   crime mobile labs, 235
   hidden error, 244
   open heart surgery, 236–238
     changes made, 238
     heart–lung pump, 236
     monitoring devices, 237
     oxygen saturation display, 237
     patient, 236
     reconstruction, 237–238
     resuscitation dummy, 237
     systemic leadership problem, 238
     videotape, 237
     workstation examination, 238
   patient found dead in his bed, 241–244
     anesthesia, 241
     anesthesiologist, 243
     catheter, 243
     communication, 243–244
     mannequin, 241
     nerve block, 243
     pain relief procedures in hospitals, 244
     peripheral nerve stimulation, 242
   summary, 244
   ticking bomb, 236
Reporting system, human factor focused (for hospital medical staff), 245–261
   administration of medication, 248
   discussion of report results, 255–256
   documentation and record keeping, 248
   equipment and instrumentation, 248
   general discussion, 257–260
     ambiguity in job definitions, 58
     daily work difficulties, 258
     expansion of the use of the proposed system, 260
     motivation to report, 259
     raising staff awareness of the impact of human engineering factors on patient safety, 259–260
     system management, 260
     system utility, 259
     validity of reporting system, 258
   human factors problem areas, 250
   major problems, 249
   method efficiency, 247
   outbreak of epidemic, 246
   physical layout, 248
   preliminary analysis used to develop forms, 248
   problem reports, 249
   results, 251–255
     ambiguity in job definitions, 254
     database, 253
     difference between systems, 251
     meetings, 254
     reports collected, 251
     type and frequency of reports, 253, 254
   self-report forms, 247
   shortcomings, 246
   specific problems, 249
   study population, 249
   study questions, 247
   study stages, 247
   topics, 248
   validation study of centrality of major problems reported by department staff, 256–257
     communication, 257
     criteria, 256
     indices, 256
     patient charts, 256–257
   work procedures and patterns, 248
   work procedure and structure report form, 252
Resident physicians' medical errors, managerial factors related to, 285–292
   active learning, 286–287
   department safety climate, 286
   employee behavior, 286
   positive attitudes, 287
   safety climate, 286
   subordinates, 287
   systematic-managerial factors, 286

autonomy, 288–289
  example, 288
  hierarchical linear models, 289
  instructions, 288
  motivational value, 288
  perceived autonomy, 289
  work outcomes, 288
fertile ground for errors, 285
method, 287–288
  active learning climate, 287
  hierarchical linear models, 288
  index and intraclass correlations, 287
  level of fatigue, 288
  treatment methods, 287
  within-department agreement, 287
organizational voice, 289–290
  assessment, 290
  hierarchical linear models, 290
  open-door policy, 289
  organizational cues, 289
  organizational silence, 289
residency, 285
standardization, 290–291
  basis for errors, 291
  checklists, 291
  negative effects of fatigue, 290
  perception of hospital safety procedures, 290
Risk management, difference from accident prevention, 229–233
  actions of risk management, 232
  blood transfusion process, 231
  data to be collected, 233
    around the bed, 233
    medical chart, 233
    personnel, 233
    risk management, 233
  example, 232
  first public conference, 229
  lawyers, 231
  operating room utilization, 232
  primary role of risk management, 230
  procedure for treating the deceased, 233
  safety personnel staff, 232
  Swiss Cheese model, 231
Rounds; *See* Physician rounds, rate of interruptions during

S

Safety; *See* Patient safety climate
Self-report forms, 247
Shared situational awareness, 208
Shift changeovers, errors made close to, 27
Spaghetti, 164
Staff briefing, 208; *See also* Operating room briefing

Stickers, designed; *See* Labeling of drugs and medical tubing, designed stickers for
Surgery
  communication problems during, 50
  elective, 33
  gynecological, 38
  incidence rate of complications, 30
  mishaps stemming from weak points, 44–48

T

Teamwork
  communication, 40
  emergency medicine department, 60
  human factor characteristics in contemporary medical systems, 8
  lessons learned, 302–304
    cognitive-functional dimension, 303–304
    dimensions of medical staff activity, 303
    shared knowledge, 303
    social organizational dimension, 304
  operating room briefing, 213
  operating room and operating process, 40–41
    absenteeism, 41
    communication, 40
    coordination, 40
    inexpert requests, 40
    team changeover, 40–41
    team discipline, 41
  role of nurses in patient safety climate, 271
  work procedures and work patterns, 8–9
Thinking patterns and communication, physicians and nurses in intensive care unit, 195–206
  analysis of physicians' and nurses' thinking patterns, 195–199
    active information intake, 198
    analogical models, 196
    calculation, 198
    characteristics of data processing, 198
    characteristics of performance, 199
    cognitive processes, tools available to investigate, 196
    comparison, 198
    data collection, 197
    data organization, 198–199
    documentation, 199
    error, 199
    factoring, 198
    functional architecture, 197
    functional components, 196
    hypothesis, 198
    input data, 198
    logical models, 196
    models of mapping, 196
    order, 199

Index

passive information intake, 198
planning, 198
problem, 199
process, 198
relationships between objects, 196
results of tests, 198
searching for information, 198
significance, 198
status, 198
study population, 197
thinking aloud, 197
treatment, 199
verbal protocol analysis method, 197
results, 199–204
  added value, 204
  character of data intake and thinking style, 203
  communication and conveying information between physician and nurse, 202–203
  deficiency, 203–204
  dominance of inner knowledge base over external information, 203
  instruction sheet, 202
  lack of systematic manner of worldview construction, 203
  level of complexity of thinking structure, 203
  miscommunication, 202
  monitoring sheet, 204
  nature of messages, 203
  picture of situation, 201
  quality of communication between the physician and the nurse, 203–204
  scope of activities in thinking process, 200
  thinking path, 199
  worldview, 199
summary, 204–205
  anchor, 205
  data collection, 205
  doctor's rounds, 205
  knowledge acquired, 205
  lack of methodology, 204
  world image, 204
Total parenteral nutrition (TPN), 165

U

Unfamiliar drugs, administration of, 88
Unique database, 142
Unisex design, operating room clothes, 190
Unit climate, 280
Unsuitability
  human factors, 309
  implications, 4
Upheaval-prone surgery, 14
U.S. Food and Drug Administration (FDA), 80
U.S. Occupational Safety and Health Administration (OSHA), 177

V

Verbal protocol analysis method, 197
Veteran Health Association (VHA) hospitals, 20
Videotape, reconstruction to investigation event sources, 237, 240

W

Work outfit, 176
Workstation, 5–7
  analysis, 24, 238
  design of devices and unit systems, 5–6
  design and layout of individual workstations, 6
  design and planning of large workspaces, 6–7
  incubator, 116
  medication, 132
  NICU, 106
  nurses', 88
  recording of information, 7
Worldview, 196, 203, 300

X

X-rays, 14
  decision on need for, 72
  equipment, 41, 42
  examination of image, 49
  methodology, 123
  radiation monitoring devices for, 181, 187
  small NICU, 109
  technicians, 129